Magnesium-Based Cementitious Materials

Magnesium-Based Cementitious Materials

Developments and Applications

YONG YANG

Shandong Meijiatu New Materials Technology Co., Ltd., Jinan, P.R. China

TINGLU SONG

Beijing Institute of Technology, Beijing, P.R. China

ELSEVIER

TSINGHUA UNIVERSITY PRESS

Elsevier
Radarweg 29, PO Box 211, 1000 AE Amsterdam, Netherlands
125 London Wall, London EC2Y 5AS, United Kingdom
50 Hampshire Street, 5th Floor, Cambridge, MA 02139, United States

ISBN: 978-0-443-54075-2

For Information on all Elsevier publications
visit our website at https://www.elsevier.com/books-and-journals

Publisher: Matthew Dean
Acquisitions Editor: Glyn Jones
Editorial Project Manager: Akansha Rathi
Production Project Manager: Fahmida Sultana
Cover Designer: Mark Rogers

Typeset by MPS Limited, Chennai, India

Contents

Preface

As a practical implementer and promoter of magnesium-based cementitious materials, over the past 20 years of my journey, I have experienced a lot. In 2014, inspired by the diverse industry processes and coinciding with my 10-year career, I compiled the book "Limestone Technology," advocating the "four-dimensional integration" technical concept, and innovatively proposed a sulfur-chlorine composite four-element reaction system beyond the three existing reaction systems of magnesium-based cementitious materials, opening up a new path for the development of limestone and providing solutions to many entrepreneurs. Five years later, I integrated the exploration experience in the cement and gypsum industries and published the book "Innovative Practice of Inorganic Cementitious Materials," further introducing the company's practical exploration in the sulfur-oxygen-magnesium system and sulfur-chlorine composite four-element system, and being highly regarded by many colleagues.

To this day, my team and I have been engaged in the field of magnesium-based cementitious materials for over 20 years. As the saying goes, "Having navigated through the world for twenty years," this book, *Magnesium-Based Cementitious Materials: Developments and Applications*, continues to introduce to industry colleagues the practical and developmental experiences of chloromagnesium, sulfomagnesium, sulfur-chlorine composite, and phosphomagnesium cementitious materials over the past 20 years.

During the 20-year development of magnesium-based cementitious materials, magnesium products have gradually evolved from low-cost agricultural and forestry products towards energy conservation and environmental protection and have gradually given rise to special products suitable for industries such as healthcare, environmental protection, energy, and special cement: the antibacterial and anti-infective magnesium purification board remains steadfast in the intensive care unit, the foamed porous magnesium filter material contributes to the sewage treatment plant, and the salt-resistant and halogen-resistant magnesium photovoltaic pipe piles shine brightly in saline-alkali areas. Over the past 20 years, the company has remained steadfast on the industry front line, contributing to industry colleagues.

This book was organized and compiled by Yong Yang from Shandong Magnesium Jiatu New Materials Technology Co., Ltd. and Prof. Tinglu Song from Beijing Institute of Technology. This book is divided into 4 chapters: Chapter 1, "Raw Materials for Magnesium-Based Cementitious Materials"; Chapter 2, "Theoretical Ratio"; Chapter 3, "Process and Related Equipment"; and Chapter 4, "Magnesium-Based Cementitious Material Products." The translation and proofreading of this book were primarily completed by Gao Lijuan, Yang Zhuoming, and Prof. Xu Fan from the University of Electronic Science and Technology of China (UESTC). During the compilation process, I also greatly appreciated the support and assistance of experts and scholars at home and abroad.

Given our limited knowledge, we request the readers to kindly criticize and correct any mistakes.

Content overview

This book focuses on the latest developments and applications of magnesium-based cementitious materials, condensing and summarizing the pioneering breakthroughs in raw material preparation processes, reaction mechanisms, and production processes. It is guided by the "four-dimensional integration" technology concept for research and production, addressing industry issues and social application demands. It is the accumulation and summary of the author's years of research and production experience.

This book is mainly for technicians in the magnesium-based cementitious materials industry for learning and use. It can also be read and referred to by relevant personnel in universities, research institutes, etc.

Introduction

Magnesium-based cementitious materials are a type of air-hardening inorganic cementitious materials. Due to their environmentally friendly, nontoxic, pollution-free, lightweight, and high-strength characteristics, they are widely used. This book is divided into 5 chapters, covering raw materials, process equipment, and various production applications of magnesium products. From the reaction mechanism, theoretical ratios, to actual production application technologies, it provides a comprehensive and systematic introduction for the majority of readers.

This book provides a detailed account of the formation process of the $5Mg(OH)_2{\cdot}MgCl_2{\cdot}8H_2O$ phase (referred to as the 5·1·8 phase) in the chloromagnesium system and discusses the influence of factors such as MgO activity, initial ratio, reaction temperature, and water usage on the final formed phase. In practical applications, some magnesite producers have not mastered the production concept of "100–10" and "details determine success," resulting in problems such as reoxidation and rusting of the products, leading to the prohibition of using magnesite materials in some areas and severe challenges for enterprises producing products mainly based on chloromagnesium. In response to this, magnesite enterprises were forced to start researching the new application of the sulfur-oxygen-magnesium system. After modification, the sulfur-oxygen-magnesium system mainly forms the new hydrated product $5Mg(OH)_2{\cdot}MgSO_4{\cdot}7H_2O$ (referred to as the 5·1·7 phase). Through experimental research, when the molar ratio of $MgO/MgSO_4$ and $H_2O/MgSO_4$ is at a reasonable ratio, $3Mg(OH)_2{\cdot}MgSO_4{\cdot}8H_2O$ (referred to as the 3·1·8 phase) gradually disappears, and the 5·1·7 phase becomes the dominant product. At this time, the mechanical strength and water resistance of the magnesium oxychloride system are high. When producing sulfur-oxygen-magnesium products under low-temperature conditions, the curing process is slow and the strength is low. Many problems occurred in the initial application, especially in the production of door core boards, where problems such as mold collapse and high scrap rates often occurred and needed to be solved urgently. In view of the above problems, the research team of the author proposed the sulfur-chlorine four-component reaction system during the research on the sulfur-oxygen-magnesium system. By replacing magnesium chloride with some magnesium sulfate, when the raw material ratio is controlled within a reasonable range, the water-hydrated product of the sulfur-chlorine four-component system has a dense microstructure, uniform crystal distribution, closely arranged, interlaced and crisscrossed, making its mechanical properties and water resistance lie between sulfur-oxygen-magnesium cement and chloromagnesium cement. The sulfur-chlorine composite system has the characteristics of fast curing and high initial strength of chloromagnesium cement and the non-reoxidation and non-rusting of sulfur-oxygen-magnesium and has been widely used in wood door mold pouring production enterprises in Sichuan and Chongqing. With the continuous practice and research of production technicians, the sulfur-oxygen-magnesium technology has become increasingly mature and widely applied.

During this period, magnesium phosphate technology also made significant breakthroughs, and its application scope expanded comprehensively. The magnesium phosphate system essentially involves a neutralization reaction between calcined magnesium oxide and phosphate salts, forming a ternary cementitious system with the structure of $Mg(M)PO_4{\cdot}6H_2O$. This system has the characteristics of rapid setting, small shrinkage upon drying, high strength, corrosion resistance, and resistance to high and low temperatures. From its relatively limited application only for quick-hardening repair of airport runways 10 years ago, to its current application in medical fields such as bone repair, it has achieved a series of breakthroughs.

As the technology of magnesium-based cementitious materials evolves, the supporting processes and equipment also continue to advance. In the molding process, it has progressed from initial manual and semi-manual/semi-mechanical methods to current mechanized and automated production, significantly increasing production efficiency and enhancing product quality. In the curing process, through in-depth research, the important influence of moisture, temperature, and time on product performance has been identified. Simultaneously, microwave heating curing has been innovatively introduced, improving product yield and shortening curing time. In the post-processing stage, treatments such as coating the product surfaces have been applied to enhance product quality and increase added value.

Magnesium-based materials are widely used, ranging from greenhouse frames in farmlands to skyscrapers in cities, with magnesium-based products found everywhere. In recent years, the application of magnesium-based materials has rapidly expanded into other fields. For instance, in the field of new energy, photovoltaic pipe piles have been implemented. In harsh ocean environments with high salt content, high humidity, and strong sea winds, steel structures and Portland cement are not resistant to corrosion and have a short lifespan. However, pipe piles made from magnesium-based materials can be compatible with seawater, resistant to corrosion, nontoxic, and pollution-free, and their strength does not decrease but instead increases. There are also exemplary prefabricated houses that can be entirely constructed using magnesium-based materials, leveraging their lightweight and high-strength characteristics. Purification boards made utilizing the antibacterial properties of magnesium-based materials have been widely applied in scenarios such as hospitals. Magnesium-based cementitious materials are also increasingly playing a significant role in handling solid waste.

No matter how magnesium-based cementitious material technology develops, production cannot be separated from the "four-dimensional integration" technical concept: the quality of raw materials is the prerequisite; reasonable proportioning is the key; scientific modification is crucial; and the process flow cannot be overlooked. The production of the products discussed in this book adheres to the "four-dimensional integration" technical concept, achieving a win-win outcome between quality and production efficiency.

This book is guided by the national advocated "dual carbon" goal and always implements the green and energy-saving environmental protection concept.

CHAPTER 1

Raw materials for magnesium-based cementitious materials

1.1 Magnesium oxide

Since the introduction of magnesium (Mg)-based cementitious materials, the corresponding production processes and mixing techniques have exhibited multiple iterations; however, magnesium oxide (MgO) has consistently remained an indispensable raw material. MgO is typically produced through the calcination of natural minerals such as magnesite. With technological advancements, various industrial by-products have also been processed into MgO products. Depending on the production process and calcination temperature, the type of MgO may also vary. During manufacturing, the appropriate specifications of MgO products need to be selected according to the different types of Mg-based products.

The quality of MgO is determined by multiple technical indicators. In production, one should not focus solely on achieving a high or low value for a single indicator. Instead, it is necessary to combine the design performance of Mg-based products and the process conditions to select MgO products with optimal comprehensive performance.

1.1.1 Source of magnesium oxide

As the core element of the crystalline structure in Mg-based cementitious materials, the source of the Mg element should match both the production process and the type of products. The commonly used MgO in production is generally obtained through the calcination of minerals such as magnesite, dolomite, and willemite. These products are widely used in the construction and insulation industries. Some enterprises also synthesize brucite from brine and then further calcine it into MgO. This method produces MgO with high purity and can be used in the pharmaceutical industry, etc. [1].

The cement factory has low added value, and its raw materials are readily available, so its radiation radius is only about 200 km. Cement production enterprises are scattered across the country. However, for the MgO production enterprises that "burn rocks," they are restricted by the

Magnesium-Based Cementitious Materials
DOI: https://doi.org/10.1016/B978-0-443-54075-2.00004-7

distribution of mineral resources and are concentrated in several major production areas. From the distribution of magnesite mineral resources, China's magnesite deposits can be mainly divided into four categories: regional metamorphic–type magnesite deposits (Liaodong and Jiaodong regions), olivine hydrothermal alteration–type magnesite deposits (Hanzhong area of Shaanxi Province), serpentine weathering leaching–type magnesite deposits (Suolun Mountain area of Inner Mongolia), and Quaternary Lake sedimentary–type magnesite deposits (Naqu area of Xizang) [2,3]. The above-listed typical magnesite deposits are the main production areas of MgO-related products. For example, the line from Haicheng to Dashiqiao to Jiaodong Fenzi Mountain is one such zone. In the area of Haicheng and Dashiqiao alone, there are several large and super-large magnesite mines such as Shengshui Temple, Jinjiabaosi, and Xiafangshen, with proven reserves of over 3.261 billion tons [4].

Dolomite ($CaMg(CO_3)_2$) is a carbonate complex salt composed of magnesite ($MgCO_3$) and calcite ($CaCO_3$). It is widely distributed in various provinces and autonomous regions of China. Magnesian dolomite with a CaO/MgO ratio lower than 1.39 is also an excellent ore for producing MgO. In addition, due to metasomatic action, metamorphism, and desilication, dolomite mines often lose a considerable amount of Mg elements. Combined with certain geological conditions, the lost Mg elements can be accompanied by the formation of magnesite deposits and halite deposits near the original solomite mines. The small-scale magnesite deposit in Tianshan, Shanshan County, Xinjiang, discovered in 2004, is a layer-controlled crystalline magnesite deposit formed by magnesium carbonate rock layers under various geological processes of calcite [5]. The magnesite deposits in the Sidaohongshan area of JinTa County are also associated with dolomite and talc.

Overall, practitioners in the southeastern coastal areas may purchase MgO-related products from the Liaoning, Hebei, and Shandong regions nearby. Meanwhile, colleagues in the inland areas may choose small-scale magnesite products from places like Hanyuan in Sichuan Province, as well as a series of products from the salt lake industrial chains in Qinghai and Xizang.

1.1.2 Production process of magnesium oxide

The production processes suitable for different Mg-containing ores vary, and the process requirements for different specifications of Mg-containing products also differ. The typical production processes for common specifications of MgO on the market are as follows.

1.1.2.1 Roasting magnesite

The main chemical component contents of the magnesite ore in the Haicheng area are as follows (Fig. 1.1): (MgO) 34.90wt.%–47.27wt.%, (CaO) 0.47wt.%–14.30wt.%, (SiO_2) 0.28wt.%–4.70wt.%, (Fe_2O_3) 0.15wt.%–0.76wt.%, and (Al_2O_3) 0.06wt.%–0.73wt.%. The composition of pure magnesite, high silicon magnesite, and high calcium magnesite is slightly different. The roasting process of magnesite in the Haicheng area generally includes the following procedures.

High-grade ores with high MgO content can be used directly for roasting after simple crushing. As environmental requirements become increasingly stringent and the magnesite industry faces pressure to transform, manufacturers are compelled to utilize low-grade ores with lower MgO content. This shift has introduced the need for complex beneficiation processes.

Low-grade ores are first crushed by jaw crushers, then ground in ball mills, and classified to achieve the appropriate particle size. Afterward, dewatering treatment is conducted. The ore is subsequently separated using a flotation machine. At this stage, reverse flotation is typically applied to remove siliceous particles, such as quartz and silicate minerals, from the pulp. In reverse flotation, the pulp pH is usually adjusted with hydrochloric

Figure 1.1 Liaoning Haicheng magnesite mine.

acid or similar reagents, followed by the addition of sodium silicate as a depressant. The flotation machine introduces air into the pulp, and with amine collectors, siliceous particles attach to bubbles and rise to the surface as a foam layer, which is then removed. Following reverse flotation, the tailings are upgraded to produce a coarse concentrate.

For ores with high calcium oxide (CaO) content, a positive flotation process is required to further remove calcium. After pH adjustment, sodium hexametaphosphate and water glass are employed as inhibitors, while sodium oleate or 2# oil serves as a frother for the flotation of MgO particles. Since magnesite is often associated with siderite, magnetic separation may be needed for concentrates with high iron content to reduce iron oxide levels.

For impurities that are difficult to remove, additional mineral processing techniques such as chemical treatment, thermal separation, or gravity separation may be combined. For instance, after drying, the ore slurry can be preirradiated at low temperature for 30 minutes. During this process, most magnesite decomposes into MgO, reducing its density. The undecomposed magnesite remains denser, enabling separation based on density differences. The purified MgO can then be used in downstream applications.

The MgO series products are generally classified according to firing temperature: light-baked MgO (700°C–1000°C), heavy-baked MgO (1000°C–1500°C), dead-baked MgO (1500°C–2000°C), and molten MgO (above 2800°C) [6,7]. The structure of light-baked MgO is fluffy and it has high activity. It is mainly used in industries such as chemical engineering, building materials, environmental protection, feed, and fertilizer. Heavy-baked MgO and dead-baked MgO have very low activity and are often used to make refractory materials and are applied in industries such as metallurgy. They are the most widely used MgO products. The MgO crystals in molten MgO are large in size and have high density. Their thermal insulation and insulation performance are also the best.

During roasting, $MgCO_3$ in magnesite decomposes at high temperatures to generate MgO and release carbon dioxide (CO_2) gas. The sintering degree and chemical activity of MgO crystals vary with roasting temperature, directly influencing the end use of the product.

Different MgO specifications are produced using various processes and equipment. Light-burned MgO is typically produced from high-grade or beneficiated magnesite in gas-fired reverberatory furnaces, though some enterprises use rotary kilns. Reverberatory furnaces often suffer from uneven temperature distribution, leading to inconsistent calcination. Undercooked material must be manually selected and reroasted, resulting in lower quality stability compared to rotary kilns. Since 2021, under policy pressure, most producers have upgraded to rotary kilns for light-burned MgO production.

As the magnesite industry continues to evolve, some larger production enterprises, under the pressure of environmental protection and in order to improve production efficiency, have begun to use suspension kilns to produce MgO. The suspension kiln can bake small particle–sized ores that traditional kilns cannot utilize and is particularly suitable for the baking of refined ores after mineral processing. When using the suspension kiln to bake MgO, the temperature control accuracy is higher, the ore is heated evenly, and there is almost no overbaking or underbaking, thus ensuring stable quality and better product activity. The finished product can be packaged and used directly after cooling, eliminating the subsequent crushing process.

Heavy-baked and dead-burned MgO products are generally produced by using coke as the fuel and are baked in a vertical kiln. The products after leaving the kiln are magnesia. Depending on the application form, they can be further processed into heavy-baked powder (Fig. 1.2).

Due to the inevitable contact between the ore and the fuel, the impurity content of magnesia is relatively high, with the MgO content

Figure 1.2 Heavy-baked magnesium oxide.

being less than 92%. Due to its excellent refractory properties and low price, this type of product is widely used in the refractory materials industry. As the market demand for high-end refractory materials expands, magnesia products have gradually extended to medium-grade magnesia and high-purity magnesia. Medium-grade magnesia uses light-baked magnesia as the raw material, which is processed by a Raymond mill, then semidry pressed into balls, and then subjected to high-temperature roasting in a vertical kiln. The purity of medium-grade magnesia can reach 95%, with a higher density, and it is an important raw material for manufacturing medium-grade magnesite bricks and magnesium spinel bricks. The production process of high-purity magnesite clinker is similar to that of medium-grade magnesite clinker, using light-baked magnesia for ball pressing and secondary calcination, but the high-purity magnesia ball pressing uses high-pressure dry pressing technology, and the roasting fuel is changed to less-impure heavy oil and natural gas. The purity of high-purity magnesia products can reach over 97%, with extremely high added value.

Molten magnesia is generally produced using an electric arc furnace, using high-quality magnesite ore or light-baked magnesia as raw materials, and the resulting magnesia has a larger particle size, which is an important raw material for manufacturing magnesium–carbon (Mg–C) bricks. Mg–C bricks combine the high slag-eroding resistance of magnesia and the high thermal conductivity and low expansion of carbon very well and are used as the inner lining of electric arc furnaces and converters.

1.1.2.2 Calcination of dolomite

Dolomite is widely distributed in various provinces and cities across the country. It is a carbonate compound salt composed of magnesite and calcite. Depending on the different CaO/MgO ratios, it is classified as calcareous dolomite and magnesium dolomite. Through chemical methods and other beneficiation techniques, the calcium content in dolomite can be reduced, thereby producing MgO products.

1.1.2.3 Natural magnesite

Magnesite is magnesium hydroxide ($Mg(OH)_2$), which often occurs in association with magnesite, dolomite, etc. It is also the precursor of lake sedimentary magnesite. When heated at high temperatures, magnesite can be decomposed into MgO and water.

1.1.2.4 Other magnesium-containing materials

1.1.2.4.1 Brine

Seawater and salt lakes contain substantial quantities of magnesium chloride. In regions such as the Qinghai and Xizang salt lakes, where potassium salt resources have been extensively exploited, the residual brine rich in magnesium chloride poses disposal challenges. Beyond its application in brine sheet production, this brine can be utilized to synthesize magnesium carbonate, which is subsequently calcined to yield high-purity MgO.

A conventional approach for brine treatment involves introducing soda ash, hydrated lime, or ammonia into purified brine. Soda ash reacts with magnesium chloride in the brine to precipitate basic magnesium carbonate. The precipitate is then washed, dried, and calcined to produce MgO. Hydrated lime is processed into a lime slurry, which interacts with the brine to form $Mg(OH)_2$ precipitate. This $Mg(OH)_2$ is filtered, washed, dried, and calcined into MgO. Alternatively, introducing ammonia gas into the purified brine can also generate $Mg(OH)_2$ precipitate, which is separated through multiple stages to obtain $Mg(OH)_2$ before calcination into MgO.

Additionally, electrolysis has been reported as a method for treating brine to produce magnesium carbonate. Electrolysis of brine generates chlorine (Cl_2) and hydrogen (H_2) gases, elevating the brine's pH value. Produce brucite precipitate near the cathode, obtain the brucite precipitate through separation, and then continue calcination to prepare magnesium oxide.

The calcination of magnesium carbonate necessitates relatively high temperatures, typically around 650°C, with a minimum calcination temperature exceeding 400°C. Other studies have explored calcination temperatures. Notably, thermal analysis, chemical analysis, X-ray diffraction (XRD), scanning electron microscope (SEM), and other techniques indicate that magnesium chloride hexahydrate can undergo stepwise decomposition at lower temperatures to yield highly reactive MgO. The decomposition steps are as follows:

$$3MgCl_2 + 3H_2O \rightarrow Mg_3Cl_2(OH)_4 \cdot 2H2O + 4HCl \tag{1.1}$$

$$Mg_3Cl_2(OH)_4 \cdot H2O \rightarrow Mg_3(OH)_4Cl_2 + 2H_2O \tag{1.2}$$

$$Mg_3(OH)_4Cl_2 \rightarrow 3MgO + 2HCl + H_2O \tag{1.3}$$

The maximum temperature of this stepwise decomposition is 360°C, which is lower than the calcination temperature required for magnesium carbonate in traditional processes, demonstrating superior energy efficiency. However, the stepwise decomposition releases corrosive hydrogen chloride gas, demanding equipment with enhanced corrosion resistance.

1.1.2.4.2 Salt sludge

Sludge is a common waste in the chlor–alkali industry. After the salt water is purified, the remaining insoluble substances are the sludge. The sludge contains a large amount of salt and alkali substances. If it is piled up in a concentrated form, it will cause serious environmental problems. Generally, it is recycled and utilized in various forms, such as preparing industrial inorganic salts. The sludge contains a large amount of $Mg(OH)_2$ precipitate. $Mg(OH)_2$ can be dissolved in ammonium salt solutions. Through further filtration and purification, the Mg element in the sludge can be extracted. First, the sludge is leached with ammonium chloride solution. During the reaction, magnesium chloride is produced and ammonia gas is generated. Magnesium chloride is purified through filtration and recrystallization, and the escaped ammonia gas is collected and added to the purified magnesium chloride solution again in the form of ammonia water to obtain $Mg(OH)_2$ precipitate. After filtration and washing of the precipitate, it can be calcined to become MgO series products. The filtrate mainly consists of ammonium chloride and can be used again to leach the sludge, achieving recycling.

1.1.3 Specifications and indicators of magnesium oxide

Depending on the type of products produced, the relevant indicators of the MgO series products also have certain differences, and the applicable standards are not very unified. Generally speaking, for the production of chlorinated MgO and sulfurated MgO cement for building materials and insulation materials, the lightweight calcined MgO used should comply with the requirements of "Lightweight Calcined Magnesium Oxide for Magnesium Limestone Products" WB/T 1019—2002; while for the production of phosphoric MgO and refractory materials, the heavy calcined MgO used needs to meet the relevant requirements of "Industrial Heavy Magnesium Oxide" HG/T 2679—2006. In addition, for the production of refractory materials such as magnesium carbon bricks, relevant indicators of "Electrically Fused Magnesia" YB/T 5266—2004 should be referred to. The following are some excerpts of the specifications of national standards and industry standards for MgO.

1.1.3.1 Standard for light-weight calcined magnesium oxide

The relevant standards for light-weight calcined MgO are presented in Tables 1.1 and 1.2.

The above standards are based on "Light-burned Magnesium Oxide for Magnesite Products" WB/T 1019—2002, which is applicable to the production of building materials, handicrafts, etc., such as chlorine oxide magnesium and sulfur oxide magnesium. The specified strength values pertain to the Mg-based cement mortar specimens. These specimens were prepared using (450 ± 2) g of lightly burned MgO, (1350 ± 5) g of standard

Table 1.1 Quantitative indicators for light-burned magnesium powder used in magnesite products.

Grade	QM-85	QM-80	QM-75
Quality level	Premium (A)	First grade (B)	Qualified (C)
Magnesium oxide/E%	≥85	≥80	≥75
Active magnesium oxide/E%	≥65	≥60	≥50
Free calcium oxide/E%	≤ 1.5	≤2.0	≤2.0
Loss on ignition/E%	4~9		≤ 12.0
Fineness (screening rate)	≤ 10% (180 mesh)		≤3% (120 mesh)
Initial setting time	≥45 min		
Final setting time	≤6 h		

Table 1.2 Strength requirements.

Grade		QM-85	QM-80	QM-75
Quality level		Premium (A)	First grade (B)	Qualified (C)
Flexural strength/MPa	3 days	6.5	5.5	4.0
	28 days	10.0	8.5	6.5
Compressive strength/MPa	3 days	35.0	28.0	21.0
	28 days	52.5	42.5	32.5

sand, and (225 ± 1) mL of a magnesium chloride solution with a density of 1.2 kg/L (24.5°Be brine). The procedures outlined in the standard "Test Method for Cement Mortar Strength (ISO Method)" (GB/T 17671—1999) were followed to fabricate standard 40 × 40 × 160 mm prism specimens. Subsequently, the specimens were cured under standard conditions until the designated testing age, at which point their strength was measured.

1.1.3.2 Standards related to refired and dead-fired magnesium oxide

The standards for refired and dead-fired MgO are shown in Table 1.3.

The above standards are based on "Industrial Heavy Magnesium Oxide" HG/T 2679—2006 and are applicable to the production of refractory materials and magnesium phosphate cement.

1.1.3.3 Standards related to molten magnesium oxide

The relevant standards for molten MgO are shown in Table 1.4.

The above indicators are derived from "Electrically Fused Magnesia" YB/T 5266—2004 and are applicable to the production of insulating and thermal insulation materials used in the steel and metallurgical industries.

1.1.3.4 Quality inspection of magnesium oxide

1.1.3.4.1 Determination of related indicators of light-burned magnesium oxide

Light-burned MgO is prone to absorbing moisture and reacting with CO_2 in the air, generating $Mg(OH)_2$ or magnesium carbonate, which is not conducive to the preservation of light-burned MgO. Consequently, MgO can be considered to have a shelf life. Under standard storage conditions, the activity of freshly procured light-burned MgO diminishes significantly after 3 months. In practical production, it is crucial to closely monitor key parameters of light-burned MgO, including active MgO content, free CaO content, fineness, loss on ignition, and setting time, as these directly influence product quality.

In principle, the appropriate type of lightweight calcined magnesia with a suitable active magnesia content should be selected based on the specific product and process requirements. Lightweight calcined magnesia with a high active magnesia content exhibits a rapid reaction rate, significant heat release during hydration, and accelerated curing of the product. The faster reaction promotes the formation of multinuclear hydroxyl magnesium ions in the slurry, and the number of crystallization sites increases geometrically.

Table 1.3 Specifications for heavy magnesium oxide.

Project		Class I		Class II	
		First-grade product	Qualified product	First-grade product	Qualified product
Magnesium oxide/ω%		≥95.0	≥93.0	≥93.0	≥90.0
Calcium oxide/ω%		≤ 1.0	≤ 1.5	—	—
Hydrochloric acid insoluble matter/ω%		≤0.6	≤1.0	—	—
Iron oxide/ω%		≤0.5	≤1.0	—	—
Loss on ignition/ω%		≤2.0	≤3.0	≤3.5	≤4.0
Screen residue (125μm sieve)		≤0.5	≤0.5	≤0.5	≤0.5
Density (g/mL)		—	—	3.30 ~ 3.43	—
Setting time	Initial setting/min	—	—	≤40	≤40
	Final setting/min	—	—	≤4.0	≤8.0
Flexural strength/MPa		—	—	≥10	—

Table 1.4 Specifications for molten magnesium oxide.

Project	Brand name					
	FM990	FM985	FM980	FM975	FM970	FM960
Magnesium oxide/ω%	≥99.0	≥98.5	≥98.0	≥97.5	≥97.0	≥96.0
Calcium oxide/ω%	≤0.8	≤ 1.0	≤ 1.2	≤ 1.4	≤ 1.5	≤2.0
Silicon dioxide/ω%	≤0.3	≤0.4	≤0.6	≤ 1.0	≤ 1.5	≤2.2
Iron oxide/ω%	≤0.3	≤0.4	≤0.6	≤0.7	≤0.8	≤0.9
Aluminum oxide/ω%	≤0.2	≤0.2	≤0.2	≤0.2	≤0.3	≤0.3
Density/ω%	≥3.5	≥3.5	≥3.5	≥3.45	≥3.45	≥3.35

Consequently, the cured slurry forms finer crystals with lower crystallinity. Although products produced under these conditions achieve high strength, their toughness is comparatively reduced.

Conversely, when the active magnesia content in lightweight calcined magnesia is low, the reaction rate of the system decreases, heat release is diminished, and the final product exhibits lower strength. However, the material demonstrates improved toughness. Under these conditions, the concentration of multinuclear hydroxyl magnesium ions in the system is reduced. When crystals of similar strength are formed, the ordered arrangement of ions leads to larger crystal sizes and a more fibrous morphology. The interconnecting crystal structure substantially enhances the toughness of the product.

When producing foaming products or products with strict strength requirements, lightweight calcined magnesia with a higher content of active magnesia may be selected for production. In the production of foaming products, the slurry prepared with high-active lightweight calcined magnesia cures quickly, is less prone to mold collapse, and has higher production efficiency; using lightweight calcined magnesia with a high content of active magnesia to produce products with strict strength requirements can significantly increase the hardness of the product. When producing products such as colored tiles and manhole covers, using lightweight calcined magnesia with a lower content of active magnesia and lower price for production can instead improve the toughness of the product, making such products less prone to breakage. In addition, colored tiles and other products often undergo stacking for curing. The hydration heat of lightweight calcined magnesia with a lower content of active magnesia is also lower, and the stacked products are less likely to burn plates, resulting in better quality of the final product than those produced with high-active lightweight calcined magnesia.

The presence of free CaO exerts a measurable influence on the quality of Mg-based products. During the flotation processing of magnesite ore, the complete removal of calcium-containing minerals proves challenging. Consequently, a small amount of free CaO inevitably becomes incorporated into light-burned MgO following calcination. Free CaO is similar to CaO and, when mixed with brine, can also undergo a hydration reaction, ultimately generating $Ca(OH)_2$, which will damage the structure of Mg-based products. When the content of free CaO is too high, it will affect the stability of light-burned MgO, causing the Mg-based products to crack.

This not only affects the appearance of the products but also has a negative impact on the performance of Mg-based products.

The particle size of light-burned MgO can generally be selected in accordance with specific requirements. The fineness of the commercially available light-burned MgO is around 200 mesh. For the same raw material of light-burned MgO, the larger the amount of mesh after processing, the greater the specific surface area and the faster the reaction in the slurry. Under normal conditions, light-burned MgO with a fineness of around 200 mesh can meet the production requirements. MgO with a high number of mesh reacts quickly and intensely with brine and magnesium sulfate solution, with concentrated heat release, and is not suitable for producing products with larger volume and higher requirements for toughness.

The main component of magnesite ore is magnesium carbonate. During the roasting process, magnesium carbonate gradually decomposes to form MgO with a larger specific surface area and higher activity. As the temperature rises, the crystallinity of the highly active MgO further improves and the crystal phase becomes more complete, while the activity gradually decreases. When the temperature is further increased, the active MgO gradually sintered and becomes a type of reconstituted MgO product. When magnesite ore is roasted at 700°C–1000°C, it has not completely decomposed and there will be a small amount of substances such as magnesium carbonate remaining. The amount of these remaining substances that can be roasted and decomposed at high temperatures affects the degree of loss of light-burned MgO. In addition, during the crushing and grinding process of the finished light-burned MgO, it will absorb moisture and CO_2 from the air, generating substances such as basic carbonate magnesium, which can also increase the loss of light-burned MgO. In fact, the amount of loss reflects the degree of calcination of light-burned MgO during the roasting process, which is the macroscopic manifestation of the size of the internal crystallinity of light-burned MgO. When the loss is large, it indicates that the roasting time of light-burned MgO is short or (and) the roasting temperature is low and the crystallinity of MgO is also low. Light-burned MgO contains a large amount of highly active and defective crystal MgO crystals, showing a fast and intense reaction. When the loss is small, it indicates that the roasting time of light-burned MgO is longer or (and) the roasting temperature is higher, and light-burned MgO mainly exists in a form with high crystallinity and more complete crystals, with a lower reaction rate in the slurry. Generally, products with different loss rates can be selected according to changes in environmental temperature. In hot weather, light-burned MgO with a lower loss rate can be selected to better control the reaction speed and prevent burn plates and burn cores. In autumn and winter, light-burned MgO with a higher loss rate can be selected to increase the reaction speed of the slurry and thereby improve production efficiency. The size of the loss has a certain influence on the reaction rate of the slurry. In actual production, even when the loss rate is not suitable for current production conditions, it can be further adjusted by adding certain additives (refer to Section 1.3 on modified materials) to regulate the reaction rate of the slurry.

The setting time of light-burned MgO is influenced by multiple factors. Under given environmental conditions and material ratios, measuring the setting time of light-burned MgO can indirectly reflect its overall characteristics, including fineness, activity, and loss on ignition. The results of setting time measurement provide valuable guidance for actual production and are closely related to production efficiency. Before implementing additives on a large scale to regulate slurry reaction speed, relevant setting time experiments should be conducted. Setting time experiments require specific conditions, such as controlled environmental settings and equipment like curing chambers. If proper curing conditions are unavailable, the setting time test can be designed as a comparative experiment with limited time efficiency and primarily reference value. Other indicators specified in the standard have a limited impact on the production of conventional Mg-based products and can be selectively tested according to specific needs.

Content of active magnesium oxide According to the relevant provisions of "Light Burning Magnesium Oxide for Magnesium Silicate Products" WB/T 1019—2002, the content of active MgO in light-burning MgO is determined by the water method.

- Reagents: distilled water.
- Apparatus: analytical balance (0.0001 g), glassware, oven, and desiccator.

Standard steps for water method to detect the content of active MgO are as follows: Accurately weigh 2.0000 g (accurately to 0.0001 g) of light-burned MgO sample and transfer it to a glass weighing bottle with dimensions of ø24×40 mm. Add 20 mL of distilled water, cover the bottle loosely (leaving a small gap), and allow it to hydrate under static conditions at (20±2)°C and (70±5)% relative humidity for 24 hours. Afterward, place the sample in an oven for hydration and predrying at 100°C–110°C. Then raise the temperature to 150°C and dry to constant weight. Remove the sample, cool it to room temperature in a desiccator, and weigh the mass of the hydrated sample [8].

The content of active MgO in light-burning MgO is calculated according to the following formula (accurately to 0.01%):

$$W = \frac{W_2 - W_1}{0.45W_1} \times 100\%$$

In the formula, W represents the content of active MgO in light-burned MgO, expressed in percentage; W_1 is the mass of the sample, expressed in grams; and W_2 is the mass of the sample after hydration, also expressed in grams.

The standard water method demands extremely high precision in experiments. Not only must the operators possess the requisite professional expertise, but the instruments must also comply with the relevant standards outlined in "Cement Chemical Analysis Methods" GB/T 176—2008. Based on the hydration method, the experimenters can simplify the experimental steps and employ the rapid water method to determine the MgO content in light-burned MgO.

- Experimental instruments: electronic balance (0.01 g), beaker (1000 mL), glass rod, oven, and desiccator.

Place the sample in a (100 ± 5)°C oven and dry for 1 hour. Then, remove it and transfer it to a desiccator to cool to room temperature. Accurately weigh 100.00 g of light-burned MgO into a predried beaker (previously weighed to constant weight and its dry weight recorded). Add 400 mL of distilled water to the beaker and stir evenly with a glass rod. During stirring, take care to avoid sample loss (any light-burned MgO adhering to the glass rod should be rinsed back into the beaker with distilled water). Transfer the sample into the oven at (100 ± 5)°C and allow it to hydrate for 2–2.5 hours. Then, raise the oven temperature to 150°C and continue drying until constant weight is achieved.

Calculation of active MgO content is as follows:

$$\text{Content of active magnesium oxide} = \frac{W_1 - 100}{45} \times 100\%$$

In the formula, W_1 represents the dry weight of the hydrated light-burned MgO after the reaction, with the unit being grams.

The results obtained from this method are largely consistent with those of the standard method. Even nonprofessionals can control the errors effectively; however, a drawback is that the detection period is relatively long. Please note that during the detection process, the beaker should not be placed directly on the electric stove for heating in an attempt to shorten the detection period. Improper operation may cause the heating temperature to become too high, leading to the decomposition of the $Mg(OH)_2$ formed by hydration and resulting in lower detection values.

Since calcium and Mg belong to the same group, free CaO can also undergo hydration to form calcium hydroxide, which interferes with the determination of active MgO content. If conditions allow, the content of free CaO can be measured after determining the active MgO content via the water method, in order to eliminate this interference.

Content of free calcium oxide Weigh 0.5 g of light-burned MgO accurately to 0.0001 g. Place it in 250 mL conical flask a. Add 10 mL of anhydrous ethanol and 20 mL of ethylene glycol to conical flask a. Heat and stir it until boiling for 5 minutes in a fume hood. Filter it hotly into 250 mL conical flask b. Conical flask a and the filter residue need to be rinsed with anhydrous ethanol three to four times. Add 1 mL of 1:2 hydrochloric acid to conical flask b and let it stand for a period of time. Add 3–5 mL of triethanolamine, 30 mL of distilled water, and about 20 mL of 0.05 mol/L sodium hydroxide to adjust the pH value to 13.0. Add a small amount of calcium red indicator. The color turns red.

Wash the burette with ethylenediaminetetraacetic acid disodium (EDTA) three times and use 0.05 mol/L EDTA to titrate the liquid in the above conical flask. Make the solution blue. Pay attention to reading the two consecutive scales of the burette. Calculate the free CaO from the three titration results and take the average value.

Data calculation is as follows:

$$\text{Free calcium oxide content} = \frac{0.0028 \times v_1}{m_1} \times 100\%$$

In the formula, v_1 represents the amount of EDTA used for each titration of free calcium, m_1 represents the mass of calcined magnesia, and 0.0028 is the proportional constant.

Total calcium oxide content Weigh 2.00 g of calcined magnesia accurately to 0.0001 g and place it in a dry 250 mL beaker. Add a small amount of distilled water and then add a small amount of 1:2 hydrochloric acid to adjust the pH value to 3.0. Let it stand for a while and then filter. Pay attention to weighing the filter paper mass in advance. First, rinse the filter residue with a small amount of 1:2 hydrochloric acid, and then rinse the filter residue with distilled water. During the rinsing process, pay attention to using a glass rod for drainage. The filtrate becomes filtrate A and filter residue B.

Transfer the filtrate to a 500 mL volumetric flask, rinse the beaker three times, and transfer the rinsing liquid to the volumetric flask as well. At room temperature, use distilled water to make up the volume. If filtrate A has a temperature, place it at room temperature before making up the volume, that is, solution C.

Take three clean 250 mL conical flasks. Use the rinsed pipette to respectively transfer 20 mL of the aforementioned solution C into each conical flask. Add 3–5 mL of triethanolamine to each conical flask, and then add 10 mL of water and approximately 20 mL of 0.05 mol/L sodium hydroxide solution (10 mL of 0.1 mol/L sodium hydroxide solution). Adjust the pH value to 13. Add a small amount of chrome black T. At this point, the solution turns pink.

Rinse the burette with EDTA three times, and use 0.05 mol/L EDTA to titrate the liquid in the aforementioned conical flasks. Make the solution blue. Pay attention to reading the two previous and current scales of the burette.

If the data of the three titrations using EDTA differ greatly, then resample and titrate; if only one of the three sets of data has a large deviation, then retitrate with a new set; if the deviations are not significant, then take the average of the three sets of data as the amount of EDTA used.

Data calculation is as follows:

$$\text{Total calcium oxide content} = \frac{0.0701 \times v_1}{m_1} \times 100\%$$

In the formula, v_1 represents the average amount of EDTA used, m_1 is the mass of the light-burned MgO, and 0.0701 is the proportion constant.

Total magnesium oxide content Take three clean 250 mL conical flasks. Using the rinsed pipette, transfer 20 mL each of the above solution C into the conical flasks. Add 3–5 mL of triethanolamine to each flask, followed by 20 mL of water, and approximately 10 mL of a buffer solution with a pH of 10. Adjust the pH to 10, and then add a small amount of chrome black T. At this point, the solution will turn pink.

Rinse the burette with EDTA solution three times. Titrate the liquid in the conical flasks with 0.05 mol/L EDTA until the solution turns blue, noting the initial and final scale readings carefully.

If the titration volumes from the three EDTA titrations vary significantly, collect a new sample for titration. If only one of the three data points shows a large deviation, perform an additional titration. If the differences are not significant, take the average of the three volumes as the amount of EDTA used.

Data calculation is as follows:

$$\text{Total magnesium oxide content} = \frac{0.0504 \times (v_2 - v_1)}{m_1} \times 100\%$$

In the formula, v_2 represents the average amount of EDTA used, v_1 represents the average amount of EDTA used in the total CaO content, m_1 represents the mass of light-burned magnesia, and 0.0504 is the proportion constant.

Loss on ignition Take approximately 10 g of light-burned magnesia and place it in a constant-temperature drying oven at (105 ± 5)°C for 2 hours. After removing it, transfer it to a desiccator until it reaches room temperature. Next, place the porcelain crucible in a muffle furnace at (450 ± 50)°C for 1 hour, then transfer it to the desiccator, and allow it to cool to room temperature. Accurately weigh about 1 g of the dried light-burned magnesia and place it into the crucible, which has been precalcined to constant weight, with the measurement precisely being 0.0001 g. Be sure to record the mass of the crucible and prepare two parallel samples.

Position the lid on the crucible at an angle, leaving a small gap, and then place it in a muffle furnace at 600°C. Gradually heat it to (1050 ± 50)°C and maintain this temperature for 1 hour. After removal, allow it to cool slightly before transferring to a desiccator and cooling to room temperature for weighing. Repeat the firing process (each cycle lasting 15 minutes) and weigh again until the weight stabilizes (the difference between two consecutive weighings is no more than 0.0004 g).

Data calculation is as follows:

$$\text{Ignition loss} = 1 - \frac{\mathrm{m}_2 - \text{the quality of the crucible}}{\mathrm{m}_1} \times 100\%$$

In the formula, m_2 is the mass after combustion and m_1 is the mass of the lightly calcined MgO.

The mean value should be taken for the parallel group data.

Fineness The fineness of the lightly calcined MgO is expressed by the sieve residue rate.

Weigh approximately 10 g of the lightly calcined MgO with an accuracy of 0.0001 g. Transfer it to a 180-mesh sieve, cover with the lid, and shake it horizontally (side to side) for about 10 minutes (avoid vertical shaking). Allow it to settle for a period of time, then remove the sieve, and weigh the sieved lightly calcined MgO.

Data calculation is as follows:

$$\text{Screening rate} = 1 - \frac{m_2}{m_1} \times 100\%$$

In the formula, m_2 represents the mass of the sieved calcined MgO and m_1 represents the mass of the calcined MgO before sieving.

Setting time Adjust the Vickers tester needle so that when it contacts the glass plate, the pointer reads zero.

After sampling 500 g of calcined MgO, mix it with magnesium chloride that complies with the "Industrial Magnesium Chloride for Magnesite Products" standard WB/T 1018—2002, at a molar ratio of MgO to magnesium chloride to distilled water of 5:1:13. The molar quantity of MgO in the calcined MgO is calculated based on the active MgO content. Before mixing, the magnesium chloride must be dissolved in distilled water to prepare the mixing brine.

According to the "Test Methods for Standard Consistency Water Content, Setting Time and Soundness of Cement" GB/T 1346—2011, prepare the pure cement paste using a cement paste mixer. The mixing pot and blades must be wiped clean in advance with a damp cloth. Pour the prepared brine into the mixing pot, and then carefully add the weighed 500 g of light-burned magnesia into the brine within 5–10 seconds to prevent splashing. Place the pot on the mixer's potholder, raise it to the mixing position, start the mixer, and follow the preset program: mix at low speed for 120 seconds, stop for 15 seconds, and during this pause, scrape the cement paste from the pot walls and blades into the center of the pot. Then, mix at high speed for 120 seconds before stopping the machine.

After mixing, immediately take an appropriate amount of cement paste and pour it all at once into the test mold placed on a glass base plate, filling it slightly above the top of the mold. Use a straight-edged knife about 25 mm wide to gently tap the excess paste above the mold surface five times to remove air bubbles. Then, tilt the knife at about a one-third angle to the mold surface and saw off the excess paste gently. Smooth the top of the mold by lightly stroking from the edge toward the center to create a level surface. During the sawing and smoothing process, take care not to compact the cement paste.

After smoothing the mold, immediately place it in a curing environment maintained at a temperature of (20 ± 2)°C and a relative humidity of

(70 ± 5)%. The time when the light-burned magnesia is completely added to the brine is recorded as the start time for setting.

The first measurement is taken 30 minutes after the sample is placed in the curing chamber and after water addition. For measurement, remove the mold from the curing chamber and place it under the test needle, lowering the needle until it lightly contacts the paste surface. Tighten the screw for 1–2 seconds, and then release it suddenly, allowing the test needle to sink vertically into the paste freely. Record the pointer reading when the needle stops sinking or 30 seconds after release. Take measurements every 5 minutes (or at shorter intervals) as the initial setting time approaches. When the test needle sinks to a distance of (4 ± 1) mm from the base plate, the paste has reached its initial setting state. The time from when all light-burned magnesia is added to the brine until the initial setting state is achieved is defined as the initial setting time of the light-burned magnesia, expressed in minutes.

After determining the initial setting time, immediately and carefully slide the test mold together with the paste off the glass plate, rotate it 180 degrees so that the larger-diameter end faces upward and the smaller end downward, and place it back on the glass plate. Then, return it to the constant temperature and humidity curing chamber for further curing. As the final setting time approaches, take measurements every 15 minutes (or at shorter intervals). When the final setting test needle sinks 0.5 mm into the specimen—that is, when the circular attachment leaves no mark on the specimen—the paste has reached its final setting state. The time from when all calcined magnesia is added to the brine until the final setting state is achieved is defined as the final setting time of the paste, expressed in minutes.

1.1.3.4.2 Determination of related indicators of recalcined magnesia

After the magnesia from the magnesite or heavy calcined magnesia powder is calcined, its activity significantly decreases. When used in refractory material products, more attention will be paid to the MgO content, impurity content, and physical indicators in the raw materials.

Total magnesium oxide and total calcium oxide content The detection methods for total MgO and total CaO content can refer to the previous section on the analysis of total MgO and total CaO in light-burned magnesia. The test sample is still digested and then titrated under a certain pH value.

Fineness The fineness of the refired MgO is determined in accordance with the aforementioned method for testing the fineness of the light-fired MgO, using a 120-mesh sieve for sieving and calculation.

For sintered magnesia, the determination can be conducted in accordance with "General Method for Sampling and Sample Preparation of Bulk Mineral Products—Grading Method by Manual Sieving" (GB/T 2007.7) and will not be elaborated further.

Iron (II, III) oxide content The iron (II, III) oxide content can be determined according to the "General Method for Determining Iron Content in Industrial Chemical Products: 1,10-Phenylphenanthroline Spectrophotometric Method" (GB/T 3049). This method requires high experimental precision and involves rather complex steps, which will not be elaborated further here.

1.1.3.4.3 Determination of related indicators of molten magnesia

When magnesite or calcined magnesia is burned in an electric arc furnace at temperatures above 2800°C, it is transformed into molten magnesia. This type of magnesia has larger crystal grains and higher density, making it an excellent refractory material. The purity of molten magnesia is often above 96%, and as the purity increases, the price becomes more expensive.

Total magnesium oxide content The method for detecting the total MgO content can refer to the previous section on the analysis of total MgO in calcined magnesia. It still involves dissolving the sample and titrating under a certain pH value.

Fineness The fineness of molten magnesia is determined using the same detection method as that for sintered magnesia, as previously described. This process is conducted in accordance with "Sampling and Sample Preparation General Principles for Bulk Mineral Products—Particle Size Determination Method—Manual Sieving Method" GB/T 2007.7, without further elaboration.

1.2 Modifier

Whether it is the ternary system of MgO, magnesium sulfide, and magnesium phosphate or the quaternary system containing sulfur and chlorine, the production of Mg-based products necessitates the use of admixtures. The designation of Mg-based cementitious materials across different systems also reflects the specific type of admixture employed. Generally, MgO reacts directly with water to yield relatively low strength, whereas in the presence of an admixture, the comprehensive performance—particularly the strength—of ternary and quaternary reaction systems is significantly enhanced compared to that of binary reaction systems. The advantages of various Mg-based cementitious materials differ; therefore different types of admixtures can be selected according to the specific performance requirements of the final products, such as magnesium chloride, magnesium sulfate, or magnesium dihydrogen phosphate.

The quality of admixture products determines the performance quality of Mg-based products. With scientifically and rationally designed formulations, the use of high-quality admixtures can yield high-quality Mg-based products. In production activities, it is essential to monitor and adjust the key indicators of the admixture and modify the production ratio accordingly based on these indicators.

1.2.1 Magnesium chloride

Magnesium chloride is a raw material in the ternary system of magnesium oxychloride cement. Commonly used magnesium chloride products include hexahydrate magnesium chloride, industrial brine, and anhydrous magnesium chloride. The main sources of hexahydrate magnesium chloride are halite flakes and halite grains.

Magnesium chloride (Fig. 1.3) is generally produced in places such as Shandong, Tianjin, and Sichuan. After seawater and salt well water is dried, the remaining old brine is boiled to produce brine sheets. The salt lake

Figure 1.3 Magnesium chloride.

water in places like Qinghai, after extracting potassium, the remaining waste liquid old brine, after evaporation, can be used to produce brine particles. The industrial brine left over from the borax industry, after treatment, can also be used as brine. However, it contains certain impurities, and some brines have a strong acidity. When using them, careful discrimination is necessary.

The sponge titanium production plant in the southwest region employs the Claus process, utilizing a reaction between titanium tetrachloride and metallic magnesium to produce sponge titanium. The by-product is molten magnesium chloride, which can be recycled through thermal decomposition for reuse. Alternatively, the plant may only recover the residual heat from the molten magnesium chloride and obtain higher-purity pure magnesium chloride products, namely, anhydrous magnesium chloride [9]. Additionally, the production of anhydrous magnesium chloride more commonly originates from the chlorination of minerals such as magnesite and dolomite, as well as the dehydration of halite and brine. After crushing magnesite and other minerals, they are mixed with tar in a specific proportion and fed into a chlorination furnace. Under high temperatures, they react with introduced chlorine gas to form anhydrous magnesium chloride of higher purity. Halite and other materials undergo dehydration and chlorination steps to yield anhydrous magnesium chloride of a certain purity [10].

Anhydrous magnesium chloride releases a significant amount of heat upon dissolution, raising the brine temperature, which is particularly advantageous for winter production. It should be noted that the heat released during dissolution can easily cause local boiling of the brine. When using it, anhydrous magnesium chloride must be added slowly to water, and care must be taken to avoid burns.

1.2.1.1 Standards related to magnesium chloride products

For the production of chloromagnesium cement products, halide sheet materials can be selected according to the relevant requirements specified in “Industrial Magnesium Chloride for Magnesium Oxide Products” WB/T 1018—2002. Specific standards are provided in Table 1.5. By purchasing common finished halide sheets with a nominal purity higher than 46.5%, production requirements can be satisfied.

When preparing the brine for magnesium chloride hexahydrate, refer to the “Comparison Table of Common Brine Ratios, Specific Gravity, Density, etc. ” in the appendix for ingredient selection. The brine should

Table 1.5 Chemical composition standards of industrial magnesium chloride.

Item	Standard/%
Magnesium chloride	≥45.00
Sodium chloride	≤1.50
Potassium chloride	≤0.70
Calcium chloride	≤1.00
Sulfate	≤3.00

be prepared at least 1 day before production to allow foam and bubbles to dissipate. Stir thoroughly during the ingredient mixing process. The brine pool must be free of any sediment from dried flakes. Simultaneously, a multilevel brine pool design should be implemented to allow impurities in the brine to settle fully. Typically, bagged magnesium chloride hexahydrate has inner lining bags. Unused magnesium chloride hexahydrate should be resealed in its inner lining bags for storage to prevent the flakes from absorbing moisture.

For products involving anhydrous magnesium chloride, there are currently no relevant domestic standards or industry experience data. During the production of anhydrous magnesium chloride products, certain impurities are often introduced due to raw materials. In the high-temperature dehydration stage, excessively high local temperatures in the reaction tank can lead to the formation of impurities such as basic magnesium chloride and even MgO. During application, when anhydrous magnesium chloride dissolves, it releases significant heat of dissolution. Excessive heat can cause incomplete hydrolysis of magnesium chloride, resulting in the formation of flocculent basic magnesium chloride and $Mg(OH)_2$. Basic magnesium chloride can convert into $Mg(OH)_2$ and magnesium chloride in an alkaline slurry. Typically, a small amount of magnesium ions in brine undergo hydrolysis to produce $Mg(OH)_2$, imparting a certain acidity to the brine. In Mg-containing slurry, a considerable amount of $Mg(OH)_2$ is also present. The flocculent basic magnesium chloride and $Mg(OH)_2$ may adversely affect the Mg-containing slurry. However, since these flocculent substances also contribute to the Baumé-specific gravity to some extent, there will be a slight deviation between the actual Baumé-specific gravity and the measured value. In practical applications, for products requiring precise ratios—such as whiskers—the anhydrous magnesium chloride solution can be filtered simply before proceeding with methods to measure chloride and magnesium ion concentrations in the brine, followed by ingredient proportioning based on the molar ratio.

1.2.1.2 Purity determination of magnesium chloride products

1.2.1.2.1 Simple detection and analysis of magnesium chloride hexahydrate products

Magnesium chloride hexahydrate, as a relatively inexpensive industrial product, generally has sufficient purity. If quality issues arise, the halogen content can be ruled out using the following methods. The primary indicator of magnesium chloride hexahydrate is purity, with the main impurities being other chlorides. When determining purity, it is generally feasible to measure Mg^{2+} and Cl^- separately.

Measuring with Mg^{2+}

1. Total calcium–Mg determination

   Weigh approximately 10 g of the sample, accurately to 0.001 g, and denote it as m_1. Place it in a porcelain crucible (use of a glass crucible is strictly prohibited) at a temperature of (450 ± 50)°C until constant mass is achieved. Then, place it in a constant temperature drying oven at (105 ± 5)°C for 2 hours. Denote this mass as m_2.

   Place the dried sample of mass m_2 into a 200 mL beaker. Add approximately 50 mL of distilled water to dissolve it. Filter the solution and transfer it to a 250 mL volumetric flask. After allowing it to stand for a period, adjust to volume and denote it as solution A.

   Prepare three clean 250 mL conical flasks. Using a rinsed pipette, transfer 20 mL of solution A into each flask. Add 3–5 mL of triethanolamine to each conical flask, followed by 20 mL of water. Then add approximately 10 mL of pH 10 buffer solution, adjust the pH to 10, and add a small amount of Eriochrome Black T indicator. At this point, the solution will turn pink.

   Rinse the acidic burette with EDTA solution three times. Then, using 0.05 mol/L EDTA, titrate the contents of each conical flask until the solution turns blue. Note and record the initial and final burette readings.

   If the EDTA volumes from the three titrations differ significantly, repeat the sampling and titration. If only one of the three data points shows a large deviation, repeat that single titration. If the deviations are not significant, calculate the average of the three titration volumes and denote it as v_1, the volume of EDTA consumed for titrating total calcium and Mg.

2. Determination of calcium ion content

Take three clean 250 mL conical flasks, use the burette washed with EDTA to respectively take 20 mL of the above solution A, and place them in the conical flasks. Add 3–5 mL of triethanolamine to each conical flask and add 20 mL of water and about 10 mL of 0.05 mol/L sodium hydroxide solution. Adjust the pH value to 13, and add a small amount of calcium red indicator. At this time, the solution is red.

Wash the acidic burette with EDTA three times. Use 0.05 mol/L EDTA to titrate the liquid in the above conical flask until it turns light-blue. Pay attention to recording the two previous and subsequent readings of the burette;

If the volume data of EDTA used in the three titrations differ greatly, then resample for titration. If only one of the three sets of data has a large deviation, then retitrate with a new set. If the deviations are not significant, take the average of the three sets of data as the volume v_2 of EDTA used for titrating calcium content.

Data calculation is as follows:

$$\text{Magnesium chloride content} = \frac{(v_1 - v_2) \times 0.05 \times 250}{1000 \times 20} \times \frac{95.211}{m_2} \times 100\%$$

In the formula, v_1 represents the average amount of EDTA used when titrating the total calcium and magnesium ions, with units of mL; v_2 represents the average amount of EDTA used when titrating the calcium ions, with units of mL; 0.05 is the concentration of EDTA, with units of mol/L (for actual calculation, the actual concentration of EDTA should be substituted); 20 is the volume of solution A in the conical flask, with units of mL; 250 is the total volume of solution A, with units of mL; 95.211 is the molar quantity of magnesium chloride; and m_2 is the total mass of the dried halogen sheet, with units of g.

Or it can be simplified as follows:

$$\text{Magnesium chloride content} = \frac{0.05 \times 1.1901 \times (v_1 - v_2)}{m_2} \times 100\%$$

In the formula, v_1 represents the average amount of EDTA used when titrating the total calcium and magnesium ions, in units of mL; v_2 represents the average amount of EDTA used when titrating calcium ions, in units of mL; m_2 represents the total mass of the dried halogenized sheets, in units of g; 0.05 is the concentration of EDTA, in units of mol/L, and in actual calculations, the actual concentration of EDTA should be substituted; 1.1901 is the proportion constant.

Measured in Cl^- Take three clean 150 mL conical flasks. Use the rinsed pipette to respectively take 20 mL of the solution A obtained from the Mg^{2+} determination test and place it in each conical flask. Add a few drops of potassium chromate solution as an indicator to each conical flask. Use the acid burette containing the preprepared 0.1 mol/L silver nitrate solution for titration. During the titration, shake the conical flask while titrating. White precipitate appears in the conical flask first. At the titration endpoint, brick-red precipitate appears and the brick-red precipitate does not fade after shaking. Record the amount of silver nitrate used.

If the quantities of silver nitrate used in the three titrations differ significantly, a new sample should be taken for titration. If only one of the three sets of data shows a large deviation, an additional titration should be performed to obtain a new set of data. If the differences are not significant, the average value of the three sets of data should be calculated and taken as the amount of silver nitrate solution used.

Data calculation is as follows:

$$\text{Magnesium chloride content} = \frac{v_3 \times 0.1 \times 250 \times 95.211}{1000 \times 20 \times 2 \times m_2} \times 100\%$$

In the formula, v_3 represents the average amount of silver nitrate solution used, in units of mL; 0.1 is the concentration of the silver nitrate solution, in units of mol/L, and in actual calculations, the actual concentration of the silver nitrate solution should be substituted; 250 is the total volume of solution A, in units of mL; 95.211 is the molar quantity of magnesium chloride; 20 is the volume of solution A in the conical flask, in units of mL; 2 is the number of Cl^- ions in one molecule of magnesium chloride; and m_2 is the total mass of the dried halide sheet, in units of g.

Or it can be simplified as follows:

$$\text{Magnesium chloride content} = \frac{v_3 \times 0.1 \times 0.5951}{m_2} \times 100\%$$

In the formula, v_3 represents the average amount of silver nitrate solution used, in units of mL; m_2 represents the total mass of the dried magnesium chloride hexahydrate, in units of g; and 0.1 is the concentration of the silver nitrate solution, in units of mol/L. During actual calculation, the actual concentration of the silver nitrate solution should be substituted; 0.5951 is the proportion constant.

Generally, there is a certain discrepancy between the purity of magnesium chloride hexahydrate calculated by Mg^{2+} and that calculated by Cl^-. Certain amounts of chlorides, sulfates, and other impurities are present in magnesium chloride hexahydrate, which affect the test results. When the results of the two are significantly different and neither meets the relevant requirements of "Industrial Magnesium Chloride for Magnesium Ore Products" WB/T 1018—2002, different situations should be considered. If the purity calculated by Mg^{2+} is higher than that calculated by Cl^-, the content of sulfate ions should be considered for measurement; if the purity calculated by Cl^- is higher than that calculated by Mg^{2+}, it indicates that the contents of potassium chloride, sodium chloride, calcium chloride, etc. may be higher. It is advisable to avoid using substandard magnesium chloride hexahydrate products.

1.2.1.2.2 Simple detection and analysis of magnesium chloride without water

Simple analysis of magnesium chloride without water Weigh approximately 5 g of the magnesium chloride sample, accurately to 0.001 g, and denote it as m_3. Follow the sample processing method in the magnesium chloride hexahydrate analysis section to process it, obtaining a sample with a mass of m_4. Dissolve the sample using the same method as in the magnesium chloride hexahydrate processing section, and add a few drops of a 1:1 nitric acid solution to inhibit the formation of basic magnesium chloride. Then, bring it to the required volume and label it as solution B.

Solution B was analyzed according to the detection method described in the halide analysis section. The calcium content, Mg content, and chloride ion content were measured. The results were expressed in terms of the amount of EDTA and silver nitrate used, and the corresponding purities were calculated using the formula.

Simple analysis of anhydrous magnesium chloride preparation When anhydrous magnesium chloride is mixed with water to form brine, approximately 10 g is accurately weighed and added to the brine, which has reached room temperature and been filtered. The weight is recorded to 0.001 g and is denoted as m_5. Then, approximately 50 mL of distilled water is added to dissolve the solution. The solution is transferred to a 250 mL volumetric flask and left to stand for a specified period before being diluted to the mark. This is referred to as solution C.

According to the detection method for the magnesium chloride hexahydrate, the calcium content, Mg content, and chloride ion content of solution C are measured. The results are expressed in terms of the amount of EDTA used and the amount of silver nitrate used. Then,

1. The moles of magnesium chloride are represented by Mg^{2+}

$$\text{Molar quantity of magnesium chloride} = \frac{(v_4 - v_5) \times 0.05 \times 250 \times \rho}{20 \times m_5}$$

In the formula, v_4 represents the average amount of EDTA used when titrating the total calcium and magnesium ions, with units of mL; v_5 represents the average amount of EDTA used when titrating calcium ions, with units of mL; and 0.05 is the concentration of EDTA, with units of mol/L. During actual calculation, the actual concentration of EDTA should be substituted; 20 is the volume of solution C in the conical flask, with units of mL; 250 is the total volume of solution C, with units of mL; ρ is the density of the magnesium chloride solution, which can be obtained from Appendix 1, "Common Magnesium Chloride Ratios and Comparisons with Potassium Magnesium, Density, etc., unit: g/cm^3"; and m_5 is the mass of the magnesium chloride sample, with units of g.

2. Represent the molar quantity of magnesium chloride as Cl^-

$$\text{Molar quantity of magnesium chloride} = \frac{v_6 \times 0.1 \times 250 \times \rho}{20 \times 2 \times m_5}$$

In the formula, v_6 represents the average amount of silver nitrate solution used, in units of mL; 0.1 represents the concentration of the silver nitrate solution, in units of mol/L, and in actual calculations, the actual concentration of the silver nitrate solution should be substituted; 250 represents the total volume of solution A, in units of mL; ρ represents the density of the magnesium chloride solution, which can be obtained from Appendix 1, "Common Magnesium Chloride Ratios and Comparisons of Salinity, Density, etc.", in units of g/cm^3; m_5 represents the mass of the magnesium chloride sample, in units of g; 20 represents the volume of solution A in the conical flask, in units of mL; 2 represents the number of Cl^- ions in one magnesium chloride molecule; and m_5 represents the magnesium chloride sample mass, in units of g.

1.2.2 Magnesium sulfate

Magnesium sulfate (Fig. 1.4) is an indispensable material in sulfur–oxygen magnesium cement. Its most common application form is in various specifications of the heptahydrate. Heptahydrate magnesium sulfate is typically produced by dissolving Mg-containing ores with industrial sulfuric acid. With technological advancements, numerous methods for preparing magnesium sulfate from by-products have also emerged.

1.2.2.1 Production process of magnesium sulfate

1.2.2.1.1 Sulfuric acid method

The sulfuric acid method uses light-burned MgO or other Mg-containing ores (such as magnesite, dolomite, and magnesite hydrate) as raw materials. After being neutralized by sulfuric acid, magnesium sulfate heptahydrate is produced. During the production process, water or mother liquor is usually first added to the neutralization tank, followed by the addition of magnesium ore powder. With continuous stirring, the measured amount of sulfuric acid is slowly added to the reaction mixture. During the production process, the reaction temperature is maintained between 80°C and 90°C to enhance solubility and the final pH is adjusted to 5–6. After the reaction is complete, the solution is filtered and crystallization is performed at a maximum temperature of 30°C. The magnesium sulfate heptahydrate crystals are separated by centrifugation and subsequently dried

Figure 1.4 Magnesium sulfate.

using a vibrating fluidized bed. Magnesium sulfate produced via the sulfuric acid method exhibits high quality but at a relatively higher cost.

Boric acid production enterprises generally use sulfuric acid to dissolve boron–magnesium ore or rich boron slag (by-product from iron smelting) to produce boric acid and borax products. After the ore powder is dissolved, it can be used in a reaction vessel to cause anhydrous magnesium sulfate to crystallize under high-temperature and high pressure, resulting in a relatively high yield [11]. The solid waste boron sludge from the boric acid industry can also be dissolved with sulfuric acid to produce magnesium sulfate heptahydrate products [12].

The spent industrial acid from the alkylation oil and titanium dioxide industries can also be utilized for producing magnesium sulfate products. After dissolving Mg-containing ore with the waste acid, the solution can be adjusted to a specific Baumé degree and directly used in the manufacture of magnesium oxysulfate cement products. Alternatively, it can be purified through methods such as recrystallization to prepare magnesium sulfate heptahydrate. The sources of the waste acid vary, and the resulting finished products contain different impurities. Therefore they should be properly identified before use.

1.2.2.1.2 Brine extraction method

After extracting potassium, mirabilite, etc. from the salt lake brine, the remaining mother liquor contains a large amount of Mg^{2+}, SO_4^{2-}, Na^+, Cl^-, and other ions. By evaporating water or lowering the temperature, magnesium sulfate can be precipitated in the form of hydrated magnesium sulfate crystals, resulting in crude magnesium sulfate that is used in agriculture and other fields. Depending on the application, this crude product can be recrystallized to obtain a purified form suitable for use in the food industry, pharmaceuticals, and other sectors.

After seawater is evaporated to yield salt, the remaining brine in winter can precipitate hydrated magnesium sulfate containing a substantial amount of sodium chloride impurities. The collected crystals can then undergo recrystallization to yield a purified product, which can be used to produce hydrated magnesium sulfate.

1.2.2.1.3 Magnesium-based desulfurization

After the MgO products are used for desulfurization, the waste liquid contains a large amount of magnesium sulfite. Through aeration or adding acid for oxidation and neutralization, it can be filtered and crystallized to produce magnesium sulfate heptahydrate [13].

1.2.2.1.4 White sodium magnesium vanadium method

White sodium magnesium vanadium is a compound salt of sodium sulfate and magnesium sulfate, which is a common by-product of salt evaporation in salt lakes and is generally used for extracting mirabilite. In the market with low mirabilite prices, white sodium magnesium vanadium and magnesium chloride can be used as raw materials to produce magnesium sulfate and sodium chloride products. After the white sodium magnesium vanadium and magnesium chloride are mixed in a certain proportion and prepared as a solution, magnesium sulfate heptahydrate is first precipitated at −5°C and the filtrate is evaporated at 55°C to separate sodium chloride. This method further processes white sodium magnesium vanadium and generates certain economic benefits [14].

1.2.2.2 Detection indicators of magnesium sulfate

When producing products related to magnesium sulfate, the magnesium sulfate raw materials used generally refer to CMMA/T 1—2015 "Magnesium-based Cementitious Materials Products Magnesium Sulfate." The specific requirements are shown in Table 1.6.

Magnesium sulfate heptahydrate is generally quite stable. During transportation, it may only lose water and become powdery due to heat, but the resulting magnesium sulfate powder has little impact on production. It can be used normally when diluted to the required specific gravity.

Table 1.6 Specifications and indicators of magnesium sulfate heptahydrate.

Project	Indicator	
	First-grade product	Qualified product
Magnesium sulfate content (after calcination)/ω%	≥97.0	≥93.0
Sodium/ω%	≤0.080	≤0.20
Potassium/ω%	≤0.015	≤0.30
Chloride (calculated as Cl)/ω%	≤0.40	≤0.40
Boric acid (H_3BO_3)/ω%	≤0.28	≤0.28
Water-insoluble matter/ω%	≤0.50	≤0.70

If there are concerns about the quality of the raw materials, the following detection methods can be referred to for a brief analysis of magnesium sulfate heptahydrate.

1.2.2.3 Relevant detection of magnesium sulfate

The hydrated magnesium sulfate delivered to the factory can generally be tested for two indicators: the magnesium content and the content of magnesium sulfate after calcination. When both of the above two indicators are qualified, the influence of other indicators on the production of Mg-based products is limited.

1.2.2.3.1 Magnesium content determination

1. Weigh approximately 10 g of the sample, accurately to 0.0001 g, and record it as m_1. Place it in a porcelain crucible (strictly prohibit using glass crucibles) at a temperature of 450 ± 50°C for constant mass. It should be dried in a constant temperature drying oven at 105 ± 5°C for 2 hours. Then transfer it to a muffle furnace at 450 ± 50°C for calcination until the mass remains constant. Record the mass of the magnesium sulfate after calcination as $m_{2.}$
2. Weigh approximately 1 g of the calcined sample, accurately to 0.0001 g, and record it as mass m_3. Place it in a 200 mL beaker. Add 20 mL of distilled water to dissolve the solution. Transfer the solution to a 500 mL volumetric flask and adjust the volume to be constant. Record this as solution A.
3. Take three clean 250 mL conical flasks. Use the rinsed pipette to add 20 mL of solution A to each conical flask. To each conical flask, add 3–5 mL of triethanolamine, followed by 20 mL of water and approximately 10 mL of pH 10 buffer solution. Adjust the pH to 10 and add a small amount of Eriochrome Black T. At this point, the solution turns pink.
4. Rinse the burette with EDTA three times. Use 0.05 mol/L EDTA to titrate the liquid in the conical flask. Make the solution turn blue. Pay attention to reading the two scales before and after the titration.
5. If the data from the three titrations differ greatly, resample and titrate. If only one of the three sets of data has a large deviation, retitrate with a new set. If the deviations are not significant, take the average of the three sets of data as the amount of EDTA used.

   Data calculation is as follows:

$$\text{Magnesium content} = \frac{v_1 \times 0.05 \times 500 \times 24.305}{1000 \times 20} \times \frac{m_2}{m_1 \times m_3} \times 100\%$$

In the formula, v_1 represents the average amount of EDTA used, 0.05 is the concentration of EDTA, 20 is the volume of solution C in the conical flask, 500 is the total volume of solution C, 24.305 is the molar quantity of magnesium, m_3 is the mass of magnesium sulfate used for preparing the solution, m_2 is the total mass of magnesium sulfate after calcination, and m_1 is the mass of magnesium sulfate heptahydrate before calcination.

Or it can be simplified as follows:

$$\text{Totalmagnesiumcontent} = \frac{0.0304 \times m_2 \times v_1}{m_1 \times m_3} \times 100\%$$

In the formula, v_1 represents the average amount of EDTA used, m_3 is the mass of magnesium sulfate used for preparing the solution, m_2 is the total mass of magnesium sulfate after calcination, m_1 is the mass of heptahydrate magnesium sulfate before calcination, and 0.0304 is the proportion constant.

1.2.2.3.2 Magnesium sulfate content (after calcination)

1. Take three clean 250 mL conical flasks. Use the rinsed pipette to respectively take 20 mL of the above solution A and place it in the conical flask. Add at least 0.09 g of dihydrate barium chloride to each conical flask, accurately to 0.0001 g. Record the mass of barium chloride added in each bottle $m4$. Add approximately 5 mL of 1:2 hydrochloric acid. After standing for a period of time, add 3–5 mL of triethanolamine, add 20 mL of distilled water, add 10 mL of pH 10 buffer solution, adjust the pH to 10, and add a small amount of chrome black T. At this time, the solution is pink.
2. Rinse the burette with EDTA three times. Use 0.05 mol/L EDTA to titrate the liquid in the above conical flask. Make the solution turn blue. Pay attention to reading the two previous scales of the burette, calculate the magnesium sulfate content separately, and take the average value. Data calculation is as follows:

$$\textit{Magnesium sulfate content} = \left[\frac{m_4}{244.27} - \frac{(v_2 - v_1) \times 0.05}{1000}\right] \times \frac{500 \times 120.3676}{20 \times m_3} \times 100\%$$

In the formula, v_2 represents the amount of EDTA used for each titration, v_1 represents the average amount of EDTA used for the magnesium content titration, 0.05 represents the concentration of EDTA, 20 represents the volume of solution C in the conical flask, 500 represents the total volume of solution A, 120.3676 represents the molecular weight of magnesium sulfate, m_3 represents the mass of magnesium sulfate used for preparing the solution, and m_4 represents the mass of dihydrate barium chloride used for each bottle.

Repeat the above experiment three times, and take the average of the results for the magnesium sulfate (after calcination).

1.2.2.3.3 Insoluble substances

1. Take 20 g of heptahydrate magnesium sulfate and place it in a 100 mL beaker. Record the mass precisely to 0.01 g as m_5. Use weighed filter paper for filtration. Rinse the beaker and filter residue with distilled water three times. Use 0.05 mol/L barium chloride solution to test the filtrate until no white precipitate is observed. This indicates that the filtration is complete.
2. Transfer the filter paper and filter residue to a preweighed clean beaker. Dry them at a constant temperature of (105 ± 5)°C in an oven until a constant weight is reached. Record it as m_6.

   Data calculation is as follows:

$$\textit{Soluble matter content} = \frac{m_6 - \textit{The mass of the beaker} - \textit{Filter paper quality}}{m_5} \times 100\%$$

In the formula, m_5 represents the mass of magnesium sulfate heptahydrate and m_6 represents the total mass of the drying-dried beaker and its contents.

1.2.3 Phosphate

In addition to common magnesium chloride and magnesium sulfate, phosphates can also combine with MgO to form cementitious materials, generating certain strength components through neutralization reactions. More precisely, the phosphates involved here need to possess a certain degree of acidity, such as phosphate hydrates and phosphate dihydroxides. Currently, the more widely studied and more effective types mainly include ammonium dihydrogen phosphate and potassium dihydrogen phosphate (Fig. 1.5). Other phosphate-based products, such as orthophosphoric acid, sodium dihydrogen phosphate, and monohydroxyl phosphate, although also capable of reacting with MgO to form certain strength components, either produce water-soluble products (as with orthophosphoric acid) or result in low crystallinity or amorphous structures (as with sodium dihydrogen phosphate). This leads to numerous micro-cracks in the product structure, compromising its performance [15].

1.2.3.1 Sources of phosphate

Phosphate products mainly come from the mining of phosphate rocks. The phosphate resources in China are mainly distributed in Yunnan, Guizhou, Sichuan, Hunan, and Hubei. These regions account for approximately 74.5% of the total phosphate reserves in the country. The main form of phosphate rocks is apatite. Depending on the differences in their constituent elements, they can be classified as fluorapatite, chlorapatite and hydroxyapatite, etc.

Phosphate rocks primarily originate from the metamorphism of sedimentary rocks. Some deposits are also formed through biochemical processes, such as the bird droppings phosphate deposit in the Xisha Islands, which exists in the form of hydroxyapatite. In fact, although China has abundant phosphate resources, the proportion of high-grade phosphate rocks with apatite as the main component is relatively low. Most phosphate resources consist of phosphorite blocks mixed with fine-grained crypto-crystalline minerals and gangue, which poses certain difficulties in the development and utilization of phosphate rocks.

Figure 1.5 Potassium dihydrogen phosphate.

1.2.3.2 Production process of phosphate

Apart from existing in the form of phosphate rock, there are also numerous deposits in the form of phosphate blocks. The phosphorus elements in phosphate blocks often exist in the form of amorphous substances or phosphorus-containing colloids, which imposes higher requirements for mineral purification.

In colloidal phosphate rock, in addition to containing target minerals such as apatite and phosphorite, the gangue also contains a large amount of carbonate minerals (such as calcite and dolomite) and silicate minerals (such as quartz, feldspar, and amorphous silicon dioxide). These impurities need to be removed during the beneficiation stage. After the raw ore is screened, ground, and deslimed, the silicate impurities are first removed using positive flotation, and then calcite, dolomite, etc. are removed through reverse flotation.

After phosphate ore is processed through beneficiation, it is mainly further treated using thermal or wet methods. The thermal phosphoric acid process utilizes silica, coke, etc., to reduce the phosphorus elements in the phosphate ore to yellow phosphorus, which is then further processed into higher-purity phosphoric acid. The wet method generally uses sulfuric acid for decomposition, generating phosphate salts and phosphogypsum. The phosphate salts are further refined and used for the production of phosphoric acid, phosphate fertilizers, organic phosphorus, lithium iron phosphate, lithium hexafluorophosphate, and fine chemical phosphate salt products.

1.2.3.2.1 Production of ammonium dihydrogen phosphate

Ammonium dihydrogen phosphate is commonly used as fertilizer, food additive, and fire retardant and is an important product in the phosphorus chemical industry. Ammonium dihydrogen phosphate is generally produced by thermal method or wet method.

Thermal method The thermal method typically involves directly reacting a specific concentration of thermal phosphoric acid with ammonia gas. After neutralizing to a certain pH value, the mother liquor is filtered to obtain a relatively pure ammonium dihydrogen phosphate product. The thermal method is characterized by high energy consumption and significant pollution, lacking cost advantages.

Wet method The wet method is carried out using predefluoride and predesulfurized wet phosphoric acid. After introducing ammonia gas into the wet phosphoric acid, it undergoes concentration and crystallization processes to produce ammonium dihydrogen phosphate.

1.2.3.2.2 Production of potassium dihydrogen phosphate

Potassium dihydrogen phosphate is an inorganic phosphate salt product that is widely used in industries such as chemical engineering, agriculture, medicine, and food. The production method is relatively mature, and it mainly employs methods such as neutralization, extraction, ion exchange, double decomposition, and direct methods [16].

Neutralization method The neutralization method typically uses phosphoric acid and potassium hydroxide or potassium carbonate to produce potassium dihydrogen phosphate through a neutralization reaction. This method is simple in process and has a high purity, but it requires the use of potassium hydroxide and potassium carbonate as raw materials, resulting in a relatively high cost.

Extraction method The extraction method mainly consists of two technical routes: inorganic extraction and organic extraction. The inorganic extraction method has a simple process and can directly utilize phosphorite powder, reacting with inorganic acids and potassium salts directly to produce potassium dihydrogen phosphate. It can also coproduce phosphoric acid, phosphate fertilizers, and other phosphorus-based chemical products.

The organic extraction method generally utilizes organic solvents to separate inorganic salts with different solubilities. In production, triethanolamine and tertiary amines are commonly used as organic solvents. The reaction between phosphoric acid and potassium chloride to generate hydrogen chloride is concentrated in the organic solvent, promoting the reaction toward the formation of potassium dihydrogen phosphate. Potassium dihydrogen phosphate can be crystallized and separated by controlling the temperature. The organic solvent can be recovered and recycled after treatment.

Ion exchange method The ion exchange method is a commonly used synthesis technique in the field of synthetic chemistry. In the production of potassium dihydrogen phosphate, it is generally categorized into two types: the cation exchange method and the anion exchange method.

The cation exchange method involves treating a potassium chloride solution with a strongly acidic styrene-based cation exchange resin, adsorbing the potassium ions onto the resin, and then eluting the resin with a potassium dihydrogen phosphate solution to replace the potassium ions with ammonium ions, thereby producing potassium dihydrogen phosphate. The purity obtained can meet agricultural standards.

The anion exchange method utilizes a weakly basic styrene anion exchange resin. First, it adsorbs phosphoric acid, and then the resin is washed with a potassium chloride solution to obtain potassium dihydrogen phosphate. The ion exchange resin that adsorbs chloride ions can also be used in subsequent transformation processes to produce ammonium chloride. This method yields potassium dihydrogen phosphate with higher purity, suitable for industrial-level applications.

Double decomposition method The double decomposition method involves the use of potassium chloride or potassium acetate, which reacts with phosphoric acid and acidic phosphates. The by-products, such as hydrogen chloride gas, acetic acid, or chlorides, are continuously separated out, promoting the reaction to proceed in the forward direction.

Direct method The direct method involves first reacting concentrated sulfuric acid with potassium chloride to produce potassium bisulfate, then mixing the potassium bisulfate with sulfuric acid, and reacting it with phosphate rock. After separation and purification, phosphoric acid dihydrogen potassium is obtained. The direct method is relatively simple in process, but the product purity is not high.

Multi-step method The multistep method can directly use wet-process phosphoric acid products to produce phosphoric acid dihydrogen potassium, eliminating the step of phosphoric acid purification. After wet-process phosphoric acid precipitates under the action of a flocculant, it reacts with carbamide to produce phosphoric urea. Then, the filtered phosphoric urea reacts with caustic potassium, and the solution is filtered to obtain high-purity phosphoric acid dihydrogen potassium. The carbamide in the filtrate can be recycled and reused. The multistep method can use phosphoric acid raw materials of lower purity, greatly reducing production costs. No organic solvents are used in the production process, and the production process is relatively simple and environmentally friendly [17].

1.2.3.3 Specifications and indicators of phosphate

Given that this book primarily focuses on introducing industries related to Mg-based product series, the phosphates discussed in this section are based on the standards for producing phosphate-based cementitious materials. Only brief introductions to the relevant indicators of commonly used ammonium dihydrogen phosphate and potassium dihydrogen phosphate are provided.

1.2.3.3.1 Indicators of ammonium dihydrogen phosphate

The indicators of ammonium dihydrogen phosphate are shown in Table 1.7.

The above standards are based on "Industrial Monopotassium Phosphate" HG/T 4133—2021. In actual production, it is recommended to use Class I monopotassium phosphate.

1.2.3.3.2 Indicators related to potassium dihydrogen phosphate

The indicators related to potassium dihydrogen phosphate are shown in Table 1.8.

The above standards are based on "Industrial Disodium Phosphate" HG/T 4511—2013. In actual production, it is recommended to use superior-grade disodium phosphate.

1.2.3.4 Quality inspection of phosphate salts

1.2.3.4.1 Determination of ammonium indicator in ammonium phosphate dihydrogen

When ammonium phosphate dihydrogen is exposed to the air for storage, a small amount of ammonia is lost. During the process of forming the strength of magnesium phosphate cement, phosphate ions and ammonia are both indispensable components. Therefore it is necessary to test the content of phosphate ions and ammonia in the raw material ammonium phosphate dihydrogen.

Table 1.7 Relevant indicators of industrial phosphoric acid dihydrate.

Test items		Standard		
		Class I	Class II	Class III
Purity (ω%)	With $NH_4H_2PO_4$/ω%	≥99.0	≥98.5	≥98.0
	With P_2O_5/ω%	≥61.1	≥60.8	≥60.5
	With N/ω%	≥12.0	≥11.8	≥11.5
Arsenic (As)/ω%		≤0.0050	—	≤0.0015
Fluoride (measured as F)/ω%		≤0.02	—	—
Sulfate (measured as SO_4^{2-})/ω%		≤0.5	≤0.9	—
Water content/ω%		≤0.2	≤0.3	≤0.5
Water-insoluble matter/ω%		≤0.1	≤0.2	≤0.2
pH value (10 g/L solution)		≤4.2 ~ 4.8		

Table 1.8 Relevant indicators of industrial phosphoric acid potassium dihydrate.

Project	Indicator		
	First-class product	First-grade product	Qualified product
Purity (based on KH_2PO_4)/ω%	≥99.0	≥98.0	≥97.0
Potassium Oxide/ω%	≥34.0	≥33.5	≥33.0
Moisture/ω%	≤0.5	≤ 1.0	≤2.0
Chloride/ω%	≤0.05	≤0.2	—
Iron/ω%	≤0.003	≤0.008	—
Arsenic/ω%	≤0.005	≤0.015	—
Heavy metals/ω%	≤0.005	≤0.008	—
Water insoluble matter/ω%	≤0.1	≤0.2	≤0.5
pH Value (30 g/L Solution)	4.3 ~ 4.7		

Purity According to the relevant provisions of "Industrial Ammonium Phosphate Dihydrogen" HG/T 4133—2021, the phosphomolybdic acid quinoline gravimetric method can be used to quantitatively detect the content of phosphate ions, and then the purity of ammonium phosphate dihydrogen can be obtained through the formula.

- Reagents: 1:1 nitric acid solution, quinolinemolybdenum solution.
- Apparatus: analytical balance (0.0001 g), glass beaker (100 , 250 mL), 500 mL volumetric flask, oven, and glass sand core crucible (filter plate pore diameter 5–15 μm).

The steps for determining the content of phosphate ions using the phosphomolybdic acid quinoline gravimetric method are as follows: Take approximately 1.0000 g of the ammonium dihydrogen phosphate sample (accurately to 0.0001 g), place it in a 100 mL beaker, dissolve it with water, and then adjust the volume to 500 mL using a 500 mL volumetric flask. Take 20 mL of the ammonium dihydrogen phosphate solution and place it in a 250 mL beaker. Add 10 mL of nitric acid solution, then add water to approximately 100 mL, add 50 mL of quinolinemolybdic acid solution, cover the surface， and place it in a water bath at 75 ± 5°C for 30 seconds. During the insulation process, it is strictly prohibited to use open flames or stir. Instead, 20 mL of distilled water is used to replace the ammonium dihydrogen phosphate solution as the blank group and the ammonium dihydrogen phosphate group, and they are treated in the same way. Stirring is used to accelerate the cooling process. A glass sand core crucible that has been dried to a constant weight at 180 ± 5°C beforehand is used for filtration. The solution in the beaker, along with the precipitate, must be transferred to the crucible completely. After the precipitate is washed, it, along with the glass sand core crucible, is placed in a 180 ± 5°C oven and dried to a constant weight.

The purity in terms of P_2O_5 can be calculated according to the following formula (accurately to 0.01%):

$$\omega_1 = \frac{(m_2 - m_1) \times 0.03208}{m(20/500)} \times 100\%$$

In the formula, ω_1 represents the purity of ammonium dihydrogen phosphate in terms of P_2O_5, expressed in percentage (%); m represents the mass of ammonium dihydrogen phosphate weighed out, expressed in grams (g); m_1 represents the mass of the phosphomolybdic acid quinoline precipitation in the ammonium dihydrogen phosphate group, expressed in grams (g); m_2 represents the mass of the precipitation in the blank group, also expressed in grams (g); and 0.03208 is the conversion coefficient between phosphomolybdic acid quinoline and pentoxide of phosphorus.

If the purity is expressed in terms of $NH_4H_2PO_4$, it can be converted according to the following formula (accurately to 0.01%):

$$\omega_2 = 1.6207 \times \omega_1 \times 100\%$$

In the formula, ω_2 represents the purity of ammonium dihydrogen phosphate in terms of $NH_4H_2PO_4$, in units of %; ω_1 represents the purity of ammonium dihydrogen phosphate in terms of P_2O_5, in units of %; and 1.6207 is the conversion coefficient between ammonium dihydrogen phosphate and pentoxide of phosphorus.

Total nitrogen content According to the distillation–back titration method in "Determination Methods of Phosphate Monoammonium and Diammonium: Total Nitrogen Content" GB/T 10209.1—2008, through distillation, the ammonia gas in ammonium dihydrogen phosphate is separated out and a certain amount of sulfuric acid is used to absorb it. Then, the excess sulfuric acid is titrated back with sodium hydroxide solution. This method has high requirements for the instrument and operators and will not be elaborated here.

1.2.3.4.2 Determination of potassium dihydrogen phosphate indicators

During the production of magnesium phosphate products, when using ammonium dihydrogen phosphate and MgO to prepare slurry, ammonia gas is easily released, which poses certain hazards to the producers. In the market, potassium dihydrogen phosphate is mostly used to replace ammonium dihydrogen phosphate to prepare magnesium phosphate slurry. Due to the stability of potassium dihydrogen phosphate, quantitative detection of phosphate groups and potassium elements can be carried out in actual production.

Purity According to the relevant provisions of "Industrial Potassium Dihydrogen Phosphate" HG/T 4511—2013, the phosphorus molybdate quinoline gravimetric method can be used to quantitatively detect the content of phosphate groups, and then the purity of potassium dihydrogen phosphate can be obtained through the formula. The specific operation steps are basically the same as those for the purity detection of ammonium dihydrogen phosphate, and attention should be paid to the change of the conversion coefficient during conversion.

The purity in terms of $NH_4H_2PO_4$ can be calculated according to the following formula (accurately to 0.01%):

$$\omega_3 = \frac{(m_4 - m_3) \times 0.0615}{m\,(20/500)} \times 100\%$$

In the formula, ω_3 represents the purity of potassium dihydrogen phosphate in terms of KH_2PO_4, with the unit being %; m_3 represents the mass of the phosphomolybdic acid quinoline precipitation in the potassium dihydrogen phosphate group, with the unit being g; m_4 represents the mass of the precipitation in the blank group, with the unit being g; and 0.0615 is the conversion coefficient between phosphomolybdic acid quinoline and potassium dihydrogen phosphate.

Apart from the weight method for determining phosphomolybdic acid, the volumetric method can also be used to measure the purity of potassium dihydrogen phosphate.

- Reagents: sodium chloride, 1 mol/L NaOH solution.
- Instruments: analytical balance (0.0001 g), glass beaker (250 mL), burette, oven, drying cabinet, pH meter (with an accuracy of 0.02 pH unit), and electromagnetic stirrer.

The steps for determining the concentration of potassium dihydrogen phosphate by the volumetric method are as follows: Preheat the potassium dihydrogen phosphate to 105 ± 5°C and dry it for 2 hours. Then, accurately weigh approximately 3.0 g (precise to 0.0002 g) and place it in a 250 mL beaker. Dissolve it with 80 mL of distilled water, and add approximately 5 g of sodium chloride. Place the beaker on an electromagnetic stirrer for dissolution. Titrate the above potassium dihydrogen phosphate mixture solution with sodium hydroxide solution. Detect the pH value change of the solution using a pH meter. The titration is considered complete when the pH value is 9.1. During the titration, set up a blank group. The blank group does not add potassium dihydrogen phosphate, and other processing methods are the same as the experimental group. Pay attention to setting parallel groups for the blank group and the experimental group. The difference in results between the parallel groups should not be greater than 3‰.

The purity of potassium dihydrogen phosphate can be calculated according to the following formula:

$$\omega_4 = \frac{(v_2 - v_1) \times c \times 0.1361}{m_5} \times 100\%$$

In the formula, ω_4 represents the purity of potassium dihydrogen phosphate in terms of $NH_4H_2PO_4$, expressed in %; v_2 represents the dosage of sodium hydroxide solution in the experimental group, expressed in mL; v_1 represents the dosage of sodium hydroxide solution in the blank group, also expressed in mL; c represents the actual concentration of the sodium hydroxide solution; m_5 represents the actual dosage of potassium dihydrogen phosphate; and 0.1361 is the conversion factor.

Oxide potassium content According to the steps mentioned in "Industrial Potassium Dihydrogen Phosphate" HG/T 4511—2013, the oxide potassium content is determined using the potassium tetraphenylborate gravimetric method.

- Reagents: 37% formaldehyde solution, 15 g/L potassium tetraphenylborate solution, 1.5 g/L potassium tetraphenylborate solution, 40 g/L EDTA solution, 10 mol/L sodium hydroxide solution, and phenolphthalein reagent.
- Instruments: analytical balance (0.0001 g), glass beaker (250 mL), 250 mL volumetric flask, 25 mL pipette, oven, and fume hood.

The steps for determining the oxide potassium content using the potassium tetraphenylborate gravimetric method are as follows: After drying the potassium dihydrogen phosphate at 105 ± 5°C for 2 hours, take approximately 1.2 g accurately to 0.0002 g and place it in a 250 mL beaker. Add approximately 150 mL of distilled water and heat to boiling for 30 minutes. Allow it to cool and use a 250 mL volumetric flask to make up to the volume. The blank group does not add potassium dihydrogen phosphate and is processed and made up to the volume using the same steps.

Using a pipette, accurately transfer 25 mL of the sample to be tested into a 250 mL beaker. Add 20 mL of EDTA solution to mask the interference from other cations. After adding a small amount of phenolphthalein reagent, slowly add sodium hydroxide solution until the solution just begins to show red. Continue to add 1 mL of sodium hydroxide solution and 0.15 mL of formaldehyde solution. If the red color disappears at this point, add sodium hydroxide solution again until the red color reappears.

Heat the solution in a fume hood for 15 minutes, then cool it down. Observe whether the red color disappears. If it does, add sodium hydroxide solution to make the red color reappear. Under stirring conditions, add tetraphenylborane solution to the solution. The amount of tetraphenylborane solution added is calculated based on the estimated mass of potassium oxide. For every 1 mg of potassium oxide, add 0.5 mL of tetraphenylborane solution. On this basis, add an additional 7 mL of tetraphenylborane solution. Let the solution stand for 15 minutes, and then transfer it to a glass sand crucible that has been dried at 105 ± 5°C until it reaches a constant weight. Use no more than 40 mL of tetraphenylborane washing solution to rinse the sediment multiple times, and then wash with 5 mL of distilled water three times. Finally, place the glass sand crucible in a 105 ± 5°C oven to dry until a constant weight is reached.

The content of potassium oxide can be calculated according to the following formula (accurately to 0.01%):

$$\omega_5 = \frac{(m_6 - m_7) \times 0.1314}{m_8\,(25/250)} \times 100\%$$

In the formula, ω_5 represents the potassium oxide content, with units of %; m_6 represents the mass of the potassium dihydrogen phosphate group's tetraphenylborate precipitation, with units of g; m_7 represents the precipitation mass of the blank group, with units of g; m_8 represents the actual sampling amount of potassium dihydrogen phosphate, with units of g; and 0.1314 is the conversion coefficient between tetraphenylborate potassium and potassium oxide.

1.3 Modified material

Modified materials play a crucial role in product manufacturing and are indispensable components of production formulations. When modification is carried out properly, the product structure becomes stable and mechanical properties are excellent; conversely, improper modification significantly impairs product performance. Modification technology is a scientific field, and the use of modified materials is just one of the methods for material modification. Broadly speaking, modification also includes ratio adjustment, process modification, and maintenance methods. The concept discussed in this book refers to the narrow sense, which involves optimizing the physical and chemical properties of the slurry and the product through the combination of specific chemical raw materials. In a certain sense, the development of modification technology is based on the progress of materials science, and the advancement of modified materials is fundamental for enhancing material optimization.

There are numerous types of modified material. In this book, the modifiers are classified into four categories based on the types of Mg-based cementitious materials: chloromagnesium modifier, sulfomagnesium modifier, sulfur–chlorine composite modifier, and phosphoric–magnesium modifier.

It can be classified into specialized modifiers and general modifiers based on their application scope. The specialized modifiers (Fig. 1.6) are divided into chlorine–oxygen–magnesium specialized modified agents, sulfur–oxygen–magnesium specialized modified agents, sulfur–chlorine composite specialized modified agents, phosphate–magnesium specialized modified agents, and crystal seed–type modified agents. Among them, the chlorine–oxygen–magnesium specialized modified agents include accelerator-enhancing agents (GX-0#), retarder antisalt agents (GX-1#), antisalt enhancing agents (GX-4#), accelerator antisalt agents (GX-8#), and toughening solvent-reducing agents (GX-12#); the sulfur–oxygen–magnesium specialized modified agents include sulfur–oxygen–magnesium modifiers (GX-15#) and neutral sulfur–oxygen–magnesium modifiers (GX-15-2#). The general modified agents have a wide range of applications and are divided into foaming agents, waterproofing agents, accelerator agents, defoaming agents, curing agents, coupling agents, and release agents (Fig. 1.7). This section will discuss the specialized modified agents and general modified agents.

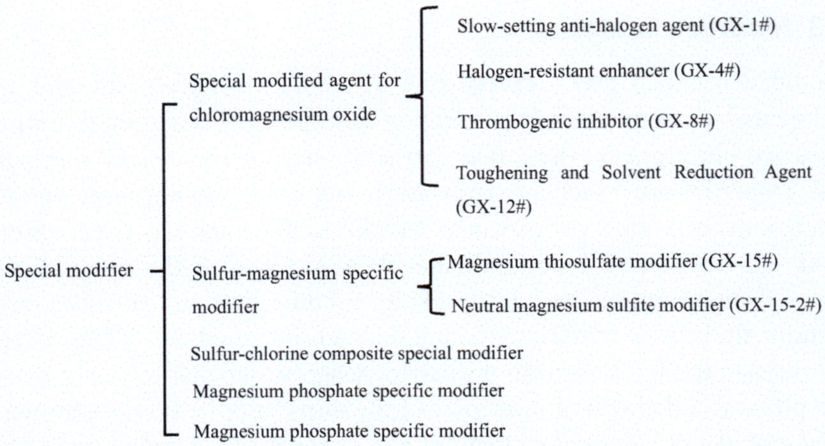

Figure 1.6 Special modifier.

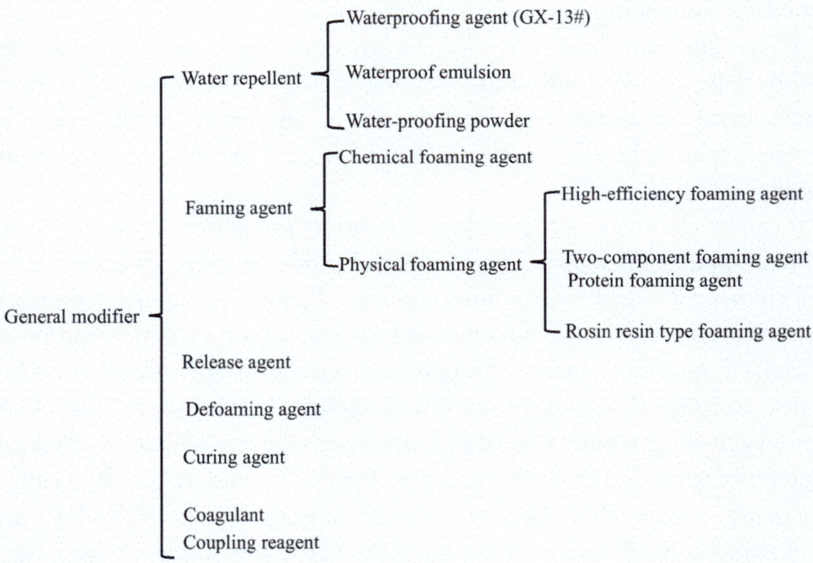

Figure 1.7 General modifier.

1.3.1 Special modified agents

Incorporating admixtures into the magnesia cement slurry can refine the structure of magnesia-based products, enhance the performance of the magnesia cement, reduce its consumption, and accelerate construction progress. This approach offers considerable economic and social benefits.

To address the aforementioned problems encountered in the practical use of magnesia material products, Shandong Meijiatu New Materials Technology Co. Ltd. (MJT) has conducted years of dedicated research and continuous innovation to develop a variety of modifier products. These products comply with the WB/T 1023—2005 standard for "Magnesia Cement Admixtures." This standard specifies in detail the technical requirements, test methods, and inspection rules for magnesia cement admixtures. The subsequent chapters also introduce the modifier products in accordance with this standard.

1.3.1.1 Special modified agent for magnesium oxychloride cement

1.3.1.1.1 Accelerator-enhancing agents (GX-0#) (Fig. 1.8)

Definition and mechanism

- Accelerating setting enhancer (flash setting and reinforcing agent): An additive that can shorten the setting time of magnesite cement materials at 0°C–10°C and enhance the early strength of the products.

There are many factors that affect the setting speed of magnesite cementitious materials. Under the same conditions, the higher the activity of MgO, the faster the reaction speed; the higher the environmental temperature and the temperature of the raw materials, the faster the reaction speed; and the higher the concentration of brine, the faster the reaction speed. The speed of the reaction of magnesia-based gelling materials has both advantages and disadvantages for the performance of magnesia products. At low temperatures or when the content of active MgO in the system is low, the reaction speed of magnesia materials will slow down, the curing time will be prolonged, resulting in an extended mold turnover time and reduced production efficiency, and even more seriously, it may affect the crystal structure of the magnesia material, causing a significant decline in the performance of the products. Appropriately accelerating the reaction speed and reducing the curing time of the products can improve the early strength of magnesia products, shorten the mold turnover time, and increase production efficiency [18]. MJT has developed a series of low-temperature early-setting products to address the issue of low curing efficiency of magnesite materials caused by low temperatures. These products can meet the production requirements of magnesite products in an environment ranging from 0°C to 10°C and can achieve demolding within 24 hours without any heating measures.

Figure 1.8 Accelerator-enhancing agents (GX-0#).

By combining thermodynamic functions in chemical thermodynamics, low temperatures affect the reaction rates at each stage, thereby influencing the formation of five phases, and the accelerator-enhancing agents (GX-0#) increase the curing speed of the products and the strength at each age. Through research, it was found that compared with the magnesium oxychloride cement (MOC) system without adding the accelerator-enhancing agents (GX-0#), the relative content of $MgCl_2$ was the same but the concentration of $[Mg_x(OH)_y(H_2O)_z]^{(2x-y)+}$ differed. This was caused by the different reaction conditions where the reaction of MgO in the $MgCl_2$ solution and the hydrolysis bridging reaction of the mononuclear complex were at different rates. During the hydration process of MgO in the $MgCl_2$ solution, the dissolved $[Mg(OH)(H_2O)_5]^+$ can be encapsulated on the surface of the MgO particles, forming a double-layer hydration shielding film. Low-temperature conditions can increase the surface tension of the hydration shielding film, thereby inhibiting the further diffusion of $[Mg(OH)(H_2O)_5]^+$. Generally speaking, the electrostatic repulsion can weaken the surface tension of the double layer, thereby changing the zeta potential of the MgO surface and increasing the rate of diffusion of $[Mg(OH)(H_2O)_5]^+$ from the MgO surface [19]. Clearly, with the addition of the accelerator-enhancing agents (GX-0#), the zeta potential of the MOC system increased and the strong repulsion promoted the diffusion of the shielding layer of $[Mg(OH)(H_2O)_5]^+$, causing the concentration of the mononuclear complex in the solution to increase rapidly, promoting the hydrolysis bridging reaction rate of the mononuclear complex, and thus accelerating the hydration reaction rate of the MOC system, playing a role in promoting coagulation. At the same time, the production of $Mg(OH)_2$ was reduced, causing the crystals to become slenderer and the microstructure denser, thus enhancing the later strength of the product.

The specific reaction process is as follows: First, under the action of the modifier, the shielding layer $[Mg(OH)(H_2O)_5]^+$ of the system reacts with the modifier through a complexation reaction; then, $[(H_2O)_5Mg(OH)]^+$ or $[Mg(H_2O)_6]^{2+}$ connect with each other in the form of "hydroxyl bridge" or "water bridge" to form a series of $[Mg_x(OH)_{2y}(H_2O)_z]^{(2x-y)+}$, releasing the modifier; then, $[Mg_x(OH)_{2y}(H_2O)_z]^{(2x-y)+}$ reacts with the Cl^- and H_2O or OH^- in the solution to form hydrated substances (five phases), and the reaction equation is as follows:

$$MgO + H^+(aq) + 5H_2O \Rightarrow Mg(OH)(H_2O)_5]^+(aq) \tag{1.4}$$

$$[Mg(OH)(H_2O)_5]^+ + GX\text{-}0\# == [[Mg\text{-}(OH)\text{-}(H_2O)_5\text{-}GX\text{-}0\#]^+ \tag{1.5}$$

$$\begin{aligned}&[(H_2O)_5\text{-}Mg-(OH)-GX-0\#]^+ + [H_2O-Mg-(OH)-(H_2O)_4-GX-0\#]^+\\ &\Leftrightarrow [(H_2O)_5Mg-OH-Mg(OH)(H_2O)_4]^{2+} + H_2O + 2GX-0\#\end{aligned} \tag{1.6}$$

$$\begin{aligned}&\left[Mg_x(OH)_y(H_2O)_{x-y}\right]^{2x-y} + mCl^- + (2x-y-m)OH^-\\ &\rightarrow mMg_3(OH)_5(H_2O)_3]^+\cdot Cl-\cdot H_2O(5-phase)\end{aligned} \tag{1.7}$$

Usage and precautions The dosage of the accelerator-enhancing agents (GX-0#) is generally 0.5%–1% of the weight of the light-burned MgO. When using, directly add it to the aged and clear brine. First, stir and disperse the modifier in the brine, then add the MgO and powder, and stir to avoid the modifier being difficult to disperse in the powder, which may affect the modification effect.

The accelerator-enhancing agents (GX-0#) is a weakly acidic liquid and cannot be used together with alkaline substances. During winter production, using both anhydrous magnesium chloride together will have a better effect.

Storage and transportation

1. Storage: Store in a cool environment below 25°C, seal, and cover. The shelf life of the accelerator-enhancing agents (GX-0#) is 1 year.
2. It is an environmentally friendly material, nontoxic and harmless, and nonflammable and nonexplosive and can be transported as general chemicals.

Product performance The experimental temperature was 10°C, and the same raw materials and raw material ratios were used to test the effect of different amounts of setting enhancer (GX-0 #) on the initial setting time and strength. The specific results are shown in Table 1.9.

In Table 1.9, it can be seen that as the dosage of the promoting and strengthening agent (GX-0#) increases continuously, the initial setting time of magnesite cement shortens and the compressive strength increases. When the dosage exceeds the recommended amount, the promoting and strengthening effects are not obvious, and even when the dosage is too high, it may have an adverse effect.

1.3.1.1.2 The retarder and halogen-resistant agent (GX-1#) (Fig. 1.9)

Definition

- Retarder: An additive that can prolong the setting time of magnesite cementitious materials.

During the hot season, magnesite cementitious materials have a faster setting speed and are prone to hardening before pouring or construction. During the setting process of magnesite cementitious materials, heat is continuously released, and when the reaction speed is fast, the products release a large amount of heat suddenly, which can cause the products to expand and subsequently crack. Adding an appropriate amount of retarder to magnesite cementitious materials can improve this phenomenon [18].

Function and principle The mainstream viewpoints regarding the mechanism of chloromagnesium retardation are the precipitation hypothesis, the complex salt hypothesis, the adsorption hypothesis, and the nucleation formation hypothesis. The formation of a protective layer on the surface of MgO particles can inhibit the hydration of

Table 1.9 Effect of accelerator-enhancing agents (GX-0#).

Addition amount/%	Initial setting time/min	1 d compressive strength/MPa	28 d compressive strength/MPa
0.0	360	55.02	105.8
0.3	180	65.32	112.48
0.6	120	71.56	115.78
1.0	180	56.37	109.74
2.0	300	46.77	89.93

chloromagnesium and slow down the condensation of chloromagnesium.

Through research on the effects of different retarders on the setting time, compressive strength, and water resistance of magnesium oxychloride cement and by analyzing the different hydration products using x-ray diffraction equipment, it was found that the setting time of magnesium oxychloride cement continuously prolonged as the amount of retarder added increased, while the compressive strength of magnesium oxychloride cement decreased. This is because the retarders affect the pore structure in magnesium oxychloride cement and the formation of phase 5·1·8 in the hydration products.

After years of in-depth research, the retarder developed by MJT not only can delay the setting speed of magnesite cement in high-temperature environments (25°C–40°C) and improve the operability of the slurry but also can enhance the various properties of the products.

At present, most of the research directions on retarders for chloro-magnesite cement mainly focus on the setting time, water resistance, and strength [20]. The retarder and halogen-resistant agent (GX-1#) first reacts with the MgO in the slurry to form a "thin film" covering the surface of the 5.1.8 crystal seed, reducing the growth speed of the 5.1.8 crystal seed, and at the same time covering the surface of the light-burned MgO to slow down the formation speed of single nuclei of MgO, reducing the peak of hydration heat and preventing core burning of the plates. The retarder and halogen-resistant agent (GX-1#) will also react with the raw halogen water in the chloro-magnesite cement to form insoluble products that fill the

Figure 1.9 The retarder and halogen-resistant agent (GX-1#).

pores of the products, reducing the porosity of the products, increasing the water resistance of chloro-magnesite cement, and at the same time inhibiting the reprecipitation and frost heaving phenomenon of chloro-magnesite cement products.

Usage and precautions The dosage is generally 0.5%–1% of the weight of light-burned MgO. When using, simply add it directly to the slurry and stir evenly.

The retarder and halogen-resistant agent (GX-1#) is an acidic liquid and cannot be mixed with alkaline substances.

Storage and transportation

1. Store in a cool environment below 25°C and seal the container tightly; the shelf life is 1 year.
2. It is an environmentally friendly, nontoxic, harmless, nonflammable, and nonexplosive material. It can be transported as a general chemical product.

Product usage effect At a test temperature of 30°C, with the same raw materials and ratios. Test the effect of different addition amounts of the retarder and halogen-resistant agent (GX-1#) on the initial setting time; after the test block is cured for 15 days, detect the free chlorine content. The results are shown in Table 1.10.

As can be seen in Table 1.10, as the dosage of the retarder and halogen-resistant agent (GX-1#) increases, the initial setting time of the slurry gradually prolongs and the content of free chloride ions gradually decreases. After testing, it was found that with the addition of the retarder and halogen-resistant agent (GX-1#), no rebleeding phenomenon was observed in the product, which is beneficial for the long-term use of the product.

1.3.1.1.3 Antihalogen enhancer (GX-4#) (Fig. 1.10)

- Antihalogen enhancer (halogen-resistant and reinforcing agent): An additive that can enhance the water resistance, antihumidification, and antihalogen readsorption properties of magnesium silicate composite products and increase the strength of the products.

Causes and solutions The main reaction products in the hardened body of chloromagnesium cement are $3Mg(OH)_2$-$MgCl_2$–$8H_2O$ (3 phase), 5Mg $(OH)_2$-$MgCl_2$–$8H_2O$ (5 phase), and $Mg(OH)_2$ phase. The mechanical strength of the hardened body mainly comes from the 5 phase and the 3 phase, especially the 5 phase. The 5 phase is an unstable phase. Under

Table 1.10 The influence of retarder (GX-1#) on the initial setting time and free chloride ion content of magnesium oxychloride cement.

Addition amount/%	Initial setting time/min	Free chloride ion content/%
0.0	90	6.0
0.3	120	3.1
0.5	155	1.48
0.7	187	1.35
1.0	210	1.2

Figure 1.10 Antihalogen enhancer (GX-4#).

humid conditions, the 5 phase undergoes a phase transformation, reducing the mechanical strength of the hardened body [21]. During the production process, to meet operational requirements, the workability of the slurry is adjusted. In this case, there is inevitably an excess of magnesium chloride and water in the mixture. Excess magnesium chloride in the mixture absorbs moisture and condenses on the surface of the product. After drying, it forms dew frost again. Excessive water causes more capillary pores in the product. The chloride salts contained in the magnesium chloride, such as sodium chloride and potassium chloride, do not coordinate or chelate with MgO during the product's condensation process. These impurities are adsorbed or encapsulated in the crystal structure in the form of free state in the solidified product. Generally, the more halite impurities there are, the more severe the moisture absorption and reprecipitation phenomenon become. At the same time, due to the existence of capillary action, water vapor penetrates into the product interior, causing excessive humidity inside the product. Long-term wetting will lead to the destruction of rhodochrosite crystals, and the internal stress generated by alternating wet and dry conditions will also exacerbate the damage to the product [18].

To enhance the water resistance of chloromagnesium cement, the following methods are currently employed: (1) Add waterproof dispersing emulsions or other waterproof fillers to the cement slurry, thereby reducing the porosity of the chloromagnesium cement, which in turn improves the pore structure and prevents water molecules from penetrating into the chloromagnesium cement, thus protecting the five phases from being damaged; [22] (2) Add hydrophobic agents, forming a film on the pores of the chloromagnesium cement, reducing the contact angle of water on the surface of the chloromagnesium cement pores, and inhibiting water penetration; (3) Add acid-resistant hydraulic binders, such as aluminate cement and composite slag, which, upon hardening, form a new strength structure network, inhibiting the erosion of water on the chloromagnesium cement and thereby preventing the destruction of the hardened body's strength. Through experimental verification, it has been found that methods (1) and (2), due to the weak natural weathering corrosion resistance of the added waterproof agents, are insufficient in improving the waterproof performance of chloromagnesium cement products; while method (3), when preparing chloromagnesium cement with the formulated materials, has a lower strength and insufficient elasticity [21]. Considering all these factors, after a large number of comparative experiments, we have developed an ideal chloromagnesium cement antihalogen agent–antihalogen enhancer (GX-4#) and accelerator anti–halogen agent (GX-8#), which not only significantly improves the water resistance of the cement but also enhances its durability.

Principle and function The antihalogen enhancer (GX-4#) is a composite modifier. Its surface-active components can adjust the surface adsorption degree of the materials and at the same time introduce some hydrophobic substances to change the five-phase lattice structure and interpenetrate and connect between the crystal phases to form a stable spatial structure network, blocking the capillary channels and reducing capillary phenomena. The dispersant component in the antihalogen enhancer (GX-4#) can enable the quick contact and reaction between light-burned magnesia and magnesium chloride, improving the dispersion of light-burned magnesia in brine, increasing the contact area between active MgO and $MgCl_2$, making the generated crystal phases finer and more closely connected, thereby enhancing the strength of magnesite cement, having a lower apparent porosity, and enabling it to have a lower water absorption rate and higher water resistance.

Functions are as follows:

1. High water resistance: By adding the antihalogen enhancer (GX-4#) to magnesite cement, its waterproof performance can be significantly improved, enabling it to have better durability in humid environments.
2. Low water absorption rate: The antihalogen enhancer (GX-4#) reduces the porosity of magnesite cement, making its water absorption rate lower, effectively preventing water penetration and damage to the structural phase.
3. High mechanical strength: Adding the antihalogen enhancer (GX-4#) can enhance the mechanical strength of magnesite cement and increase its compressive strength.
4. Good thermal stability: Adding the antihalogen enhancer (GX-4#) to magnesite cement can improve its thermal stability, enabling it to maintain better performance in high-temperature environments.

Usage and precautions The dosage is generally 0.6%–0.8% of the weight of the light-burned magnesia. The specific dosage needs to be adjusted according to the activity of the light-burned magnesia, production temperature, product requirements, etc. When using, add it to the well-mixed slurry and stir evenly.

The antihalogen enhancer (GX-4#) is an alkaline liquid and cannot be mixed with acidic substances.

Storage, transportation, and precautions It is suitable for storage in a sealed manner in a cool, dark, and shaded environment at 0°C–30°C for a shelf life of 1 year.

It is not a hazardous product and can be transported as a general item. If it comes into contact with the skin or splashes into the eyes, immediately rinse with plenty of water; if ingested, seek medical examination immediately.

Product usage effect At 20°C conditions, using the same raw materials and ratios, test the effect of different dosages of the antihalogen enhancer (GX-4#) on compressive strength, softening coefficient, and free chloride ion content. The results are shown in Table 1.11.

In Table 1.11, it can be seen that after adding the antihalogen enhancer (GX-4#), the compressive strength of the sample block significantly increased and the 28-day softening coefficient rose to approximately 0.9. The content of free chloride ions decreased as the dosage of the modifier increased. This indicates that the sample blocks with the addition of the antihalogen enhancer (GX-4#) have a significant effect in improving the water resistance of the products.

1.3.1.1.4 Coagulation-promoting and halogen-resistant agent (GX-8#) (Fig. 1.11)

Definition and mechanism

The coagulation-promoting and halogen-resistant agent (GX-8#) enhances the water resistance of the product, increases the surface smoothness of the product, eliminates air pores, and inhibits cracking, moldiness, and discoloration in magnesite products. The coagulation-promoting and halogen-resistant agent (GX-8#) not only accelerates the formation of crystal seeds, increases the curing speed of magnesium cement, but also changes the entropy of the system and promotes chemical reactions toward the formation of five-phase crystals, making it particularly suitable for the production of magnesite products in environments with an average daily temperature of around 10°C. Moreover, the substances in the coagulant antihalogen agent can form a water-repellent film on the surface of magnesite crystals and in the capillary channels, protecting the crystal structure in the product from being eroded by water, significantly improving the halogen resistance of the product.

Usage method and precautions The dosage of the coagulation-promoting and halogen-resistant agent (GX-8#) is generally 0.5%–1% of the mass of the light-burned MgO. The exact dosage can be adjusted based on factors such as the activity of the light-burned MgO. When using, add it evenly and slowly to the uniformly mixed slurry.

The coagulation-promoting and halogen-resistant agent (GX-8#) is an alkaline liquid and cannot be stored together with acidic substances.

Storage and transportation

1. Store in a cool environment below 25°C, and seal the container tightly. The shelf life of the coagulation-promoting and halogen-resistant agent (GX-8#) is 1 year.
2. It is an environmentally friendly, nontoxic, harmless, nonflammable, and non-explosive material. It can be transported as a general chemical product.

Product performance At an experimental temperature of 10°C, with the same raw materials and raw material ratios, the effect of different addition amounts of the setting coagulation-promoting and halogen-resistant agent (GX-8#) on the initial setting time was tested and standard test blocks were

Table 1.11 Effect of antihalogen enhancer (GX-4#) usage.

Addition amount/%	Compressive strength/MPa		Soaking in water for 28 days, softening coefficient	Free chloride ion content/%
	7 days	28 days		
0	65.03	70.21	0.78	6
0.3	74.45	80.6	0.89	2.8
0.6	82.34	91.2	0.93	1.45
1.0	80.33	87.15	0.91	1.45

Figure 1.11 Coagulation-promoting and halogen-resistant agent (GX-8#).

prepared. After 15 days of curing, the salt resistance was tested and the results are shown in Table 1.12.

As can be seen in Table 1.12, under low-temperature conditions, as the amount of the coagulation-promoting and halogen-resistant agent (GX-8#) added increases, the initial setting time gradually shortens and the halogen resistance performance improves. When the addition amount exceeds 0.5%, good halogen resistance performance can be achieved.

1.3.1.1.5 Toughening and solvent reduction (GX-12#) (Fig. 1.12)

- Toughening agent: An additive that can enhance the toughness of magnesia-based cementitious materials.

Causes and solutions Magnesia-based cement has the advantages of quick setting and hardening, high strength, good fluidity, and easy operation. However, its brittleness severely limits its application. Based on the research on improving the toughness of magnesia-based cement conducted both domestically and internationally, the main methods of toughening modification include adding polymer emulsions, short-cut fibers, glass fibers, fine steel fibers, and fiber fabrics. When modifying magnesia-based cement with polymer emulsions, it can play an inhibitory role during crack propagation, improving the deformation of the product. However, the dosage is large and the final strength of the product will also decrease and it will also affect the workability of the slurry. When using

Table 1.12 Effect of the coagulation-promoting and halogen-resistant agent (GX-8#).

Addition amount/%	Initial setting time/min	15-day halogen resistance test
0.0	360	Water droplets on the surface
0.3	290	Slightly damp on the surface
0.5	200	Surface drying
0.8	145	Surface drying
1.0	110	Surface drying

Figure 1.12 Toughening and solvent removal agent (GX-12#).

short-cut fibers, glass fibers, fine steel fibers, and fiber fabrics for toughening, the fracture of the fibers can consume the stress of the product cracking and can also prevent the expansion of microcracks through the bridging and other effects of the fibers, improving the adhesion, crack resistance, and impact resistance of the magnesia-based system [23]. Different modulus fibers have significant differences in the toughening effect on magnesia-based cement. It is necessary to select corresponding types and specifications of fibers according to actual needs.

As a brittle material, the toughening modification of magnesite cement can refer to ceramic toughening technology, such as increasing the fineness of MgO particles, using MgO with uniform particle size, good chemical purity, and single structure, can inhibit abnormal crystal growth, and make the crystal structure closely arranged. However, considering factors such as feasibility and cost, these methods are difficult to be applied in actual engineering [23]. Therefore MJT synthesized a toughening and solvent-reducing agent (GX-12#).

Function and working principle The toughening and solvent-reducing agent (GX-12#) can form a continuous network structure inside the magnesite products, filling in between each crystal phase. This not only increases the tightness of the combination of each crystal but also improves the softening coefficient of the product.

Usage method and precautions The dosage of the toughening and solvent-reducing agent (GX-12#) is generally 0.5%–1% of the weight of the light-burned MgO. The exact dosage can be adjusted slightly according to the on-site conditions. When using, add it to the uniformly mixed Mg-based slurry and stir evenly.

Before using the toughening and solvent-reducing agent (GX-12#), it must not be mixed with other modifiers. Otherwise, it will react with other substances and lose its effectiveness.

Storage and transportation Store in a dark environment in a sealed container. Avoid contact with strong acids and strong bases. The shelf life is 1 year.

It is an environmentally friendly, nontoxic, harmless, nonflammable, and nonexplosive material. It can be transported as a general chemical product.

Product usage effect At a test temperature of 20°C, using the same raw materials and ratio, test the effect of different dosages of the toughening and solvent-reducing agent (GX-12#) on the flexural strength of the product. The results are shown in Table 1.13.

As can be seen in Table 1.13, the flexural strength of the test block increases with the increase in the amount of the modifier. When the addition amount exceeds 1%, the enhancement effect slows down.

1.3.1.2 Special modified agent for magnesium oxysulfide

- Strengthening agent (fortifier): An external agent that can enhance the mechanical properties of magnesia-based cementitious materials.

1.3.1.2.1 Difficulties and solutions in the application of magnesium oxysulfide cement

The chloride–magnesium cement products have potential problems such as moisture absorption and reoxidation, frost reformation and alkali effusion, and poor durability. To address these issues, magnesium oxysulfide cement was invented. Magnesium oxysulfide cement has many advantages similar to those of chloride–magnesium cement, such as light weight, fire resistance,

Table 1.13 Toughening effect of toughening and solvent-reducing agent (GX-12#).

Addition amount/%	0.0	0.3	0.5	0.7	1.0	2.0
7-day flexural strength/MPa	12.4	13.5	14.4	15.2	15.8	16.2
28-day flexural strength/MPa	11	13.2	14.3	15	15.6	15.5

high-temperature resistance, heat insulation, low carbon, and environmental friendliness. In addition, magnesium oxysulfide cement also has advantages such as being less prone to corrosion of metals and reoxidation. These provide a new direction for the development of Mg-based cementitious materials [24]. The production of magnesium oxysulfide cement has low energy consumption. Not only can the sulfate–magnesium additive be recycled from industrial waste, but the calcination temperature of MgO is also lower than that of cement. MgO-based products can also permanently adsorb CO_2, which is in line with the national carbon reduction plan. Overall, magnesium oxysulfide cement is a material with excellent performance, green, and low-carbon characteristics.

The main phases of magnesium oxysulfide cement are 5·1·3 phase, 3·1·8 phase, 1·2·3 phase, and 1·1·5 phase. After research, the main stable strength phase of modified magnesium oxysulfide cement is 5·1·7 phase. The solubility of 5·1·7 phase in water is relatively low and it is a stable phase in water. This is the reason why the water resistance of modified magnesium oxysulfide cement is higher than that of chlorinated magnesium cement.

In practical use, magnesium oxysulfide cement has disadvantages such as low mechanical strength and easy cracking. The MJT technology has discovered through research that by reducing the capillary porosity of the hardened cement stone and thereby reducing the internal pores of the material, the material's compactness can be improved and the strength of magnesium oxysulfide cement can be enhanced. Based on the above principle, a series of modified magnesium oxysulfide agents have been developed. This section mainly introduces the modified magnesium oxysulfide agent (GX-15#) (Fig. 1.13) and the neutral modified magnesium oxysulfide agent (GX-15-2#) (Fig. 1.14).

1.3.1.2.2 Function and action mechanism

When the sulfur–magnesium modifier is added to the slurry, first of all, the modifier reacts with the $[Mg(H_2O)_xOH]^+$ hydration layer in the cementitious system to form an organic–magnesium complex layer

Figure 1.13 Magnesium oxysulfide agent (GX-15#).

Figure 1.14 The neutral modified magnesium oxysulfide agent (GX-15-2#).

[$CA^{n-} \rightarrow Mg(OH)(H_2O)_{x-1}$]. The positive charge on the surface of the MgO hydration layer decreases, reducing the surface energy of [Mg (H_2O) $_x$OH]$^+$ and inhibiting the rate at which it reacts with water to form Mg $(OH)_2$. At the same time, the organic–magnesium complex layer continuously adsorbs SO_4^{2-} and Mg^{2+}, and in the SO_4^{2-} solution, it generates a low-crystalline phase nucleus 5·1·7. After the new strong phase is formed, the sulfur–magnesium modifier is rereleased and continues to complex [Mg (H_2O) $_x$OH]$^+$ to form the organic–magnesium complex layer. The 5·1·7 phase exists in the form of needle-like crystals or whiskers, making the structure of the sulfur–magnesium cement specimen more dense, which further improves it and significantly enhances the water resistance of the sulfur–magnesium cement. Secondly, the sulfur–magnesium modifier can extend the initial setting time of the sulfur–magnesium cement, reduce the shrinkage rate of the sulfur–magnesium cement products, and thereby inhibit the cracking of the sulfur–magnesium cement in the air.

1.3.1.2.3 Usage method and precautions

The addition amount of the sulfur–magnesium modified agent is 0.5%–1% of the light-burned MgO by weight. The specific dosage needs to be adjusted according to the activity of the light-burned MgO, production temperature, product requirements, etc. When using, the modified magnesium oxysulfide agent (GX-15# or GX-15-2#) should be added to the magnesium sulfate solution and dispersed evenly.

The modified magnesium oxysulfide agent (GX-15#) and the neutral modified magnesium oxysulfide agent (GX-15-2#) significantly improve the strength, softening coefficient, and crack resistance of the products. However, the effects of the two modified agents on the reaction rate are different. The modified magnesium oxysulfide agent (GX-15#) has a more obvious retardation effect and is suitable for use in summer when the temperature is high; the neutral modified magnesium oxysulfide agent (GX-15-2#) is suitable for use when the temperature is lower.

1.3.1.2.4 Storage and transportation

The sulfur–magnesium modified agent is suitable for storage in a cool, dark environment at a temperature range of 10°C–30°C. The shelf life is 1 year. The magnesium thiosulfate modifier is not a hazardous product and can be transported as a general item. In case of contact with the skin or splashing into the eyes, thorough rinsing with large amounts of water should be carried out immediately; if ingested, medical examination should be sought immediately.

1.3.1.2.5 Product usage effect

The setting time and strength of the standard mortar prepared in the laboratory and the standard test blocks were measured. The specific test results are shown in Table 1.14.

Table 1.14 Effect of magnesium sulfate modifier (GX-15#).

Addition amount/%	Initial setting time/min	3 days compressive strength/MPa	28 days compressive strength/MPa
0.0	135	*	*
0.3	195	17.6	47.36
0.5	230	20.7	49.42
0.8	290	33.2	52.38
1.0	325	41.75	56.26
1.5	480	35.7	56.25

*Indicates cracking.

- Conclusion: As shown in Table 1.14, the retardation effect of the modified magnesium oxysulfide agent becomes more pronounced with the increase in its dosage. The specimens without the modifier will develop numerous cracks around 3 days, significantly affecting the strength of the specimens; as the dosage of the modifier increases, the cracks in the specimens decrease and the compressive strength increases; when the dosage of the modified magnesium oxysulfide agent exceeds 1%, the improvement in performance is limited.

1.3.1.3 Special modified agent for sulfur–chlorine composite materials

1.3.1.3.1 Causes and solutions

The tetra-component cement system (sulfur–chlorine composite magnesium cement) has both the advantages and disadvantages of the three-component systems. Chloride–magnesium cement has the advantages of light weight, high strength, heat insulation, and noncombustibility, but it also has the disadvantages of poor water resistance and corrosion of reinforcing bars. Sulfur–oxygen magnesium cement has good water resistance, but its mechanical properties are low and it is prone to cracking. To further modify Mg-based cementitious materials, a Mg-based cementitious material composite system was adopted. Regarding the problems of water resistance and cracking in sulfur–chlorine composite magnesium cement, the MJT technology department, from the perspective of the hydration mechanism of sulfur–chlorine composite magnesium cement, developed sulfur–chlorine composite modifiers to improve the mechanical properties and water resistance of sulfur–chlorine composite magnesium cement.

1.3.1.3.2 Principles and functions

After the sulfur–chlorine composite modifier is added to the slurry, first of all, the modifier forms an organic magnesium complex layer by complexing with the water–magnesium ions in the cementitious system, reducing the encapsulation effect of water–magnesium ions on MgO. The organic magnesium complex layer will directly react with Cl^-, SO_4^{2-}, OH^-, and Mg^{2+} in the solution to form phases 5·1·8 and 5·1·7, increasing the generation rate of strength phases. Phases 5·1·8 and 5·1·7 have a competitive relationship and simultaneously hinder the generation of $Mg(OH)_2$, reducing the volume expansion rate of the specimen, adjusting the construction of crystal phases, and reducing the capillary porosity of the hardened cement stone, thereby making the structure of the sulfur–-chlorine composite magnesium cement specimen more dense and subsequently improving the strength, thereby significantly enhancing the water resistance of sulfur–chlorine magnesium cement. Moreover, the components in the sulfur–chlorine composite modifier react with the ions of this four-component system to form substances that block the capillary channels, improving the water resistance of the product.

1.3.1.3.3 Usage method

The dosage of the sulfur–chlorine composite modifier is 0.5%–2% of the weight of the light-burned MgO. The specific dosage needs to be adjusted according to the activity of the light-burned MgO, the production temperature, and the requirements of the product. When using, it should be directly added to the mixed slurry.

1.3.1.3.4 Storage, transportation and precautions

The sulfur–chlorine composite modifier is suitable for sealed storage in a cool, dark environment at a temperature range of 10°C–30°C. The shelf life is 1 year.

The sulfur–chlorine composite modifier is not a hazardous product and can be transported as a general item. If it comes into contact with the skin or splashes into the eyes, it is necessary to rinse with a large amount of water immediately; if accidentally ingested, medical examination should be sought immediately.

1.3.1.3.5 Usage effect

The softening coefficient and strength of the standard test block were measured by preparing standard paste in the laboratory. After the test block was immersed in water for 28 days with 1% of the sulfur–chlorine composite modifier added, the softening coefficient was 0.94. After tracking, it was found that it has long-term water resistance. And as the dosage of the modifier increases, the compressive strength increases but the improvement in effect is limited when the dosage exceeds 2%.

1.3.1.4 Special retarder for magnesium phosphate cement

1.3.1.4.1 Causes and solutions

Magnesium phosphate cement is a fast-setting cement with high early strength, possessing excellent mechanical properties, durability, fire resistance, wear resistance, cementability, and low permeability. It is commonly used in rapid repair and biological restoration fields. The hydration reaction of magnesium phosphate cement is intense, and the setting speed is too fast, making engineering operation difficult. Therefore MJT, from the perspective of the hydration mechanism of magnesium phosphate cement, conducted research on the retarder for magnesium phosphate cement and the retardation mechanism and developed a magnesium phosphate retardant, which slows down the magnesium hydration reaction while improving the mechanical properties of magnesium phosphate cement.

1.3.1.4.2 Principle and function

The magnesium phosphate retarder can combine with MgO to form amorphous or low-crystallinity mineral precipitates, reducing the consumption of magnesium in the solution and compensating for the corresponding negative charges of Mg^{2+} and K^{+} in the solution to maintain charge balance. The presence of amorphous minerals during the hydration process stabilizes the pH value of the system and slows down the crystallization and growth of the hydration products. At the same time, the hydroxyl groups of the magnesium phosphate retarder form hydrogen bonds with the hydroxyl groups on the surface of alkaline oxides, resulting in strong adsorption. The formed polymer adsorbs on the surface of the hydrated products, hindering the dissolution of MgO and the crystallization of the hydration products [25]. The presence of polymer on the surface of the hydrated crystals is conducive to strengthening the intercrystalline bonding force, reducing the average size of the hydrated crystals, and improving the performance of magnesium phosphate cement.

1.3.1.4.3 Usage and precautions

The dosage is generally 1.5%–8% of the weight of the calcined magnesia. It should be directly added to the mixed slurry during use.

The magnesium phosphate retarder is an acidic liquid and cannot be mixed with alkaline substances.

1.3.1.4.4 Storage and transportation

1. Store in a cool environment below 25°C, seal, and cover. The shelf life of the magnesium phosphate retarder is 1 year.
2. It is an environmentally friendly, nontoxic, harmless, nonflammable, and nonexplosive material. It can be transported as a general chemical product.

1.3.1.4.5 Product performance

The setting time of magnesium phosphate increases with the increase of the dosage of the magnesium phosphate retarder: from 15.25 minutes of the zero-dosage group to 1 hour of the 6% dosage group. The results of hydration heat detection have the same trend as the change in setting time. The time when the first exothermic peak is reached in the hydration heat release rate curve gradually delays with the increase of the dosage of the magnesium phosphate retarder, and the time of the second exothermic peak also delays with the increase of the dosage of the magnesium phosphate retarder. The compressive strength of the magnesium phosphate cement at each age shows a trend of first increasing and then decreasing. The compressive strength at 28 days reaches the maximum of 65.1 MPa when the dosage is 6%, which is higher than the compressive strength of 60.2 MPa of the blank group at 28 days. Combining with the microscopic morphology of magnesium phosphate, it can be known that the magnesium phosphate retarder makes magnesium phosphate generate more amorphous noncrystalline products in the early stage and then transform into bird droppings stone crystals in the later stage, increasing the hydration products in the system and the compressive strength.

1.3.1.5 Crystal seed-type modifier

According to the classical nucleation theory, during the solidification process of Mg-based materials, the hydrate phase needs to form large clusters first and then generate crystal nuclei, and finally, the particles in the system arrange in an ordered manner based on the crystal nuclei and gradually solidify and form a structure. Therefore the generation speed of crystal nuclei in the system restricts the solidification speed of the Mg-based system. The faster the crystal nuclei are generated, the faster the reaction rate of the Mg-based slurry. To increase the generation speed of crystal nuclei in the system, one can artificially add crystal-like substances to the system, thereby eliminating the spontaneous generation process of crystal nuclei in the Mg-based system. These artificially added crystal-like substances are usually called crystal seed–type modifiers.

To increase the solidification speed of the Mg-based system, corresponding types of crystal seed–type modifiers can be selected based on the different Mg-based systems, such as chloromagnesium oxide crystal seed–type modifier, sulfur–magnesium oxide crystal seed–type modifier, and sulfur–chlorine composite crystal seed–type modifier. For example, adding a crystal seed-type modifier to the chloromagnesium oxide system can accelerate the solidification speed of the slurry and promote the formation of five phases. The addition of chloromagnesium oxide crystal seed–type modifier enables the chloromagnesium oxide slurry to grow rapidly with the crystal seed as the core, which not only speeds up the solidification speed of the slurry but also promotes the formation of ordered crystal phases based on the crystal seed, increasing the proportion of the five-phase structure in the slurry and improving the strength of the Mg-based product.

Furthermore, by controlling the production conditions, products with a high aspect ratio of length to diameter can be produced, which can be used as substitutes for certain fibers. For example, adding a certain amount of sulfur–magnesium crystal seeds to the chlorine–magnesium handicrafts can enhance the water resistance of the chlorine–magnesium handicrafts, thereby preventing the cracking of the handicrafts.

1.3.1.5.1 Function of crystal seed modifier

Acceleration effect According to the test method of "Standard Clinker Water Content, Setting Time, and Soundness Test Method" GB/T 1346—2011, the setting time of the Mg-based paste is measured. The blank group of paste is prepared by mixing light-burning MgO with brine in a ratio of 1:1, and the experimental group adds an additional amount of synthetic crystal seeds. The active content of light-burning MgO in the paste is 60%, and the brix of the brine is 27. The specific ratio is shown in Table 1.15.

After adding synthetic seed crystals to the chloromagnesium paste, the setting speed of the paste significantly increased. In subsequent

Table 1.15 Experimental ratio and setting time of crystal seed–type modifier for accelerated setting.

Group	Light-burned magnesium oxide/g	Brine/g	Admixture/g	Initial setting time/min	Final setting time/min
Blank group	300	300	0	230	350
Experimental group 1			3	210	340
Experimental group 2			9	180	330
Experimental group 3			15	160	305

experiments, it was found that the effect of the seed crystals on the setting of the chloromagnesium paste had a certain limit and the specific amount of addition needed to be determined based on the product and the process.

Enhancement effect According to the China Magnesite Association Group Standard 7—2019 standard, chloromagnesium test blocks were made. The raw materials of the blank group were commercially available 60% active light-burned MgO and 27°Be brine. The experimental group was composed of the raw materials of the blank group with a certain amount of synthetic seed crystals added, as shown in Table 1.16.

The test blocks were respectively cured under standard conditions for 1, 7, and 28 days. According to the T/CMMA-2019 standard, the strength of the blocks at each age was tested and the results are shown in Table 1.17.

It can be clearly seen in Table 1.17 that the crystal seed–type modifier has a significant reinforcing effect on the products at 1 and 7 days and has no negative impact at 28 days.

Toughening effect According to the T/CMMA 7—2019 standard, chloromagnesium oxide test blocks were made. The raw materials of the blank group were commercially available active 60% light-burned MgO and 27°Be brine. The experimental group was composed of the raw materials of the blank group with a certain amount of sulfur–magnesium crystal seed–type modifier added, as shown in Table 1.18.

The test blocks were respectively cured for 1, 3, and 7 days under standard conditions, and the strength of the blocks at each age was tested according to the T/CMMA 7—2019 standard. The results are shown in Table 1.19.

It can be clearly seen in Table 1.19 that the seed-type modifier has a significant toughening effect on the products at 1, 3, and 7 days.

1.3.1.5.2 Usage method and precautions

The seed-type modifier is in suspension form and should be stored in a sealed container. When using, add it to the slurry at a concentration of 1%–5% based on the mass of light-burned MgO.

1.3.2 General modified agent

1.3.2.1 Waterproofing agents (GX-13#, waterproof emulsion, waterproof powder)

- Waterproofing agent: Also known as antiseepage agent. Through physical and chemical reactions, it reduces the formation of internal

Table 1.16 Ingredient schemes for the blank group and the experimental group.

Group	Light-burned magnesium oxide/kg	Brine/kg	Admixtures/g
Blank group	5	4.2	0
Experimental group 1			50
Experimental group 2			150
Experimental group 3			250

Table 1.17 Compressive strength by group and age group.

Group	Compressive strength/MPa		
	1 day	7 days	28 days
Blank group	24.9	55.2	90.2
Experimental group 1	27.6	66.2	91.4
Experimental group 2	29.5	70.5	92.6
Experimental group 3	28.3	67.3	91.9

Table 1.18 Ingredient schemes for the blank group and the experimental group.

Group	Light-burned magnesium oxide/kg	Brine/kg	Admixtures/g
Blank group	5	4.2	0
Experimental group 1			50
Experimental group 2			150
Experimental group 3			250

Table 1.19 Compressive strength by group and age group.

Group	Flexural strength/MPa		
	1 day	3 days	7 days
Blank group	9.66	10.10	10.31
Experimental group 1	11.81	12.42	12.60
Experimental group 2	13.08	13.36	13.55
Experimental group 3	13.76	14.15	14.23

pores, blocks, and cuts off capillary channels, thereby enhancing the seepage resistance of the external agent.

Waterproofing agents are classified by their components into inorganic waterproofing agents, organic waterproofing agents, and composite waterproofing agents. Inorganic waterproofing agents mainly include chlorinated salt waterproofing agents, ferric chloride waterproofing agents, sodium silicate waterproofing agents, and inorganic aluminum salt waterproofing agents. Organic waterproofing agents mainly include organic silicone waterproofing agents, metal soap waterproofing agents, emulsion waterproofing agents, and composite waterproofing agents [26].

Magnesium cement is a porous material. The pores and cracks affect the strength of the material, and moisture in the air will seep into the hardened body and thereby affect the service life of the product [27]. To solve this problem, waterproofing agents are usually sprayed on the surface of the product or mixed internally.

Through extensive experimental research, it has been found that 2.0% sodium silicate can improve the water resistance of MgO cement; the fillers also affect the water resistance of MgO cement handicrafts. Especially, the MgO cement handicrafts with 15% active fillers and 1.5% admixtures have the best water resistance, with a softening coefficient of 0.89 [28]. Different waterproofing agents can also enhance the water resistance of MgO cement and improve its mechanical properties. Therefore through continuous efforts, MJT has developed an ideal waterproofing agent (GX-13#), waterproof emulsion, and waterproof powder.

1.3.2.1.1 Principle and function

The waterproofing agent (GX-13#) (Fig. 1.15) is mainly applied to waterproof Mg-based cementitious materials. Due to its hydrophilic and hydrophobic effects, it forms a hydrophobic layer around the pores of

Figure 1.15 Waterproofing agent (GX-13#).

magnesium cement, simultaneously improving the crystal structure of magnesium cement. Through chemical modification, the main strength crystals of magnesium cement, namely, 5.1.7 or 5.1.8 crystals, become finer and the direct combination of crystals becomes more compact. This reduces the porosity and decreases the internal infiltration and erosion damage of water, thereby increasing the water resistance of Mg-based cementitious material products.

The waterproof emulsion (Fig. 1.16) is a waterproof material developed specifically for inorganic cementitious materials, with the emulsified polymer being the main component. It has excellent solubility in inorganic cementitious materials and excellent compatibility. The silanol groups in the molecular structure of the waterproof emulsion react with the free hydroxyl groups in the cementitious materials through dehydration, forming an excellent hydrophobic layer. At the same time, it generates reverse capillary pressure in the inorganic cementitious material system, increasing the compactness of the cementitious material and endowing it with excellent penetration crystallization properties and outstanding resistance to water and alkali.

The waterproof powder (Fig. 1.17) reacts with the calcium and magnesium ions in the light-burned MgO to form precipitates and gels, reducing or blocking the pores. The tiny particles with hydrophobic properties aggregate together, and the tiny pores between them can generate reverse capillary pressure, thereby balance the external water

Figure 1.16 Waterproof emulsion.

Figure 1.17 Waterproof powder.

pressure, and achieve the waterproof effect. The waterproof powder can effectively reduce the water absorption rate of the product, improve the antifreezing effect, and extend the service life of the product.

1.3.2.1.2 Usage and precautions

The waterproofing agent (GX-13#) should be added to the prepared slurry and stirred evenly. The dosage is generally 0.5%–1% of the light-burned MgO by weight.

The waterproofing emulsion should also be added to the prepared slurry and stirred evenly. The dosage is generally 0.5%–1% of the light-burned MgO by weight.

The waterproof powder and dry powder materials can be compounded with the slurry or directly added to the slurry, with a dosage of 0.5%–1% of the light-burned MgO by weight.

1.3.2.1.3 Storage and transportation

The waterproof agent should be stored in a cool environment below 25°C, sealed, and covered, with a shelf life of 1 year. The powder should be stored directly in an airtight container. Both belong to environmentally friendly, nontoxic, harmless, nonflammable, and nonexplosive materials. They can be transported as general chemicals.

1.3.2.1.4 Product performance

Adding waterproofing agent (GX-13#), waterproof emulsion, and waterproof powder can, to a certain extent, enhance the mechanical properties and water resistance of the magnesia cement samples. Among them, the compressive strength of the sample with 1% waterproofing agent (GX-13#) is 7.58% higher than that of the sample without adding the waterproofing agent. The reason is that the internal crystals of the hardened body change after adding the waterproofing agent (GX-13#), thereby increasing the compressive strength of the sample. Adding waterproof emulsion and waterproof powder can significantly increase the contact angle of the surface, reduce the water absorption rate of the product, and thereby increase the softening coefficient of the product. These two types of waterproofing agents have excellent water repellency. They can form a film in the internal pores of the hardened body, reducing the water absorption rate, reducing internal seepage and water erosion, and thereby improving the waterproof performance of the product. The softening coefficient of the prepared calcium magnesium aluminate cement specimens after soaking in water for 28 days is 0.94, and the compressive strength is higher than 85 MPa. This broadens the application field of calcium magnesium aluminate cement.

1.3.2.2 Foaming agent

Foam concrete is an improved form of ordinary concrete. Traditional concrete has a relatively heavy weight. By adding a foaming agent to the concrete to create foamed bubbles, a lightweight, high-strength, and highly insulating foam concrete can be formed.

A foaming agent is a substance that can enable an object to acquire uniformly dispersed bubbles, thereby achieving lightness. In terms of the method of obtaining bubbles, foaming agents can be divided into two major categories: chemical foaming agents and physical foaming agents.

- Chemical foaming agent: In theory, any substance that can undergo a chemical reaction to release gas can be used as a chemical foaming agent. However, in practice, the selection of chemical foaming agents is constrained by various factors such as reaction conditions, stability during transportation and storage, and gas production rate. Currently, the commonly used chemical foaming agents include hydrogen peroxide, aluminum powder, calcium carbide, and ammonium salts, among which hydrogen peroxide is the most commonly used, followed by aluminum powder [29,30].

  Physical foaming agent refers to an additive that can reduce the surface tension of a liquid, generate a large amount of uniform and stable foam, and be used to produce foamed cement.
- Chemical foaming: In recent years, as the national requirements for the fire resistance performance of building materials have gradually increased, traditional organic insulation materials have failed to meet the fire resistance requirements. Therefore in China, chemical foaming to prepare foam concrete has gradually emerged, and many research scholars have successfully prepared low-density foam concrete [31]. Silicate cementitious materials use hydrogen peroxide as a chemical foaming agent and produce ultra-light foam concrete with a dry volume density lower than 250 kg/m^3. With a water–cement ratio of 0.44–0.52 and adding 6%–7% hydrogen peroxide, the foamed concrete has excellent macroscopic properties and pore structure.

Hydrogen peroxide has a high decomposition rate in the alkaline environment of cement, so using hydrogen peroxide as the foaming agent for cement is feasible [32]. And as the water temperature increases, the foaming speed accelerates, while the foaming time will be prolonged with the increase of the foaming agent dosage. However, as the dosage of hydrogen peroxide increases, the dry apparent density, compressive strength, and thermal conductivity of the foam concrete all show a downward trend, while the porosity gradually increases, and the water absorption rate shows a trend of increasing first and then decreasing. Therefore many scholars have studied stabilizing foaming agents and improving the pores of foamed cement [32]. On this basis, ultra-light foamed cement insulation boards with a dry density of 100–180 kg/m^3 were prepared using the chemical foaming method with hydrogen peroxide.

In addition, using both aluminum powder and hydrogen peroxide as foaming agents can also result in products with a higher porosity. However, the porosity of the aluminum powder foaming products is higher than that of the hydrogen peroxide foaming products. Initially, by adding lime to the cement and using aluminum powder for foaming, a foam concrete with a compressive strength of 11.61 MPa could be prepared and its thermal conductivity only increased by 6% [33]. Later, by adding water glass and NaOH activator, using fly ash as the cementitious material, and using 30% hydrogen peroxide as the foaming agent, a foam concrete with a dry apparent density of 400–600 kg/m^3 was prepared. The compressive strength of this concrete could reach up to 3.4 MPa, and the thermal conductivity was as low as 0.0826 W/(m ·K) [34].

After research, it was found that the mechanical properties and thermal insulation performance of chloromagnesium foam concrete are superior to those of silicate foam materials. The water–cement ratio has a significant impact on the performance of foam concrete. It was discovered that when the water–cement ratio is less than 0.81, the pore size is smaller, the shape is irregular, and the connectivity is strong; when the water–cement ratio is greater than 0.8, the pores are circular or expanded, and the pore diameter distribution range expands. However, as the water–cement ratio increases, the number of fine pores decreases while the number of large pores increases, thereby causing the porosity of foam concrete to increase and the strength to decrease. In addition, foam concrete has excellent thermal insulation performance. When the dry density is 1000–1200 kg/m^3, the thermal conductivity is 0.23–0.42 W/(m·K). However, this thermal conductivity still cannot meet certain application requirements. To solve this problem, the chemical foaming method using hydrogen peroxide as the foaming agent was adopted to prepare foam concrete and its thermal properties and other properties were evaluated. With the increase of hydrogen peroxide, the number of microbubbles generated by rapid foaming and the microclosed bubbles in the cement slurry increases, thereby causing the volume of foam concrete to increase and the dry apparent density to decrease.

Changing the mole ratio of chloromagnesium cement foam concrete and adding different doses of hydrogen peroxide, fly ash, and antihalogen agents can significantly alter its internal pore structure, thereby affecting its mechanical properties, water resistance, and thermal insulation performance. The mole ratio mainly affects the proportion of the 5·1·8 strength phase, thereby influencing the strength performance of chloromagnesium cement foam concrete; adding hydrogen peroxide is beneficial for improving the thermal insulation performance of chloromagnesium cement foam concrete, but it will affect the mechanical properties and water resistance. Specifically, hydrogen peroxide acts as a foaming agent. As the dosage of the foaming agent increases, more tiny bubbles are generated during the rapid foaming process, and then more closed tiny bubbles are produced in the cement slurry, thereby increasing the volume of chloromagnesium cement foam concrete and reducing the dry density. The lower the dry density of chloromagnesium cement foam concrete, the lower the solid proportion, the smaller the proportion of heat transfer through conduction, and the lower the heat transfer efficiency. Therefore the thermal conductivity will decrease, and the thermal insulation performance will improve. However, the increase of hydrogen peroxide will cause changes in the internal pore structure of chloromagnesium cement foam concrete, with thinner pore walls and a looser pore structure, increasing the contact area between the material and water, leading to an increase in water absorption and accelerating the decomposition rate of hydration crystal products in contact with water. Thus the strength will decrease, and the water resistance will deteriorate; adding fly ash helps improve the thermal insulation performance of chloromagnesium cement foam concrete, but excessive fly ash will affect its mechanical properties. Adding antihalogen agents mainly improves the water resistance of chloromagnesium cement foam concrete [35].

By combining SEM analysis with the study of the structural characteristics of different types of chloromagnesium cement foam concrete, it was found that the vertical part has the characteristic of bubble stratification. This is because the hydrogen peroxide foaming agent mixed into the slurry undergoes a chemical reaction and releases gas, generating gas pressure in this area. The gas pressure causes shear stress, and when the shear stress is greater than the limit shear stress of the slurry, the bubble volume increases and different-sized bubbles form in the slurry. The additional pressure between different-sized bubbles is different, and their internal gas pressures are also different. When they meet, they will exhaust gas from the small bubbles to the large bubbles. When the small bubbles merge and form a large bubble, the large bubble has greater buoyancy and the flowing cement slurry continuously sinks along the bubble wall under the action of gravity. At this time, the large bubble continues to move upwards. In addition, due to the difficulty in synchronizing the release speed of the foaming agent with the strength development of the gas pores, the chemical foaming method results in MOC concrete with uneven bubble sizes in the upper and lower layers. However, in the physical foaming method, the gas bubbles themselves as a component of the slurry have a relatively relaxed requirement for the slurry setting time and it is easier to precisely control the integrity of the foam concrete. Although in the physical foaming method, due to the mechanical stirring force causing small bubbles to combine and form a small number of large bubbles, mechanical stirring also causes gas pores to be mixed into any position inside the foam concrete. Therefore the physical foaming method can obtain chloromagnesium cement foam concrete with uniform gas pores more easily [36]. However, there are still a few Mg-based chemical foaming products on the market. For example, the artificial footstone developed by MJT is a Mg-based cementitious material chemically foamed and has a density of about 600 kg/m^3. The reference ratio of large and small pore diameters is shown in Tables 1.20 and 1.21.

- Physical foaming: In the early 1950s, China developed two foaming agents, rosin soap and rosin thermopolymer, and applied them to mortar and foamed cement. These were the first generation of foaming agents in China. After 1980s, with the rise of China's surfactant industry, synthetic surfactant–based foaming agents became the main type in the foaming agent series, which was the second generation of foaming agents in China. At the end of the 20th century, high-performance protein-based foaming agents from developed countries such as Japan and South Korea entered the Chinese market. Due to their high stability, they have been widely applied. Animal protein foaming agents and plant protein foaming agents are the third generation of foaming agents in China. Comparison of the three generations of foaming agents is as follows: Rosin-based foaming agents have relatively low foaming power and stable foaming properties. Synthetic foaming agents have good foaming power but poor stable foaming properties. Protein-based foaming agents have good stability but low foaming power [37]. Therefore the fourth generation of composite foaming agents emerged. There are four research methods for composites: (1) complementary method, when multiple components are combined, each has its own advantages, which can achieve

Table 1.20 Large aperture reference ratio (unit: kg).

Light calcined magnesia	Brine (25°-Be)	Talcum powder	Slow-setting antihalogen agent (GX-1#)	50% hydrogen peroxide
100	120	25	0.5 ~ 1	5

Table 1.21 Small-diameter reference ratio (unit: kg).

Light calcined magnesia	Brine (27°Be)	Slow-setting antihalogen agent (GX-1#)	50% hydrogen peroxide
100	85	0.5 ~ 1	5

complementary advantages and remedy the shortcomings of a single component to improve the incomplete performance; (2) synergistic method, although multiple components cannot complement each other, they can collaborate. Originally, the performance of two or three components was not good in a certain aspect, but when multiple components are combined, a "1 + 1 + 1 > 3" synergistic effect can be generated and the effect is greatly enhanced; (3) enhancing method, when the performance of a single-component foaming agent is not good in a certain aspect, an enhancing component can be used to strengthen it, making it improve from inferior to superior; (4) adding the function method, when the function of some foaming agents is insufficient and does not meet our demand for a certain function, an external additive can be added to increase the function in that aspect [38]. In this regard, MJT has developed the following several composite foaming agents.

1.3.2.2.1 High-efficiency foaming agent (GX-7#)

Introduction MJT has developed a high-efficiency foaming agent (GX-7#) through adjustment and research, which has the advantages of high foaming ratio, good foam stability, and low water discharge (Fig. 1.18). It can be used in the production of foamed magnesite cement, with low cost and being environmentally friendly. During production, due to the stable foaming system, raw material consumption can be saved, and

Figure 1.18 High-efficiency foaming agent (GX-7#).

the board density can be reduced. At the same time, the foam moisture content is low, which helps shorten the demolding and curing time of foamed fireproof boards and other products.

Usage method and precautions

1. Accurately weigh the high-efficiency foaming agent (GX-7#), and then add 80–160 times the amount of water for dilution, to prepare the foaming liquid. When the temperature is low, appropriately reduce the dilution ratio.
2. Determine the dosage of the foaming agent for each cubic product based on the requirements of the product. The dosage of the foaming agent increases or decreases according to the bulk density of the product. The higher the designed bulk density of the product, the less the dosage of the foaming agent; the lower the designed bulk density of the product, the more the dosage of the foaming agent.
3. Use the foaming machine to foam the foaming liquid through it. Add the foam to the mixer according to the set dosage and mix evenly.

When using the high-efficiency foaming agent (GX-7#) to produce foaming products, it is necessary to pay attention to the control of factors such as raw material quality, dynamic ratio, and production process.

The high-efficiency foaming agent (GX-7#) is an alkaline liquid and cannot be mixed with acidic substances.

Technical parameters

1. High foaming multiple: 15–27 times

2. High dilution multiple: 80 to 160 times
3. Low foam bulk density: 50–60 g/L
4. Good foam stability: 1-hour sedimentation distance ≤ 5 mm
5. Low leakage volume: 1 hour ≤ 40 mL
6. Low consumption: For every 1 m^3 of fireproof door core board, the high-efficiency foaming agent (GX-7#) consumption is approximately 0.5 kg (bulk density is 300 kg/m^3)

Storage and transportation

1. Store in a cool environment at 25°C and seal, with a shelf life of 1 year.
2. The product is a green and environmentally friendly, nontoxic, harmless, nonflammable, and nonexplosive material. It does not hurt hands and can be transported as a general chemical product.

1.3.2.2.2 Two-component foaming agent

Introduction The two-component foaming agent consists of a foaming agent and stabilizing agent (Fig. 1.19). According to the requirements, it is diluted by water in a ratio of foaming agent:stabilizing agent = 9:1 or 8:2 for use.

The foaming agent is made from imported high-quality natural plant polymer materials and is produced through a series of complex chemical reactions. The product is a dark-red–brown liquid. This foaming agent is not sensitive to hard water, is nontoxic and odorless, and has no precipitates. It has no corrosive effect on cement and metals and does not cause pollution to the environment. The product can be stored for a long time and is not prone to deterioration. If there is some sedimentation, the sediment is the effective component. Stirring evenly before use is sufficient.

The stabilizing agent can control the structural stability of the bubble liquid film, allowing the surface-active agent molecules to be orderly distributed in the bubble liquid film, giving the foam good elasticity and self-repairing ability, and is an active substance that prolongs and stabilizes the long-term performance of the foam.

Usage method and precautions

1. Accurately weigh the foaming agent and stabilizing agent, and then add 30–50 times of clean water for dilution to prepare the foaming liquid. When the temperature is low, appropriately reduce the dilution ratio.
2. Determine the dosage of each cubic of foaming agent according to the product requirements. The dosage changes with the increase or

Figure 1.19 Two-component foaming agent. (A) Foaming agent. (B) Foaming stabilizer.

decrease of density. When the density increases, reduce the dosage; when the density decreases, increase the dosage;

3. Use the foaming machine to turn the foaming liquid into foam through the foaming machine. Add the foam to the mixer in the required amount and mix evenly.

When using a two-component foaming agent to produce foaming products, it is necessary to pay attention to the control of raw material quality, dynamic ratio, production process, etc.

Main features

1. Adopting an advanced closed-cell foam process, the foaming performance is excellent and the stabilizing performance is very good. The half-dissipation time of the foam is more than 120 minutes, the foam is fine with a diameter of 0.1–1 mm, and the size is uniform.
2. It can be used as the foaming material for magnesite cement and also for cement concrete.
3. The dilution ratio is high, and it can be diluted by 30–50 times, reducing the production cost.
4. It has exceptionally strong water repellency; high closed-cell ratio, resulting in a lower water absorption rate; and better insulation and heat preservation performance.

5. It has characteristics of green and environmentally friendly, nontoxic and noncorrosive, and safe for use.
6. Each cubic unit adds 0.3–0.7 kg of foaming agent, which varies with density. When the density increases, it decreases; when the density decreases, it increases.

Storage and transportation

1. Store in a cool environment at 25°C and seal, with a shelf life of 1 year.
2. The product is a green and environmentally friendly, nontoxic, harmless, nonflammable, and nonexplosive material, does not hurt hands, and can be transported as a general chemical product.

Product usage effect In some regions, the magnesium sulfate purchased by door core board manufacturers is mostly agricultural grade. During the reaction process, ammonia gas release affects the stability of the foam, causing collapse; in addition, some manufacturers use soft pipe pumps to pump the foaming slurry, resulting in an increase of more than 15% in the breakage rate. Ordinary single-component foaming agents cannot meet such strict usage requirements. However, the two-component foaming agent successfully solves the problem of pumpability stability by adjusting the ratio of A and B components and the dilution ratio. Currently, this foaming agent has been successfully used in sulfur–magnesium mold-forming fireproof door cores, mold-forming poured partition boards, etc., for foaming products with high stability requirements.

1.3.2.2.3 Protein foaming agent

Introduction The protein-based foaming agent (Fig. 1.20) has inferior foaming performance compared to anionic surfactants, but it has the advantages of stable foam and excellent properties, enabling it to have a place in the market. Protein-based foaming agents are mainly divided into plant protein foaming agents and animal protein foaming agents. Plant protein foaming agents mainly include the tea saponin type and glycyrrhizin type; animal protein foaming agents use selected animal (cow, sheep) keratin proteins as the main raw materials, which are subjected to a series of hydrolysis reactions, heated dissolution, dilution, filtration, and high-temperature shrinkage to form. It is produced using imported advanced equipment and the latest manufacturing techniques, and the production process implements strict quality management. The product is a dark-brown viscous liquid with low impurity content, a mild

Figure 1.20 Protein foaming agent.

odor, uniform quality, and good consistency in quality and has excellent foaming performance and excellent foam stability [39].

Usage method and precautions

1. Accurately weigh the animal protein foaming agent, and then add 30–40 times the amount of water for dilution to prepare the foaming solution. When the temperature is low, the dilution ratio should be appropriately reduced.
2. The dosage of the foaming agent for the combined product should be determined based on the volume per cubic meter. It varies with the increase or decrease of the density. When the density rises, the dosage should be reduced; when the density drops, the dosage should be increased.
3. Use the foaming machine to turn the foaming liquid into foam through the machine. Then, add the foam to the mixer in the prescribed amount and mix them evenly.

When producing foaming products using protein foaming agents, it is essential to pay attention to the control of factors such as the quality of raw materials, dynamic ratio, and production process.

Performance advantages

1. Due to the high surface strength and stability of the foam produced by this foaming agent, the foam products made with it have bubbles that are independently closed and not interconnected, resulting in good antipermeability.

2. Compared with foam products made with other foaming agents at the same density, its sealing and insulation properties are better and the strength is higher.
3. The excellent foam stability enables the foaming cement slurry to be poured to a height of over 1.5 m at one time without collapsing, making it suitable for the production of foam bricks, lightweight panels, lightweight walls, and magnesite products.

Storage, transportation and precautions

1. Store in a cool and ventilated warehouse at a temperature of 25°C. If exposed to direct sunlight for a long time or at excessively high or low temperatures, the foaming effect will be affected.
2. This product is a green and environmentally friendly material. It is nontoxic, harmless, nonflammable, and nonexplosive and does not cause hand injuries. It can be transported as a general chemical product.

Application scope This foaming agent is mainly used in the production of edgeless Mg-based fireproof door cores, cement fireproof door core boards, cement foamed partition boards, foamed insulation boards, and exterior wall self-insulation blocks. It can also be used for foaming cement backfilling.

1.3.2.2.4 Rosin resin–based foaming agent

The rosin resin–based foaming agent is one of the earliest types of foaming agents. Due to the continuous development of foaming agents, it has gradually become a lower-grade foaming agent in the market. It is mainly made from rosin containing aromatic hydrocarbons, aromatic alcohols, aromatic aldehydes, etc. [40] The rosin resin foaming agent was initially developed and applied as an air-entraining agent for concrete mortar and later expanded to be used as a foaming agent for foam concrete. It is the earliest type of traditional foaming agent and also the most common low-grade foaming agent in the market. The production cost of rosin thermopolymer is relatively high. As the raw materials involve phenol, there are some safety issues, so its application is relatively rare, while the production process of rosin soap is simple and the cost is low and the foam stability is better. This is because the carboxyl functional group in rosin soap reacts with Ca^{2+} in cement to form an insoluble calcium salt, thereby improving the stability of the foam. Through experiments, it is known that the amount of rosin and alkali plays a decisive role in the saponification reaction and the effect is best when the reaction is carried out under the liquid boiling state [40]. The foaming agent obtained has excellent performance. When the purity of alkali is above 80% and does not contain other strong electrolytes, the purity has little effect on the foam performance. In addition, adding a small amount of dodecyl sulfate sodium to the foaming liquid will greatly improve the foam stability. Adding gelatin and dodecyl benzene sulfonate can effectively increase the foaming volume and foam stability time [41]. Comparing the rosin resin–based foaming agent with the hair protein foaming agent, it is found that the protein-based one performs better in actual application to foam concrete [42]. The concentration of alkali also affects the reaction degree. The optimal alkali concentration is 20%–25%; analyzing the influence of nonionic surfactants on rosin soap foaming agents, it is found that the addition of nonionic surfactants can produce fine bubbles, the water solubility of the solution is well solved, and the precipitates in the solution can be uniformly dispersed during the dissolution process, which is conducive to foam stability [43].

1.3.2.3 Hemostatic agent (hemostatic quick-drying powder) (Fig. 1.21)

The MgO promoting setting and quick-drying powder can all promote crystal development, increase reaction speed, and shorten demolding time. The MgO promoting setting and quick-drying powder is a new type of MgO cement modifier, which contains a large amount of highly active substances. It can rapidly generate a large number of free OH^- ions in the Mg-based slurry, increase the pH value of the chloromagnesium cement material system, and thereby promote the hydration reaction of the light-burned MgO. At the same time, the active substances in the Mg-based promoting setting powder combine with brine, releasing a large amount of heat, and increasing the slurry temperature. As is known to all, the chloromagnesium reaction is an exothermic reaction. Once the reaction is triggered, it continuously releases heat, thus forming a positive feedback loop, further promoting the curing speed of Mg-based materials, and achieving the effect of promoting setting and quick drying. At the same time, it also improves the initial strength and later strength of chloromagnesium cement.

1.3.2.3.1 Usage method and precautions

The dosage of magnesia accelerator for quick drying powder is generally 3%–10% of the mass of light-burned MgO. The specific dosage can be adjusted slightly based on factors such as the activity of light-burned MgO. When using, add it evenly and slowly to the mixed and uniform slurry.

Figure 1.21 Accelerating coagulation quick-drying powder.

The magnesia accelerator for quick-drying powder is an alkaline powder and cannot be stored together with acidic substances. During storage, attention should be paid to moisture prevention and prevention of water leakage.

1.3.2.3.2 Storage and transportation

1. Store in a cool environment below 25°C, seal, and cover well. The shelf life of the magnesia accelerator for quick-drying powder is 3 months.
2. It is an environmentally friendly, nontoxic, harmless, nonflammable, and nonexplosive material. It can be transported as general chemicals.

1.3.2.3.3 Product usage effect

The test temperature was 13°C. With the same raw materials and raw material ratios, the influence of different addition amounts of magnesite-accelerating and quick-drying powder on the initial setting time was tested. The results are shown in Table 1.22.

As can be seen in Table 1.22, as the dosage of magnesia promoting setting and quick-drying powder continuously increases, the initial setting time of magnesia cement gradually shortens. It is worth noting that one should not blindly pursue an overly fast setting speed, because if the setting is too fast, the pores left after the hydration of the cementitious materials may be too numerous and large and an overly fast hydration rate may cause cracks in the slurry during the curing process, all of which can easily have adverse effects on the later performance of the material. The specific dosage needs to be adjusted according to the requirements of related products.

Table 1.22 Magnesium cement promoting setting and quick-drying effect.

Addition amount/%	0	4	10	15	20
Initial setting time/min	300	272	250	230	193

1.3.2.4 Defoamer

- Defoaming agent (Fig. 1.22): An additive that can reduce the bubbles and pores in magnesite cementitious materials, thereby increasing the density and strength of the products.

When magnesite cementitious materials are mixed, mechanical bubbles are easily generated. These bubbles form pores in the products after hardening. These pores are defects for most magnesite products and can affect the strength of the products, as well as their water resistance [18]. Especially for magnesite handicrafts, these products have very high requirements for appearance.

The methods to eliminate bubbles include physical and chemical methods. The physical method is to increase the permeability of gas at both ends of the liquid film and the drainage of the bubble film, so that the attenuation factors of foam are greater than the stabilizing factors, thereby reducing the number of bubbles. This mainly includes placing obstacles such as baffles or filters, mechanical stirring, high-speed centrifugation, high-frequency vibration, ultrasonic waves, static electricity, instantaneous discharge, freezing, heating, radiation exposure, pressure reduction, and pressure increase. The common drawback of these methods is that they are subject to strong constraints by environmental factors and have low defoaming rates. However, their advantages lie in being environmentally friendly and having a high reusability rate [44].

The chemical methods mainly include the chemical reaction method and the method of adding defoaming agents. The chemical reaction method refers to adding chemical reagents to react with the foaming agent or foaming promoter, generating products that are insoluble in water. However, this method has disadvantages such as the uncertainty of the composition of the foaming agent and the generation of insoluble substances that may cause harm to the system and equipment. Currently, the most widely used method in various industries is adding defoaming agents. The advantage of this method lies in its fast defoaming speed, low dosage, and good chemical stability. However, it is necessary to find

Figure 1.22 Defoamer.

suitable and efficient defoaming agents based on the usage environment [44].

Foaming agents can be classified into several major categories based on their form of existence: solid particle type, emulsion type, oil type, and paste type. According to the chemical structure and composition of the foaming agent, they can be divided into alcohol type, mineral oil type, fatty acid and fatty acid ester type, phosphate ester type, organic silicon type, amide type, polyether type, and polyether modified poly siloxane type [44]. The commonly used types in the market are mineral oil type, organic silicon type, and polyether type. Mineral oil–type foaming agents are usually composed of carriers and active agents. The carrier is a substance with low surface tension; the role of the active agent is to inhibit and eliminate foam. The disadvantage of organic silicon–type foaming agents is poor solubility and poor thermal stability, but they have the advantage of fast foaming elimination speed. Organic silicon–type foaming agents generally include polydimethylsiloxane, etc. Polyether-type foaming agents have the advantages of fast foaming elimination, good foam suppression effect, and good thermal stability. Polyether-type foaming agents include polyoxypropylene oxide ethylene glycol ether, etc.

1.3.2.4.1 Principle and function

The defoamer (GX-5#) adopts advanced international surface antipollution treatment technology and incorporates a large amount of active emulsifiers.

It can disperse quickly and uniformly in the slurry, penetrate the liquid film layer, and replace the solvents on the surface of the original liquid film. This reduces the surface tension locally, resulting in uneven forces within the liquid film. The area with reduced surface tension is strongly pulled and stretched in all directions, eventually leading to rupture. It eliminates or significantly reduces pores in the product, improving the density, which in turn enhances the mechanical properties and softening coefficient of the product. The product becomes more esthetically pleasing, and the color becomes more durable.

1.3.2.4.2 Usage method and precautions

The defoamer (GX-5#) is directly added to the slurry and stirred evenly. The dosage is generally 0.3%–0.5% of the light-burned magnesia mass. The specific addition amount should be determined through experimentation based on the specific situation. The defoamer product can be used directly or diluted for use.

1.3.2.4.3 Storage and transportation

The defoamer (GX-5#) needs to be stored at a temperature range of 10°C–30°C. It should not be placed near heat sources or exposed to direct sunlight. When not in use, it must be sealed to prevent bacterial contamination and demulsification. This product may freeze at low temperatures. If it does, after thawing, demulsification will occur. Therefore it should be stored at the recommended storage temperature for a shelf life of 1 year.

During transportation, the defoamer (GX-5#) must be sealed to prevent moisture, strong alkali and strong acid, and impurities from mixing in.

1.3.2.4.4 Usage effect

Many magnesite products have high requirements for surface pores. After adding the defoamer (GX-5#), the surface pores significantly decrease and the density increases. Taking glass magnesite plates as an example, after adding 0.5% of the defoamer (GX-5#), the density increases by 10% and the surface pores decrease from the original 15 per 10 cm^2 (with pore diameters ranging from 0.5 to 1 mm and the maximum 2 mm) to 1 per 10 cm^2 (with the maximum pore diameter of 0.5 mm).

1.3.2.5 Curing agent

- Curing agent (curing agent) (Fig. 1.23): It refers to a type of substance or mixture that enhances or controls the curing reaction.

1.3.2.5.1 Causes and solutions

During the production of magnesite products, in order to reduce production costs or improve production efficiency, the amount of brine or magnesium sulfate admixtures is increased. However, this will lead to an increase in the fluidity of the slurry, a slowdown in the setting speed, and the occurrence of slurry dripping during the curing process. In response to this, MJT, through continuous efforts, developed the curing agent (GX-16#).

1.3.2.5.2 Principles and functions

- Improving flexural strength: The internal structure of the products with added curing agent has a strong membrane effect. The three-dimensional network membrane structure combines with the hydration products to form a polymer sealing layer, which fills into the capillary pores between the aggregate matrix and the cement particles, making the paste structure denser and improving the toughness and bonding performance of the filling body.
- Preventing slurry dripping: The molecular structure of the curing agent contains hydrophilic and hydrophobic groups, which can effectively mix water-based and oil-based substances together. At the same time, through the interaction between the polymer chains, a three-dimensional network structure is formed, increasing the viscosity and consistency of the slurry and preventing the slurry from dripping during the curing and curing process.
- Improving water resistance: When the curing agent is added to the calcium magnesium aluminate cement, it can be evenly dispersed through stirring. As the hydration reaction proceeds, the moisture in the product gradually decreases and the polymer particles will aggregate together to form a network-like membrane, preventing the infiltration of moisture, thereby improving the water resistance of the calcium magnesium aluminate cement samples.

1.3.2.5.3 Usage method and precautions

The addition amount is 1%–5% of the light-burned MgO by weight. Add it to the slurry and stir evenly.

Figure 1.23 Curing agent.

1.3.2.5.4 Storage and transportation

The curing agent (GX-16#) should be stored at a temperature range of 10°C–30°C. It must not be placed near heat sources or exposed to direct sunlight. When not in use, it needs to be sealed to prevent bacterial contamination and demulsification. This product will freeze at low temperatures. If it does freeze, the demulsification phenomenon will occur after melting. Therefore it should be stored at the recommended storage temperature. For the rest, it should be stored in accordance with general chemical product methods. The shelf life is 1 year.

During transportation, the curing agent (GX-16#) should be sealed to prevent moisture, strong alkali and strong acid, and impurities from being mixed in.

1.3.2.5.5 Usage effect

Taking the sulfur–magnesium permeable board as an example, it can significantly improve the viscosity of the slurry and enhance the bonding force with expanded polystyrene (EPS) particles. As the amount of curing agent added increases, the viscosity of the slurry gradually increases, so that after the permeable board is placed on the frame, there will be no bottom drip phenomenon. After the product is fully cured, the blistering rate of the board can reach over 80%. If no curing agent is added, after the substrate is permeated by the slurry, the slurry is easily seeped out from the substrate pores, resulting in problems such as low density, poor structural strength, and easy breakage of the product.

1.3.2.6 Coupling agent

1.3.2.6.1 Definition

Coupling agent (Fig. 1.24) is a surface modifier that can improve the interfacial interaction between organic and inorganic substances. As a molecular bridge between organic and inorganic substances, the coupling agent enables better combination of organic and inorganic components in the mixed system.

Coupling agents generally include several major categories such as organic chromium, silanes, titanates, and aluminum acid. Commonly used agents in the market include silane-based KH and DL series coupling agents, phthalate-based coupling agents, etc.

1.3.2.6.2 Function and principle

The coupling agent itself possesses two different types of functional groups. Among them, the inorganic-reactive functional groups can react with inorganic substances in the system, while the organic-reactive functional groups can react with organic substances in the system or form hydrogen bonds. After the functional groups in the coupling agent react with the organic and inorganic substances in the mixed system, they act as a bridge, connecting the two substances together, making their combination tighter and improving the overall performance of the mixed system.

1.3.2.6.3 Usage method and precautions

The coupling agent produced by MJT can be added to the Mg-based slurry at a rate of 0.5%–2% based on the weight of the calcined MgO. This coupling agent is suitable for use in insulation boards such as Mg-based permeable boards and Mg-based homogeneous boards, enhancing the bonding ability between the slurry and the polystyrene particles, thereby improving the bending strength of the boards.

1.3.2.7 Demolding agent

1.3.2.7.1 Definition

- Demolding agent (Fig. 1.25): Generally used at the interface between the mold and the product as a coating to facilitate the separation of the product from the mold. Based on the classification of active ingredients, commonly used demolding agents in the building materials industry

Figure 1.24 Coupling agent.

Figure 1.25 Demolding agent.

include silicon-based, fluorine-based, and oil (wax)-based demolding agents. Among these, oil (wax)-based demolding agents include water-based and oil-based types.

During the production of Mg-based products, waste engine oil, lubricating oil, and edible oil are commonly used as demolding agents. Although these oils provide good demolding effects, their high viscosity and thick coating result in a relatively high cost when using oil-based demolding agents. Utilizing waste engine oil and other recycled oil products as demolding agents not only affects the appearance of Mg-based products but also causes environmental pollution. Environmental protection authorities in various regions are also relatively strict regarding the circulation and use of waste engine oil.

Water-based oil (wax)–type release agents are generally made from high-quality industrial oils and are environmentally friendly. The viscosity of the water-based release agent system is relatively low, and the usage cost is also lower. The release agent developed by MJT not only has the original efficacy of the release agent but also can eliminate air bubbles on the surface of the product, forming a dense layer, thereby further improving the surface strength and appearance of the product.

1.3.2.7.2 Function and principle

Commonly used release agents are generally applied to the surface of the mold. Once applied, the release agent forms a cohesive layer that acts as an interlayer. At this point, a sandwich structure is created between the mold and the product. During demolding, if separation occurs within the cohesive layer of the release agent, the product's structure remains intact and no product slurry adheres to the mold. In this case, the release agent performs effectively. If separation occurs between the product and the cohesive layer of the release agent, material exchange takes place, leading to contamination of the mold surface with dust, residue, or even larger product fragments. Under these conditions, the release agent performs poorly. Alternatively, if separation occurs between the mold and the cohesive layer of the release agent, the mold is not adequately saturated and the product surface may be contaminated with substantial amounts of release agent-like material, compromising its esthetic appearance.

1.3.2.7.3 Usage method and precautions

The water-based demolding agent produced by MJT can be diluted up to a ratio of 1:8 according to the type of product and process. The diluted demolding agent can be applied to the surface of the mold by spraying or painting. This product is suitable not only for traditional Mg-based products but also for the demolding process of Mg-based foamed products. When used in the production of Mg-based foamed products, this product can form a dense and thin layer on the surface of the foamed product, which not only reduces the demolding difficulty but also improves the appearance of the foamed product. In the subsequent processing of Mg-based foamed core boards, it is necessary to apply glue to the surface of the board and stick decorative panels. Using this demolding agent to produce core boards will result in a surface that is dense and free of air holes, which can save the amount of glue used.

1.4 Reinforced materials

Mg-based cementitious materials are widely utilized as inorganic binders. Depending on the application, they can be used to produce various products, such as fireproof boards, door core boards, insulation boards, and handicrafts. In the manufacturing of these products, MgO and admixtures serve as the primary components of Mg-based cementitious materials. However, these two materials alone are not sufficient. Additional reinforcing materials are required to enhance mechanical properties, including strength, in the final products. Common reinforcing materials include glass fiber mesh, polypropylene short fibers, bamboo fibers, jute fibers, and steel wire mesh frames.

1.4.1 Glass fiber mesh sheet

1.4.1.1 Definition and characteristics

Glass fiber mesh fabric (Fig. 1.26) is made from medium-alkali or nonalkali glass fiber woven fabric and is treated with an alkali-resistant coating. It possesses excellent alkali resistance, flexibility, and high tensile strength in both warp and weft directions. It also possesses the qualities of resistance to alkali, acid, water, cement erosion, and other forms of chemical corrosion, along with properties that prevent mold growth, repel insects, provide fire resistance, and offer thermal, acoustic, and electrical insulation.

This grid fabric is widely used in the anticracking of exterior and interior walls of buildings, as well as in enhancing the production of Mg-based materials. After undergoing high-temperature heat-setting treatments with alkali-resistant liquids and enhancers, the glass fiber grid fabric becomes more stable and durable. It plays an important structural role in the building materials product system, preventing the occurrence of cracks. Due to its excellent resistance to acids and alkalis, as well as high tensile strength in both directions, it can evenly distribute stress within the building materials product system, preventing deformation of the entire structure caused by external impacts and compression. This provides the product structure with high impact resistance strength, while also being

Figure 1.26 Glass fiber mesh fabric.

easy to construct and allowing for quality control. It functions as a "soft reinforcing bar" in the building materials product system.

1.4.1.2 Common specifications

1. Mesh area: 5 × 5 , 4 × 4 , 3 × 3 mm.
2. Mass per square meter: 80–160 g.
3. Length per roll: 50 , 100 , 200 m.
4. Width per roll: 0.5–1.5 m.
5. Colors: white, blue, green, or other colors.
6. External packaging: each roll is individually wrapped in a plastic film, and four or six rolls are packed in a carton.

1.4.1.3 Guidelines for the application of glass fibers

To achieve the best reinforcing effect of glass fibers in magnesia cement, the following requirements should be followed:

1. The more glass fiber fabric there is, the better the reinforcing effect. Generally speaking, the more layers of glass fiber fabric there are, the greater the impact force or load required to cause damage or fracture of the magnesia cement. That is to say, the more the quantity of reinforcing materials (glass fiber fabric, bamboo fibers, etc.), the stronger the bearing capacity of the magnesia cement.
2. The higher the breaking strength of the glass fiber itself, the better the reinforcing effect.

   As an reinforcing material, the glass fiber should have a better breaking strength. The higher the breaking strength, the greater the load it can withstand. The breaking strength of the glass fiber is closely related to the alkali content. Generally, the order of breaking strength of glass fibers is as follows: nonalkali glass fibers > medium-alkali glass fibers > high-alkali glass fibers. However, high-alkali glass fibers are brittle and prone to powdering, so magnesia cement mostly uses medium-alkali glass fibers or nonalkali glass fibers as reinforcing materials.

3. Distribution direction
The distribution of reinforcing materials such as glass fibers or bamboo strands in magnesium cement should be as parallel as possible to the surface of the magnesium product and perpendicular to the force direction, in order to achieve the best reinforcing effect.
4. Adequate soaking of glass fibers
When laying the glass fiber fabric onto the magnesium cement, it is necessary to perform soaking, that is, gently press the glass fiber fabric with a trowel to ensure that it is fully soaked in the magnesium cement slurry, to prevent the phenomenon of glass fibers losing their slurry.
5. Distribution position

The distribution of the glass fiber fabric within the magnesium cement should be uniform and symmetrical, be positioned as close as possible to the outer surface of the magnesium product, and preferably achieve a semiexposed effect. This means that the fabric's texture should be visible to the naked eye, but the fabric itself should not be felt when touched. Taking the magnesium insulation board as an example, the number of glass fiber layers in the magnesium surface layer on the upper and lower surfaces should be the same and their distribution positions should be symmetrical to prevent uneven thickness of the magnesium surface layers on both sides, which could lead to warping deformation.

1.4.2 Polypropylene short fibers

1.4.2.1 Definition

Polypropylene short fibers (Fig. 1.27) are high-strength braided single-filament fibers made by special composite technology using polypropylene and various organic and inorganic materials. They can form a three-dimensional anisotropic network structure in magnesium products, providing a supporting function. This prevents the microcracks formed during the initial hardening stage of magnesium products from further

Figure 1.27 Polypropylene short fibers.

development, thereby enhancing their fracture toughness and improving crack resistance, water resistance, impact resistance, and seismic resistance.

1.4.2.2 Characteristics

1. Higher strength and elastic modulus, which enhance the mechanical properties of magnesium cement.
2. Strong adhesion, easy to combine with magnesium cement.
3. Excellent dispersibility, does not aggregate, effectively ensuring the exertion of its crack prevention performance.
4. Stable chemical properties of fibers, extremely strong acid and alkali resistance.

1.4.2.3 Technical parameters

Technical parameters are shown in Table 1.23.

1.4.3 Bamboo strips and bamboo meshes

The bamboo house stands gracefully against the wind, embodying the elegance of a gentleman. The history of using bamboo in architectural art in our country dates back thousands of years. As the ancients said "Rather go hungry than live without bamboo."

Mechanical tests conducted by scientists on bamboo have demonstrated that its contraction is minimal, while its elasticity and toughness are exceptionally strong. The compressive strength along the grain of bamboo is about 600 kg/cm^2; the tensile strength along the grain per square centimeter can support 1800 kg; among them, the tensile strength along the grain of the hard bamboo can reach 2833 kg/cm^2, earning the

Table 1.23 Technical parameters of polypropylene short fibers.

Material	Polypropylene	Fiber type	Bundled single fiber
Density	0.91	Tensile strength	>358 MPa
Acid and alkali resistance	Extremely high	Elastic modulus	>3.5 GPa
Melting point	>165°C	Fiber diameter	18~48 μm
Safety	Nontoxic material	Hygroscopicity	None
Heat conductivity	Extremely low	Tensile limit	>15%
Low temperature resistance	Strong	Specification	19 /9 / 6 /3 mm

reputation of "plant steel" [45]. Bamboo possesses exceptional bending resistance. For example, the hollowness ratio of mature bamboo can reach 0.85, and its bending resistance is more than twice that of a solid pole of the same weight. Our country enjoys the reputation of being the "Kingdom of Bamboo" with extremely abundant bamboo resources. Bamboo not only helps improve air quality and beautify the environment but will also become an indispensable low-carbon and environmentally friendly building material in the pursuit of sustainable economic and social development.

Depending on the different production processes and performance requirements, adding bamboo fibers or weaving bamboo screens (Fig. 1.28) into Mg-based cementitious materials can enhance the performance of these materials.

First, the addition of bamboo fibers can enhance the crack resistance of Mg-based products. During the hardening process of Mg-based cementitious materials, cracks are prone to occur due to the evaporation of internal moisture and the impact of external loads. However, the incorporation of bamboo fibers can effectively hinder crack propagation and improve the crack resistance of the products. This is attributed to the formation of a stable cavity grid by bamboo fibers within the Mg-based cementitious materials. When cracks develop in Mg-based products, this cavity grid can absorb the crack energy and prevent its further expansion.

Secondly, the addition of bamboo fibers can enhance the compressive strength. The compressive strength of Mg-based cementitious materials mainly depends on the bonding force between the aggregates and the Mg-

Figure 1.28 Bamboo screen.

based slurry. Adding bamboo strips can increase the internal frictional force within the cementitious material, strengthen the bonding between the aggregates and the Mg-based slurry, and thereby improve the compressive strength.

In addition, bamboo fibers also offer environmental and economic advantages. Bamboo is a renewable resource, and its use can reduce dependence on other materials, contributing to sustainable resource utilization. Moreover, the relatively low cost of bamboo fibers helps lower product expenses and improve economic efficiency.

However, despite its many benefits, bamboo fiber reinforcement technology still faces some practical challenges. While bamboo exhibits excellent toughness and bending resistance, its compressive and tensile strengths are relatively low. To enhance overall load-bearing capacity, bamboo is typically woven into bamboo screens. Additionally, as a natural material, bamboo varies in quality and size. It is essential to select bamboo that is dry, pest free, crack free, and knot free and then cut it into bamboo strips and slices of specific lengths and widths. The surface wax layer should be finely polished and roughened. Presoaking processed bamboo fibers or screens in a mixed solution can also improve the bond with Mg-based slurry.

Reinforcing Mg-based materials with bamboo fibers is not a matter of random addition. Generally, bamboo fibers or screens should be evenly dispersed and oriented at an angle to the horizontal direction to enhance adhesion to the Mg-based cementitious material and improve tensile resistance. They should also be placed as close to the product surface as possible while retaining a protective layer to prevent exposure to air, corrosion, and cracking. Polishing the smooth outer surface of the bamboo further improves bonding with the Mg-based slurry. Bamboo fiber reinforcement can be effectively used in products such as Mg-based manhole covers and prefabricated partitions.

1.4.4 Brown silk

Brown coir fiber (Fig. 1.29) is derived from the leaf sheaths and coconut shells of palm plants. It is a high-performance natural fiber commonly used in the production of mattresses, shoe insoles, ropes, etc. In building materials, the product often referred to as "brown thread" is typically made from polyethylene terephthalate fiber, also known as polyester, which is a synthetic fiber. It is formed by the connection of various chain segments in the macromolecular chain through ester bonds. It has high tensile strength and elastic modulus, moderate resilience, excellent heat setting effect, good heat resistance and light resistance [46]. These fibers are cost-effective and larger in size compared to polypropylene fibers, while also offering relatively high tensile strength. In the production of products such as Mg-based partition panels, door core boards, and fireproof boards, brown fibers can be used to enhance tensile and flexural strength.

The brown fibers mainly play a crack-resisting role in Mg-based cementitious materials. The crack-resisting effect they exert can be roughly divided into two stages: the crack-resisting effect in the plastic slurry and the crack-resisting effect in the hardened slurry [47].

- Crack-resisting effect in the plastic slurry: The brown fibers are uniformly and randomly dispersed in the Mg-based cementitious materials in a three-dimensional manner, forming a three-dimensional network, which supports the aggregates and prevents the sinking of coarse and fine aggregates, effectively preventing and inhibiting the segregation tendency on the surface of the slurry. Meanwhile, the brown fibers can eliminate the water transport channels in the Mg-based slurry, significantly reducing the water loss on the surface. Thus it reduces or completely prevents the generation of surface sinkage cracks in the slurry. The strength of the plastic state Mg-based cementitious materials is extremely low. When the water evaporates, due to shrinkage, tensile stress is generated and the brown fibers in the slurry can withstand this tensile stress, reducing and preventing the generation and development of cracks [47].
- The crack-blocking effect in the hardened slurry: The shrinkage cracks of the Mg-based cementitious materials are usually quite small in the

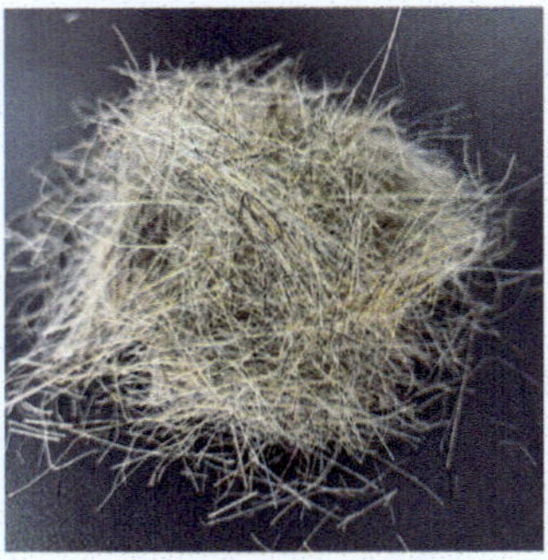

Figure 1.29 Brown silk.

initial stage, with a width generally not exceeding 0.05 mm. Due to the presence of numerous three-dimensional randomly distributed polyester fibers in the slurry, when the cracks develop in any direction, they will encounter a fiber at the farthest point not exceeding the average center distance of the fibers. At this point, the concentrated tensile stress at the crack tip is transferred to this fiber and the development of the crack is thus blocked [47].

In addition, the evenly distributed brown fibers in the slurry play a certain supporting role in the Mg-based cement slurry in the form of a microframework, reducing the shrinkage deformation of the Mg-based slurry. Microscopically, it restrains the adhesive cracks in the slurry; macroscopically, it prevents the development of plastic shrinkage cracks and drying shrinkage cracks in the slurry [47].

The crack-blocking effect of the brown fibers in the Mg-based cementitious materials is essentially to limit the initial cracks in the slurry to a harmless stage at the microscopic level, preventing them from developing into macroscopic harmful cracks [47].

The fiber's crack-retaining effect mainly depends on the average spacing s of the fibers and the number n of fibers per unit volume of the Mg-based slurry. Under the condition of uniform dispersion of the fibers, the smaller the average spacing s, the larger the number n and thus the better the crack-retaining effect of the fibers in the slurry [47]. The lower the dispersion rate of the fibers, the fewer the independent fiber roots that can produce the crack-retaining effect and the worse the crack-retaining effect.

Therefore to achieve an effective crack retention effect, the fibers must possess good dispersibility. The brown fibers utilized in Mg-based cementitious materials, after undergoing special surface treatment, enhance the repulsive forces between individual fiber strands. This treatment improves their dispersibility in the slurry, allowing them to be fully and uniformly dispersed through a conventional mixing process. Additionally, the special surface treatment strengthens the bonding force between the fibers and the hydrated MgO in the slurry, thereby facilitating effective stress transfer within the material [47].

1.4.5 Wire mesh frame

Wire mesh frames (Fig. 1.30) are mesh structures fabricated from high-strength steel wires or stainless-steel wires in various specifications and shapes according to requirements. In this book, the wire mesh frame is employed as a reinforcing material and combined with Mg-based cementitious materials to enhance the compressive and flexural strength of magnesium products. Its inherent strength is significantly superior to that of bamboo and wood (Table 1.24). However, when the wire mesh frame is integrated with magnesium materials, the presence of chloride ions can corrode the wire mesh frame, compromising its service life and the mechanical properties of the magnesium products. Therefore how to perform anticorrosion and rust prevention treatments on the surface of the wire mesh frame has become a critical aspect.

The rust prevention methods for steel wire mesh frames include hot-dip galvanizing, vacuum galvanizing, and electrogalvanizing. Currently, surface galvanizing is the most commonly used method, often combined with on-site construction or preembedding techniques to integrate with other materials for meeting production and design requirements.

The primary structure of general hot-dip galvanized steel wire mesh frames is formed by welding sturdy steel grids. It features high load-bearing capacity, a lightweight structure, and ease of hoisting. The hot-dip galvanizing surface treatment provides considerable corrosion resistance and an attractive surface luster. It is widely used in fields such as construction, transportation, and environmental engineering.

1.4.5.1 Specifications and parameters of wire mesh frame

The specifications and parameters of a wire mesh frame typically include aspects such as material, mesh size, wire diameter, specifications, and weight.

Figure 1.30 Steel wire mesh structure.

1.4.5.1.1 Material

Wire mesh frames can be fabricated from a variety of materials, including low-carbon steel wire, stainless steel wire, and aluminum alloy wire. Each material offers distinct properties in terms of corrosion resistance, strength, and cost.

1.4.5.1.2 Mesh size

Mesh size refers to the size of the mesh holes in the wire mesh frame. Common units are millimeters and inches. The mesh size determines the filtering effect and ventilation performance of the wire mesh. Different engineering applications require the selection of appropriate mesh sizes based on specific requirements.

1.4.5.1.3 Wire diameter

The wire diameter refers to the diameter of a single steel wire in the steel mesh. Common units include millimeters and inches. The wire diameter determines the strength and load-bearing capacity of the steel mesh. Generally, the larger the wire diameter, the higher the strength.

1.4.5.1.4 Specifications

The specifications of the steel mesh are usually expressed by length and width. Common specifications include 1000 × 2000 and 1220 × 2440 mm. According to specific engineering requirements, different

Table 1.24 Comparison of strength of bamboo, wood, and steel.

Type		Tensile strength/kg/cm^2		Compressive strength/kg/cm^2	
		Strength	Average	Strength	Average
Bamboo	Thorny bamboo	1948.3	2138.8	640.1	487.8
	Phyllostachys viridis	2833.6		540.5	
	Henon bamboo	1821.7		359.5	
	Dendrocalamus latiflorus	1951.3		411.2	
wood	Cedarwood	772.2	1073.4	406	444
	Pinus koraiensis	981.2		328	
	German oak	1432.1		577	
	Polish wood	1108.2		465	
Steel	Soft steel	3780 ~ 4250	5170 ~ 5638 Above all	—	—
	Semisoft steel	4400 ~ 5000		—	
	Medium-hard steel	5200 ~ 6000		—	
	Stainless steel	7300 and above		—	

specifications of steel meshes can be selected or the steel mesh can be customized as needed.

1.4.5.1.5 Weight

The weight of the steel mesh depends on factors such as material, mesh size, wire diameter, and specifications. Common units include g/m^2 and kg/m^2. Weight is an important indicator for measuring the density and durability of the steel mesh.

1.4.5.2 Guide for selecting and using steel wire mesh frames

The specifications and models of steel wire mesh frames are important reference indicators when selecting and using steel wire mesh. By understanding the basic specification parameters of the wire mesh and referring to the specification and model table, one can choose the appropriate steel wire mesh frame model for specific engineering requirements. When purchasing and using steel wire mesh, factors such as the usage environment, load-bearing capacity, installation requirements, lifespan, and maintenance need to be considered to ensure effectiveness and safety.

1.4.5.3 Application and development

1. A prefabricated steel wire mesh frame insulation board includes polystyrene board and steel wire mesh frame. Its structure features an insulation mortar layer and a facing mortar layer positioned on the outer side of the Mg-based cementitious material permeable board. The steel wire mesh frame consists of steel wire mesh sheets, U-shaped reinforcing bars, and reinforcing insertion rods. The steel wire mesh sheets are embedded within the insulation mortar layer. The base of each U-shaped reinforcing bar is fixedly connected to the steel wire mesh sheets, while its two sides are inserted into the inner side of the polystyrene board. One end of each reinforcing insertion rod is fixedly connected to the steel wire mesh sheets, and the other end extends to the inner side of the Mg-based cementitious material permeable board. The steel wire mesh frames are integrated into the Mg-based cementitious material permeable board to enhance the strength of the insulation board.
2. In the external wall insulation system using large-formwork construction, the steel wire mesh panel is also referred to as the cast-in-place concrete external wall insulation system. This system utilizes Mg-based cementitious materials to form the permeable board. Its front side is

processed into the required toothed groove and is equipped with 2.0 mm steel wire mesh sheets (with a mesh size of 50 × 50 mm). Steel wires are inserted diagonally through the core board (6–8 cm) and welded together to form a three-dimensional insulation panel. To prevent corrosion of the steel wire mesh and to enhance bonding between the surface of the insulation board and the external leveling mortar, interface agents are sprayed onto the surface of the insulation board and the steel wire mesh frame [48].

This system offers advantages including sound insulation, vibration isolation, waterproofing, fire resistance, excellent weathering resistance, energy efficiency, structural strength, and strong overall performance.

As market demands for corrosion prevention increase, the anticorrosion treatment technology for steel wire mesh coatings has also advanced accordingly. MJT has introduced a new anticorrosion treatment technology: Graphene Nanoparticle Anticorrosion Technology. By leveraging the unique properties of graphene nanoparticle materials, this technology enhances the performance of anticorrosion coatings. Graphene boasts outstanding electrical conductivity, thermal conductivity, mechanical strength, and chemical stability, making it an ideal additive for anticorrosion coatings. The nanosize effect of graphene enables it to fill in the tiny defects in the coating, forming a denser barrier layer that effectively blocks corrosive media such as water and oxygen. Graphene can react with the metal substrate to form a protective film, improving the corrosion resistance of the metal. The excellent mechanical properties of graphene can enhance the wear and scratch resistance of the coating, prolonging its service life.

Therefore graphene nanoparticle anticorrosion coatings possess excellent corrosion resistance, weather resistance, water resistance, oil resistance, and chemical resistance, among other characteristics. They can protect the steel wire mesh framework from corrosive damage.

1.5 Fillers and additives

In the manufacturing of magnesium cement products, fillers such as sawdust, fly ash, sand, and gravel are often added. Additionally, tensile-resistant materials like glass fibers or glass cloth, along with reinforcing materials such as steel bars, are used to further enhance the product strength.

Chloromagnesium cement has excellent bonding ability with various fillers (such as gravel, sand, marble powder, asbestos, wood particles, and expansive soil). Using chloromagnesium cement combined with fillers to produce various products not only results in fast curing speeds but also provides high early strength. Common filling materials mainly include sawdust (which can also be crushed to an appropriate fineness, such as straw, cotton stalks, and rice husks), fly ash, sand, and gravel. Their functions are as follows: first, to adjust the material properties of Mg-based wood substitutes, such as improving toughness, reducing weight, and enhancing processing performance; second, to reduce the consumption of light-burned MgO and lower costs by adding fillers [49].

According to the chemical composition of the fillers, fillers can be classified into organic materials and inorganic materials and different types of fillers have different regulating effects.

1.5.1 Organic materials

1.5.1.1 Sawdust-based plant straw filler material

In the new era, there is a growing demand for the environmental friendliness of building materials. Many manufacturers, both domestic and international, have utilized agricultural waste such as straw and hemp as fillers and applied them on a large scale in cement-based composite materials, generating certain social benefits [50].

Sawdust (Fig. 1.31) is the residue from wood processing and possesses excellent physical properties. It is lightweight and exhibits a loose, porous structure with outstanding water retention and air permeability. In the actual production of magnesite products, different types of sawdust can be selected based on the availability of raw materials. During selection, attention should be paid to indicators such as moisture content and fineness. Sawdust is a highly absorbent material. Without hydrophobic coating treatment, it significantly increases the brine content in Mg-based slurry, which can easily lead to the reabsorption of brine by Mg-based products. Therefore for such absorbent filler and aggregate materials, hydrophobic coating treatment should be applied before use to render them largely nonabsorbent or minimally absorbent, thereby reducing the likelihood of brine reabsorption in the products.

The functional and esthetic requirements of the product determine the amount of filler added. For example, in handicrafts, depending on the moisture content of the filler, the proportion of filler generally accounts for

Figure 1.31 Sawdust.

10%–15% of the total mass of the slurry, with a maximum of 20% of the total mass. If the amount is too large, the strength and other properties of the product are difficult to guarantee. The amount of sawdust used in Mg-based fireproof boards is much higher than that in traditional handicrafts. Depending on the process requirements, the addition of sawdust can reach up to 50%–60%, endowing it with good fire resistance while also providing certain flexibility. Therefore in Mg-based products, the rational utilization of low-cost sawdust can not only turn waste into treasure but also effectively reduce production costs and improve product performance.

In addition to sawdust, rice husk (Fig. 1.32), rice bran, etc. are also common agricultural waste materials. These materials can all be used as fillers to partially replace Mg-based raw materials in the production of Mg-based products. The properties of rice husk and rice bran differ from those of sawdust. The surfaces of rice husk and rice bran are rich in waxes, making them less absorbent and prone to floating on the surface of the slurry rather than mixing in. When using them, the ratio needs to be adjusted according to the actual situation.

The utilization of resources represents a trend and direction for the effective use of this type of agricultural waste, aligning with the requirements of sustainable agricultural development and the growth of the green circular agriculture industry. At the same time, rice husks and sawdust, as biomass energy raw materials, can yield significant economic benefits if utilized scientifically. In recent years, with continuous improvements and the promotion of resource utilization technologies, rice husks and other plant fibers, after being burned in special incinerators to remove organic matter, leave behind inorganic ashes rich in SiO_2, among other components. These ashes can then be ground by a grinder to obtain ash with a high active SiO_2 content. Currently, rice husk ash and straw ash have been widely applied in cement products, high-performance concrete, and other

Figure 1.32 Rice husk.

areas. As an active biomass silicon source admixture suitable for cement-based materials, its application in chloromagnesium cement systems is also increasing [51].

1.5.1.2 Polystyrene granules

Insulation materials are collectively referred to as thermal insulation materials and heat insulation materials. Generally, materials that prevent the loss of indoor heat are called thermal insulation materials and those that prevent outdoor heat from entering are called heat insulation materials. Thermal insulation materials are typically those with a thermal conductivity coefficient of less than or equal to 0.12 W/ (m·K). Their role in the current entire construction industry is extremely significant.

Polystyrene granules are a type of polymer. Their full name is EPS foam granules (Fig. 1.33), and they were developed by BASF in 1950. Because of their excellent insulation properties, polystyrene granules have been widely used in the construction industry. Subsequently, owing to their high impact resistance, they have also become extensively used in the packaging industry.

Polyethylene granules are produced by foaming polyethylene resin. After the polyethylene resin is mixed with foaming additives, it is heated and softened in a steam environment under certain pressure. The decomposition of the foaming additives generates gas, causing the polyethylene resin to expand into a rigid, closed-cell foam plastic structure. This uniform and sealed cellular structure endows EPS granules with the characteristics of low water absorption, good insulation, and light weight. The density of EPS granules is determined by the expansion ratio of

Figure 1.33 Expanded polystyrene foam particles.

polyethylene granules during the molding process, which typically ranges from 10 to 45 kg/m^3. The apparent density of EPS granules used in engineering applications generally falls between 10 and 30 kg/m^3 and can be customized according to customer requirements [52]. With the increasing pressure of the building materials market competition, many colleagues have already controlled the bulk density of polyethylene granules at around 5 kg/m^3.

Graphite-modified polyethylene particles (Fig. 1.34) are prepared using imported graphite-modified polystyrene as the raw material. The nanoscale sheet–like graphite particles dispersed within the polystyrene have an infrared ray–reflecting effect, which can effectively reduce long-wave radiation and further lower the thermal conductivity of the insulation board. The thermal conductivity of ordinary polystyrene foam boards ranges from 0.033 to 0.044 W/(m·K). In contrast, the insulation capacity of graphite-containing polyethylene boards is at least 30% higher than that of ordinary EPS, with a thermal conductivity of 0.030–0.032 W/(m·K). Since the raw materials for graphite-modified polyethylene particles are almost entirely imported, their price is high. Consequently, many merchants in the market use dyed polyethylene particles to pass them off as graphite-modified polyethylene particles for sale. Therefore when purchasing, it is necessary to distinguish carefully.

Currently, the Mg-based lightweight insulation boards used in the construction industry primarily achieve weight reduction, energy efficiency, and insulation effects by incorporating polystyrene particles. Examples include glass–magnesium purification boards and prefabricated fireproof insulation boards. However, in actual production, polystyrene particles also present certain unavoidable issues, such as a low heat

Figure 1.34 Graphite-modified polyphenyl particles.

deformation temperature. When the internal temperature of products containing polystyrene particles exceeds 60°C, the particles tend to shrink and deform, significantly affecting both the performance and appearance of the products. This phenomenon is commonly referred to as "core burning." Moreover, polystyrene products are manufactured from non-renewable petroleum resources, which is detrimental to the sustainable development of natural resources. There is an urgent need in the insulation industry for superior alternatives to replace such products.

1.5.2 Inorganic materials

1.5.2.1 Active materials

1.5.2.1.1 Silica fume

Silica fume (silica fume) (Fig. 1.35) is the fly ash captured from the flue during the smelting of ferrosilicon alloys or industrial silicon. It is mainly composed of amorphous silica and has a particle size ranging from 0.1 to 0.2 μm, with a specific surface area of 15–20 m^2/g. Depending on the different impurities it contains (such as carbon and iron oxide), the color of silica fume can be white, off-white, gray, etc. The original ash directly captured has a bulk density of 150–300 kg/m^3. After being densified, the density can reach 800 kg/m^3.

According to the differences in chemical composition and physical properties, the standards for silica fume grades are usually divided into grades one, two, and three. Among them, grade one silica fume has a silicon dioxide content greater than 75%, with high activity and strong bonding force, and is suitable for high-strength concrete, refractory

Figure 1.35 Silica ash.

materials, and other fields; grade two silica fume has a silicon dioxide content ranging from 65% to 75%, with moderate activity, and is suitable for ordinary concrete, cement products, and other fields; grade three silica fume has chemical components and physical properties that do not meet the requirements for construction material production, with a silicon dioxide content ranging from 45% to 65%, and low activity and is suitable for filling materials, etc.

Silica fume can enhance the water resistance of chlorinated magnesium cement, and as the addition amount increases, the water resistance of chlorinated magnesium cement further improves. After entering the chlorinated magnesium cement system, silica fume, under the stimulation of the alkaline environment, generates a large number of silicate ions, which combine with the multinuclear water hydroxyl magnesium ions in the system to form hydrated magnesium silicate gel, filling the pores of chlorinated magnesium cement and significantly improving the water resistance of chlorinated magnesium cement.

1.5.2.1.2 Slag

Slag generally refers to the waste residue produced during the metallurgical process. Its composition can be adjusted by adding appropriate amounts of fluxes (such as lime, quartzite, and fluorite). During the smelting process, by controlling the composition and properties of the slag, the products of gangue and oxide impurities can be separated from the molten metal or sulfur smoothly, harmful impurities in the metal can be removed, nonmetallic inclusions in the liquid metal can be absorbed without being directly polluted by the furnace gas, and useful metal oxides can be enriched. In electric furnace smelting, it also serves as a resistor heating body. Slag plays a decisive role in ensuring the smooth operation of smelting, the quality of smelting products, and the recovery rate of metals, and in steelmaking operations, there is a saying that "a good slag can lead to a good steel."

Depending on the different metallurgical processes, slag can be classified as smelting slag, refining slag, and synthetic slag; according to the properties of slag, there are alkaline slag, acidic slag, and neutral slag. Many slags have important uses. For example, blast furnace slag can be used as a cement raw material; high phosphorus slag can be used as fertilizer; slag containing vanadium and titanium can be used as raw materials for extracting vanadium and titanium, etc. Some slags can be used to produce slag cement, slag bricks, slag glass, etc.

Since slag (blast furnace slag) and steel slag belong to by-products in the metallurgical industry process, the chemical components of different by-products are different. They can be added according to the performance requirements of the products, not only achieving the purpose of by-product consumption but also reducing the production cost of the products.

1.5.2.2 Inert materials

1.5.2.2.1 Stone powder

Limestone powder (Fig. 1.36) is a type of powder made from various minerals. There are numerous types, classified based on the mineral composition. For instance, the main component of limestone is calcium carbonate and its forms as stone powder include heavy calcium powder and light calcium powder, which are both types of stone powder. Mica powder and quartz powder are also part of stone powder. Many rocks with different mineral compositions can be ground into powder and used as stone powder for various processes and purposes. This book only introduces commonly used stone powder related to Mg-based materials.

With the development of ecological civilization in society, the concept that "green mountains and clear waters are as valuable as gold and silver" has become increasingly deeply rooted in people's hearts. To protect the environment, the method of directly making stone powder from natural minerals will decrease. To alleviate the contradiction between supply and demand, in addition to the stone powder produced during the processing of aggregates, using industrial waste as stone powder will become a future

Figure 1.36 Limestone powder.

trend. This can solve the pollution caused by the original industrial waste to the environment.

Functions

1. When used within a certain proportion, fly ash can replace part of the fine aggregate, which can achieve the effect of filling and compacting, as well as the microaggregate effect. It can improve the workability of the Mg-based slurry, enhance the strength and impermeability of the Mg-based cementitious materials, and also reduce the amount of light-burned MgO and lower the reaction peak temperature, reduce temperature stress, and improve the crack resistance of the cementitious materials.
2. Different hardness of fly ash can be used to produce boards (or products) with different hardness requirements, such as artificial marble.
3. The current process uses Mg-based materials as the binder, adds a certain amount of MJT series modifiers, and mixes with different pigments (or coated with colored minerals) through special processes to produce high-end decorative inorganic quartz stone with excellent mechanical properties, beautiful decoration, resistance to aging, high-temperature resistance, not being prone to deformation, and easy to lay and energy-saving and environmentally friendly characteristics (Fig. 1.37).
4. The products made with Mg-based cementitious materials as the inorganic binder, combined with polystyrene foam particles and fine stone powder, not only save magnesium materials but also improve the comprehensive performance of Mg-based cementitious materials. For example, adding 300–500 mesh quartz powder in a certain proportion

Figure 1.37 Inorganic quartz stone.

to the homogeneous plates of the prefabricated structure can enhance the fire resistance and increase the strength.

5. Adding an appropriate amount of stone powder to some Mg-based products can increase the viscosity and operability of the slurry and reduce costs. For instance, adding an appropriate amount of talc powder in the production of glass–magnesium boards can achieve a good effect.
6. To improve the surface fineness and whiteness of Mg-based products, a certain amount of high-white fine talc powder can be added to the surface slurry of the products. Using low-cost talc powder to replace part of titanium white powder also works well.
7. To enhance the quality and reduce the cost of magnesium permeable insulation boards, some heavy calcium powder can be used to replace some lightweight MgO.

The above are only the applications of some common stone powders in Mg-based products. In actual production, other stone powders are also used in the products, but due to the limited space, we will not elaborate on them here.

Outlook If the stone powder can be processed slightly and used as an admixture instead of the relatively expensive silica powder or slag, it will have significant practical significance in solving the shortage of raw materials in actual engineering projects, reducing project costs, and protecting the environment [53].

1.5.2.2.2 Solid waste

All solid or semisolid substances produced during human activities that no longer have any use value for the owner are generally referred to as solid waste. This includes solid particles separated from wastewater and waste gas. Generally, solid waste generated in various production activities is called waste residue, while solid waste generated in daily life is called garbage [54]. Currently, the main cities in China produce 190 million tons of domestic garbage annually, and there are 6 billion tons of industrial solid waste piled up. These waste residues not only damage the environment and endanger the physical and mental health of the public but also occupy a large amount of land resources.

The term "waste" in "solid waste" only refers to the original owner. Most solid waste can be transformed into production raw materials through certain technical processes, and some harmless solid waste can even be directly used. Scholars from various countries are studying solid waste disposal to achieve the goals of harmless, reduced, and resource-oriented treatment and to resolve the contradiction between solid waste pollution and economic development, resource recycling, and environmental protection.

Current status of solid waste management at home and abroad The main experience of solid waste management in foreign countries is as follows: First, the efficient utilization of resources is a key trend. These countries have largely achieved the conversion of waste into resources through methods such as incineration for power or heat generation, composting, recycling, and reuse. Second, a well-designed incentive and restraint mechanism is essential. Economic incentives and legal regulations play a significant role in encouraging residents to participate in waste sorting. Third, a clear and operable waste classification system is crucial. Well-defined and practical sorting guidelines help residents effectively implement waste separation in daily life and reduce the cost of final treatment [55].

Examples of solid waste utilization At present, China's management of industrial solid waste is at a mature stage. While there are still technical breakthroughs to be made, the development prospects for related industries are very favorable. Methods such as compaction, crushing, sorting, solidification, incineration, and biological treatment are among the most widely used and highly effective approaches. The following will briefly introduce the treatment and disposal of solid waste using Mg-based cementitious materials.

Mg-based cementitious materials are the preferred binding agents for the reuse of waste residues and waste. They can be mixed with both organic and inorganic waste residues and waste to achieve cementation effects, and they allow for a larger amount of admixtures while maintaining strength, thereby enabling the digestion of more waste residues. Consequently, products made from Mg-based cementitious materials are more diverse, including handicrafts, grinding blocks, counterweight blocks, biogas digesters, trash cans, wall panels, flue ducts, and fireproof boards.

1. Utilization of fly ash in Mg-based cementitious materials
   Adding fly ash to Mg-based fireproof boards and partition boards can reduce the amount of cementitious material, lower density, enhance performance, and increase strength. In cement products, the incorporation of fly ash improves the workability of the slurry, reduces bleeding and segregation, and lowers the heat of hydration. Fly ash can be used in the production of Mg-based blocks, serving as both an admixture and a fine aggregate, with a high dosage potential. Rational use of fly ash can not only lower production costs but also improve product performance. Fly ash not only serves as physical filler but also acts as an active filler in the Mg-based cementitious material system. Its addition alters the microstructure system and optimizes the product performance. By adding an appropriate amount of fly ash, the 5·1·8 phase or 3·1·8 phase in the slurry is encapsulated on the surface of the fly ash particles. Over time, these small particles aggregate and undergo a series of hydration reactions, forming an artificial stone structure similar to talc powder [56], namely,

$$3MgO + 4SiO_2 + H_2O\text{—}\rightarrow 3MgO{\cdot}4SiO_2 + H_2O$$

   As for the fly ash used as a filler, the requirement is that the loss on ignition should be less than 8%, the content of active SiO_2 should be greater than 5%, and the clay content should be less than 1%. When the loss on ignition is greater than 8%, it will cause the residual carbon that has not combined with the cementitious materials to float on the surface of the product, affecting the secondary plastering processing and also affecting the strength of the product; moreover, the water requirement of fly ash is generally relatively high, usually 0.95–1.15 times its own weight and excessive addition will make the slurry and workability worse under specific solution usage conditions [56].

2. Application of waste building formwork in Mg-based cementitious materials
   In construction, wooden building formwork is often used. After multiple uses, the original wooden formwork becomes waste and cannot be used any longer. These discarded formworks can be first cut into different-sized strips, and then the Mg-based cementitious materials can be used to process these discarded formworks into wall panels. This not only provides fire-resistant, tough, and high-strength wall materials but also reduces pollution and saves energy.
3. Industrial solid waste and the application of Mg-based cementitious materials (a processing device and method for manufacturing handicrafts using waste materials, patent application number: CN202111201224.5)
   These can process construction solid waste into different-sized fragments and use Mg-based cementitious materials to make handicrafts or practical appliances of various shapes.
4. Other solid waste and the application of Mg-based cementitious materials
   Manufacturers of door core boards generate waste or scraps during production. One method of handling this waste is to directly crush it and add it to the door core board slurry. Another method is to collect the waste or scraps separately and place them in a mixing tank, adding a certain proportion of hydrochloric acid (or sulfuric acid) to produce brine (or magnesium sulfate solution).

This section only introduces several traditional technologies for combining solid waste with magnesium. With the continuous advancement of technology, more new technologies and methods for solid waste treatment are sure to emerge.

Outlook on solid waste treatment technologies The traditional solid waste treatment methods mostly use cement and other materials as the main components of the cementitious materials. During the cement production process, a large amount of CO is produced. Through the research and development of MJT, another solid waste treatment technology—solid waste-based cementitious materials—has emerged.

Solid waste-based cementitious materials are a new type of material with broad application prospects. They can effectively handle solid waste and can also be used as building materials and road base materials, etc. With the continuous advancement of technology and the expansion of application fields, solid waste–based cementitious materials will become an important means for solid waste treatment and resource utilization.

Through the synergistic effect of steel slag, mine slag, and desulfurization gypsum, most of the active substances in the mine slag are released to replace the clinker.

The raw materials for this technology are all made from industrial solid wastes (such as iron tailings, by-product gypsum, metallurgical slag, etc.). The concrete aggregates are all composed of iron tailings and waste rocks. According to the principle of "dual-synergy optimization of particle matching and activity," industrial waste residues are used as cementitious materials and are optimally combined with high-performance water reducers (or superplasticizers) to prepare fully solid waste–based concrete with lower heat of hydration and higher durability, completely replacing cement.

In summary, solid waste–based cementitious materials are a new type of material formed by mixing solid wastes with Mg-based materials, cement, lime, gypsum, etc. The preparation process of solid waste–based cementitious materials is simple, the raw materials are easy to obtain, and the cost is low, with broad application prospects.

1.5.2.2.3 Perlite

Perlite is a type of volcanic rock formed during ancient volcanic eruptions. The hot acidic magma erupted by volcanoes, upon cooling, contracts in volume and forms a glassy substance with a special pearl-like fissure structure, which is known as perlite [57]. China has an extremely abundant reserve of perlite, and the ore deposits have a high-grade. After being subjected to the calcination process, perlite expands, featuring low density and good fire resistance. It can be used in construction, petrochemical industry, horticulture, etc. In the field of construction, expanded perlite (EP) can be used as a raw material for cementitious materials such as magnesia cement, providing better sound insulation and heat insulation effects for walls and other structures.

Perlite mine The formation of perlite mines in China originated from large-scale volcanic eruptions during the Jurassic and Cretaceous periods. The entire mine belt runs from north to south across China, with perlite deposits distributed from Heilongjiang to the southern coast. The perlite mine belt can be divided into three subbelts: the Daxing'anling–Yanshan subbelt, the northeastern northern part–Shandong subbelt, and the southeastern coastal subbelt. Certain quantities of perlite deposits are distributed in various provinces and cities in the central and eastern parts of the country. The proven reserves are the highest in Henan Xinyang (Fig. 1.38).

Perlite ore consists of perlite, obsidian, and pitchstone. These three types differ significantly in appearance and moisture content: perlite features distinctive arc-shaped perlitic cracks with a moisture content of 2%–6%; obsidian itself exhibits a vitreous luster and conchoidal fracture, generally with a moisture content below 2%; pitchstone displays a characteristic pitchy luster and has a moisture content of 6%–10%.

The global reserves of perlite are extremely abundant, reaching an impressive 770 million tons. Aside from China, countries such as the United States and Greece are also major producers of perlite. The expansion process for perlite is straightforward, and the resulting product holds high added value. Both domestically in China and internationally, perlite production continues to increase year by year.

Production process of expanded perlite EP is commonly referred to as perlite, and it is a porous lightweight granular insulating material formed by the calcination and expansion of natural acidic volcanic ash–based glass rock. Perlite is mainly composed of 70%–76% amorphous silica and a small amount of metal oxides (such as Al_2O_3, Fe_2O_3, K_2O, and Na_2O), and the different contents and types of metal oxides result in different colors of the raw perlite ore. The commonly available EP on the market is mostly white or grayish white. Depending on the expansion ratio, the bulk density of EP

Figure 1.38 Shangtiantie pearl rock mine in Xinyang City, Henan Province.

generally ranges from 40 to 180 kg/m^3. At room temperature, the thermal conductivity of perlite is approximately 0.048 W/ (m·K), making it an excellent insulating material.

According to different heating production processes, perlite is divided into two major categories: open-cell perlite and closed-cell perlite. The main difference lies in the different firing time and temperature. Open-cell perlite uses a gas-fired high-temperature expansion furnace, which expands rapidly under high-temperature conditions, with a relatively short residence time at high temperatures above 800°C; while closed-cell perlite uses an electrically heated high-temperature expansion furnace, through precise control of the gradient heating of perlite ore and the precise control of the stagnation time, heating to above 1200°C, causing the surface of the perlite ore to melt, closing the pores, and forming a honeycomb-like structure inside. Closed-cell perlite overcomes the drawbacks of traditional EP, such as high water absorption rate, low strength, poor fluidity, and high thermal conductivity. It extends the application fields of EP [58]. Therefore the performance and usage of open-cell perlite and closed-cell perlite are different. Open-cell perlite has a light bulk density, weak strength, and strong water absorption, with a specific gravity of 80–110 kg/m^3, and the water absorption is generally greater than 200%; while closed-cell perlite has a high bulk density, good strength, and poor water absorption, with a bulk density of 100–150 kg/m^3, and the water absorption is generally less than 50%.

It should be noted that the expanded glassized microbeads used in production are closed-cell perlite. As defined in the standard, glassized microbeads refer to irregular spherical particles with a surface-baked and closed structure and internal porosity, formed by the high-temperature roasting and vitrification of vitreous volcanic lava sand. This definition does not include ordinary EP, hydrophilic modified ordinary EP, expanded vermiculite, sepiolite, and similar materials.

As a glassy substance, perlite has no fixed melting point. It generally softens at around 700°C. Therefore when expanding perlite, it is necessary to heat it above its melting point. The expansion principle of perlite is similar to that of popcorn. During the heating process of the raw perlite ore, the free water contained in the perlite will evaporate and disperse. When the heating temperature reaches above 700°C, the perlite softens, accompanied by the evaporation and expansion of the internal bound water, causing the perlite to expand and form a porous structure (Fig. 1.39).

By controlling the expansion temperature and time, the expansion ratio and porosity of perlite can be adjusted.

Hygroscopicity treatment process of expanded perlite EP possesses stable properties, is fire resistant, and is nonflammable. Owing to its high porosity of approximately 90%, its thermal conductivity coefficient is excellent, making it widely used in the production of various insulation and thermal preservation boards. In practical applications, however, the high porosity of EP results in a high-water absorption rate. This not only increases the water demand of the EP slurry (e.g., affecting the quality of the final product) but also makes the formed products difficult to dry, thereby reducing production efficiency. The high-water absorption rate severely limits the application scenarios of EP and renders hydrophobic treatment highly necessary.

Currently, there are three main methods for hydrophobic treatment of EP:

1. Coating method. A layer of hydrophobic agent is sprayed on the surface of the prepared ordinary EP products to form a hydrophobic film. While this method is straightforward, if the product's surface is damaged during construction, its hydrophobic properties will be compromised [59].
2. Penetration method. The prepared ordinary EP products are immersed in the hydrophobic agent for penetration treatment, making the interior of the products have certain hydrophobic properties. However, this process requires secondary drying and calcination, making it complex, and the surface of the products is prone to damage, resulting in high production costs and potential incomplete penetration, causing the

Figure 1.39 Physical image and microscopic structure image of expanded perlite.

products to lose their complete hydrophobic properties both inside and out [59].

3. Mixing method. The waterproofing agent is directly mixed into the slurry and then formed and cured to produce the product. This process is relatively simple. Since the surface of the raw material particles is all covered with hydrophobic agents to form a hydrophobic film and it also has good hydrophobicity in the cross-section, the mixing method is a commonly used hydrophobic treatment method in actual production [59].

Pumice rock can be used as insulation material for walls, roofs, ceilings, and other enclosure structures. It is also employed in preparing lightweight aggregate concrete, prefabricating various lightweight concrete components, and producing expanded pumice rock insulation and thermal preservation products, such as sodium silicate–expanded pumice rock, cement-expanded pumice rock, phosphate-expanded pumice rock, magnesite pumice rock, and other related products.

1.5.2.2.4 Ceramic granules

Ceramic granules are granular materials made of pottery. The main appearance features of ceramic granules are mostly round or elliptical spheres, and some imitation crushed stone ceramic granules are in the shape of irregular crushed stones [60] (Fig. 1.40).

The shape of the clay particles varies depending on the manufacturing process. The surface of the clay particles is a hard outer layer, which is either clay based or glazed. This layer has the function of waterproofing and gas retention, giving the clay particles higher strength [60].

Because the raw materials for producing silica balls are numerous, the varieties of silica balls are also diverse. The color of fired silica balls is mostly dark-red or ochre-red. Some special varieties are grayish-yellow, grayish-black, grayish-white, or bluish-gray. Nonfired silica balls have different colors depending on the type of solid waste used. They are generally grayish-black, have no luster on the surface, and are not as smooth as fired silica balls [60].

The particle size of the ceramic particles is generally 5–20 mm, with the maximum size being 25 mm. Ceramic granules are usually used to replace the crushed stones and pebbles in concrete.

Lightness is the most important aspect among the numerous excellent properties of ceramic particles, and it is also the main reason why they can replace heavy sand and gravel. The internal structure of ceramic particles is

Figure 1.40 Ceramic granules.

characterized by a fine honeycomb-like network of micropores. These micropores are all closed, rather than connected, and thus have better thermal insulation properties [60].

At the beginning of the invention and production of ceramic particles, they were mainly used in the field of lightweight partition boards for building materials. With the continuous development of technology and the deeper understanding of the properties of ceramic particles by people, the application of ceramic particles has long exceeded the traditional scope of building materials and has continuously expanded to new fields.

1.5.2.2.5 Aerogel

Aerogel is an emerging material with a nanoscale porous network structure. It was first prepared by American scientist Samuel Stephens Kistler in 1931. There are various types of aerogels, and they can be classified into silicon-based aerogels, carbon-based aerogels, sulfur-based aerogels, metal oxide-based aerogels, etc. Unless otherwise specified, the aerogels mentioned in this book refer to the common silicon-based aerogels (Fig. 1.41).

As the solid with the lowest density in the world, the density of aerogel can be as low as 3 kg/m^3 or less. It is called "solidified smoke." Due to the Rayleigh scattering effect, aerogel generally appears white with a slight bluish tint and is thus also called "blue smoke." The interior of aerogel contains a large number of nanometer-sized pores, where the gas is difficult to circulate, resulting in an extremely low heat conduction rate. Its thermal conductivity can be as low as 0.013 W/(m·K), making it a new favorite in the current insulation and thermal preservation industry.

Figure 1.41 Silica-based aerogel.

The development history of aerogels The development history of aerogels is a process where technological progress and application expansion mutually promote each other. Since the birth of aerogels in 1931, they have undergone four industrializations (Fig. 1.42). The first wave of industrialization was spearheaded by Monsanto Company in the United States but ultimately failed due to high costs and lagging application development. The second wave occurred in the 1980s, ending in failure as well, caused by high-temperature supercritical explosions and poor management of new technologies. In the third industrialization process, AspenAerogel from the United States successfully commercialized the aerogel and applied it in the fields of aerospace, military industry, and petrochemicals, which was highly favored by the market. Currently, aerogels are undergoing a fourth wave of industrialization, led by China, with new production capacity being added and their applications further expanding into civilian fields such as batteries and transportation.

Preparation process of silicon-based aerogels The preparation of aerogels generally involves two steps: the preparation of wet gel and the drying of the wet gel. When preparing silicon-based aerogels, the sol-gel method can be adopted. In a solvent composed of acetone and anhydrous ethanol, tetraethyl orthosilicate is used as the precursor and trimethylchlorosilane is added for hydrophobic modification to prepare the wet gel. After the wet gel ages, it undergoes solvent replacement treatment and is dried using methods such as supercritical drying, freeze drying, atmospheric pressure drying, and vacuum drying to obtain the finished aerogel [61]. Regardless of the drying method used, the key point is to minimize the damage to the gel structure caused by the surface tension

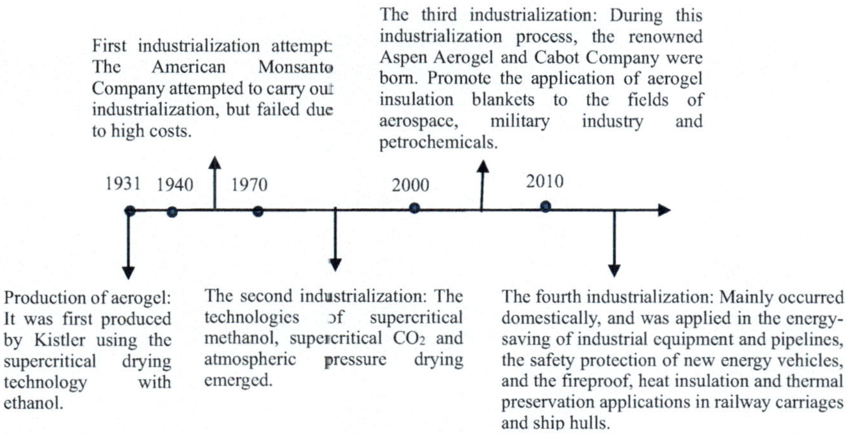

Figure 1.42 The fourth industrialization of aerogels.

of the solvent and ensure the integrity of the finished product (Fig. 1.43).

Application of aerogels in insulation materials At present, the most common and extensively studied aerogel materials are oxide-based aerogels. Aerogels used in the construction industry are mostly SiO_2 aerogels. SiO_2 aerogels possess the characteristics of heat insulation, light weight, chemical inertness, and reusability. They were the first to be produced and are also the ones that have been studied for the longest time, with the most mature sol-gel mechanism and the most perfect preparation process among all aerogels [62]. SiO_2 aerogels have excellent heat insulation and insulation performance. Various different products are constantly emerging on the market, including single-component aerogel powders, aerogel pastes, composite aerogel insulation and heat preservation mortar, aerogel insulation boards, aerogel felt, aerogel decorative integrated panels, and aerogel coatings. Test results show that the 1-mm-thick aerogel insulation coating has better heat insulation performance than the 8-cm-thick polystyrene board. Therefore applying aerogels to building construction can not only achieve energy conservation and emission reduction and promote sustainable development but also significantly save building space. The following briefly introduces several types of aerogel products.

1. The aerogel paste (Fig. 1.44) is a paste product obtained by compounding and modifying the aerogel powder with other substances.

Due to its low density and hydrophobic nature, aerogel powder is difficult to disperse in water-based systems. Moreover, during use, the

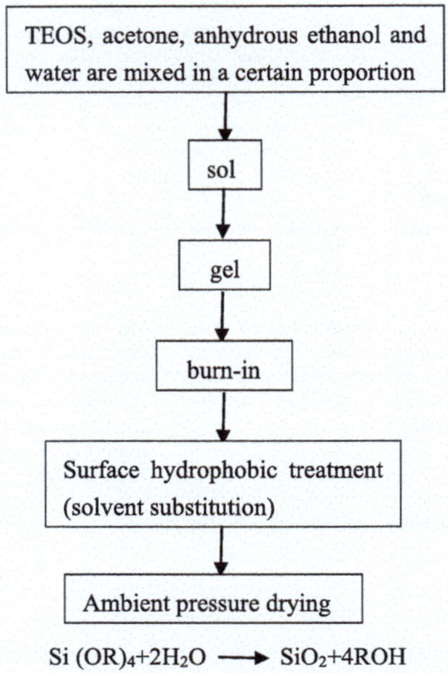

Figure 1.43 The production process and reaction equation of silica aerogel.

Figure 1.44 Silica aerogel paste preparation.

floating powder generates significant amounts of harmful dust. As a result, aerogel powder is challenging to apply in existing gelation systems. Once modified and processed into a paste, aerogel can be uniformly dispersed in water, effectively addressing the aforementioned problems encountered in practical production. It can be incorporated as a functional material into insulation and thermal barrier products, further reducing their thermal conductivity. The relevant product parameters are shown in Table 1.25.

2. Aerogel thermosetting composite polystyrene board (Fig. 1.45).

Table 1.25 Physical and chemical properties of SiO_2 aerogel paste.

Product model	Cream
Ingredients	SiO2 aerogel
Appearance	White paste like
Thermal conductivity/W/(m·K)	0.015 ~ 0.018
Temperature resistance range/°C	−60 ~ 600
Solid content/%	10 ~ 30
Density/g/cm^3	0.40 ~ 0.60
VOC content	0
pH value	6 ~ 8
Chloride	Not available

Figure 1.45 Aerogel thermosetting composite polystyrene board.

The currently prevalent internal and external insulation boards are typically manufactured using a process that integrates cementitious materials with lightweight insulating materials. However, their thermal conductivity often fails to meet the specifications required for homogeneous fireproof and insulation boards of grade 0.50 (i.e., thermal conductivity in the range of 0.045–0.050 W/[m·K]). This has led to the development of aerogel thermosetting composite polystyrene boards. In production, aerogel is commonly employed either as a substitute for lightweight aggregates or in combination with them. By incorporating SiO_2 aerogel into ordinary silicate cement or magnesium cement, the porosity of the system is increased, thereby improving its thermal and acoustic insulation properties, while also enhancing the strength and durability of the product. For instance, in the widely used Mg-based thermosetting composite polystyrene board, adding silica aerogel during the traditional manufacturing process reduces the product's thermal conductivity to below 0.045 W/(m·K), significantly improving its insulation and thermal performance (Table 1.26). Additionally, aerogel coatings can be applied to the product's surface to improve its surface properties.

However, compared with traditional insulation materials, the price of aerogel insulation and heat preservation materials is higher, which to some

Table 1.26 Performance parameters of aerogel thermosetting composite polystyrene board.

Test items	Standard request T/CIEEMA 002–2020 (050 grade)	Aerogel composite noncombustible insulation board	Testing standards
Density/kg/m^3	120 ~ 180	90 ~ 120	GB/T 5486—2008
Thermal conductivity/ W/(m·K)	0.045 ~ 0.050	≤0.042	GB/T 10294—2008
Tensile strength perpendicular to the plate surface/MPa	≥0.10	≥0.10	JG/T 536—2017
Compressive strength/MPa	≥0.15	≥0.15	GB/T 5486—2008
Dry shrinkage rate/%	≤0.6	≤0.8	JG/T 536—2017
Volume water absorption rate/%	≤8	≤5	GB/T 5486—2008
Flexural strength/MPa	≥0.2	≥0.22	GB/T 5486—2008
Softening coefficient	≥0.7	≥0.7	JG/T 536—2017
Fire grade	A (A2) grade	A (A2) grade	GB/T 14402—2007

extent increases the construction cost. In 2024, the price of ordinary building homogeneous insulation boards per cubic meter was 200–300 yuan. For example, some manufacturers in Xinjiang and Sichuan have already reduced the price to 220 yuan per cubic meter. However, in Anhui region, some manufacturers produce high-end homogeneous insulation boards, with a price as high as 1500 yuan per cubic meter, which is about seven times higher. Generally speaking, the customer base for high-end boards is smaller, the demand is lower, and the sales volume is relatively limited. If aerogel is used to improve the performance of the boards, it is recommended to add 0.5%–1% of aerogel powder to each cubic meter of the board. The market price of aerogel powder is over 200 yuan per kilogram. According to the reference dosage, the cost of homogeneous insulation boards with a density of 80–110 kg/m^3 will increase by more than 100 yuan per cubic meter. Therefore the cost issue is the main reason why this product has not been widely promoted and applied.

3. Aerogel water-based coatings and insulation mortar (Fig. 1.46). These two products have already undergone industrial production and need to be customized according to the downstream production products. Currently, they are mainly used for pipeline insulation and building wall insulation. Aerogel coatings can reach the building fire resistance grade A, which is higher than that of traditional insulation materials. In addition, aerogel coatings have strong crack resistance, preventing the cracking and even detachment of insulation materials and exterior finishes caused by thermal expansion and contraction.
4. Aerogel (Fig. 1.47) is generally composed of SiO_2 aerogel as the main material and is combined with carbon fibers or ceramic glass fiber cotton or preoxidized fiber through special processes to form a flexible insulation. Its characteristics include a low thermal conductivity, certain tensile, and compressive strength, while it can reduce the thickness of the insulation layer. It has excellent water repellency and fire resistance and is easy to install. It belongs to a new type of insulation material. The thermal conductivity is usually 0.013–0.025 W/(m·K), and the density is 3–250 kg/m^3.
5. New insulation material—transparent SiO_2 aerogel. Solar energy is an inexhaustible renewable resource. Traditionally, solar energy is utilized by means of reflection and absorption to collect solar radiation and convert it into thermal energy for utilization. Therefore the efficiency of solar energy conversion is a particularly important parameter for evaluating the performance of solar devices.

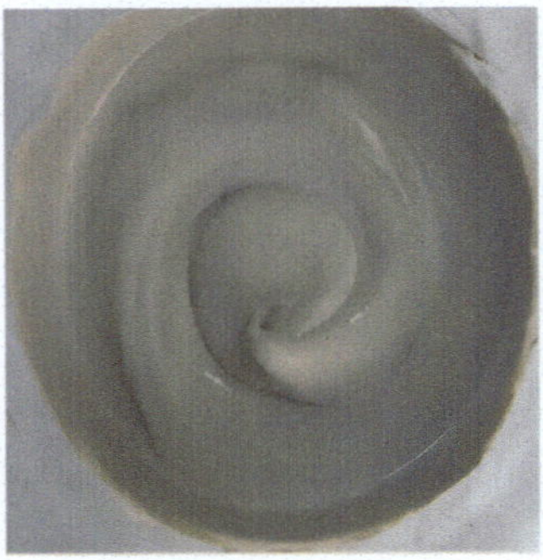

Figure 1.46 Aerogel insulation mortar.

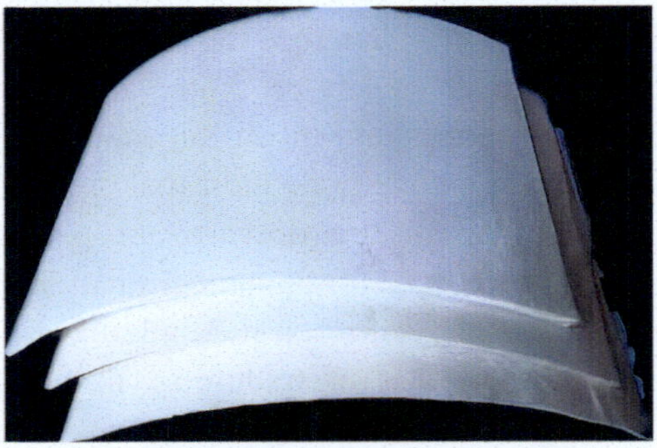

Figure 1.47 Aerogel blanket.

Common solar power generation devices have an efficiency of only about 20% in utilizing solar energy, while solar water heaters can achieve an efficiency of up to 60%. A research group from the Massachusetts Institute of Technology developed a new transparent aerogel and applied it to a solar thermal converter. Under natural light conditions without concentration or vacuum, the photothermal converter equipped with the transparent aerogel (Fig. 1.48) heated up to 265°C indoors and 220°C outdoors, significantly surpassing the performance of traditional photothermal converters. The role of the transparent aerogel is similar to that of a greenhouse, effectively trapping solar radiation and significantly increasing the utilization rate of solar energy.

Thanks to its low thermal conductivity, aerogel is well-suited for producing insulation and thermal retention materials. Various types of

Figure 1.48 Transparent SiO_2 aerogel.

aerogel-containing boards have already been used in the construction industry. In addition to traditional exterior wall insulation panels, transparent aerogel can also be applied in laminated glass production as a replacement for conventional insulating glass, providing better energy-saving performance in buildings. Although many manufacturers in the market have introduced insulation board products based on aerogel, their prices are higher than traditional EPS particle boards, making them less suitable for residential applications. Cost remains one of the key factors limiting the large-scale adoption of aerogel products.

With the in-depth promotion and implementation of the "carbon neutrality" strategy, aerogel was included in the emerging industry of building materials in June 2018. In September of the same year, the first national standard for aerogel materials, "Nanopore Aerogel Composite Insulation Products" GB/T—34336 (2017), was released and implemented. In December 2019, the National Development and Reform Commission issued a document encouraging the research and production of aerogel energy-saving materials. By November 2020, Shanxi Province adopted the "Technical Standard for Aerogel Insulation and Heat Preservation Coatings System," rectifying the chaotic market of aerogel coatings. The civilianization process of aerogel materials still requires continuous policy guidance and continuous promotion by enterprises.

References

[1] Shutao M, Yuezhong D, Jianping P, et al. Progress in the preparation of magnesium hydroxide from chlorite magnesite in Qinghai Salt Lakes. Salt Lake Research 2024;32(3):123–32. 1-10 [2024-04-28].

[2] DZ/T 0202-2002, Geological exploration specifications for bauxite and metallurgical magnesite ore. Industry Standard of the People's Republic of China.
[3] Junyan L, Yang Z, Zhiru W. Changes in mineral industry indicators from the additional edition of the mineral resources industry requirements manual. Land and Resources Intelligence 2013(6). 10+54-56.
[4] Haiyang B. Potential evaluation and development status of magnesite resources in Southern Liaoning Province. Mineral Protection and Utilization, 2016;(1):24–29.
[5] Xianglian F, Fanghong P, Zhongfeng T. Geological characteristics and genesis of Jianshan magnesite deposit in Shanshan County, Xinjiang. Xinjiang Nonferrous Metals 2012;35(S1):9–12.
[6] Unluer C, Al-Tabbaa A. Impact of hydrated magnesium carbonate additives on the carbonation of reactive MgO cements. Cement and Concrete Research 2013;54:87–97.
[7] Unluer C, Al-Tabbaa A. The role of brucite, ground gaulalfacsassa in the carbonation and performance of MgO cements. Construction and Building Materials 2015;94:629–43.
[8] Jinmei D, Hongfa Y, Ying L, et al. Comparative study on standard methods for quantitative analysis of active MgO. Salt Lake Research 2011;19(02):29–33.
[9] Guanghui L, Junxiang L, Qingbo Y. Research on the recycling and utilization of liquid magnesium chloride as a byproduct in sponge titanium production. Progress in Titanium Industry 2021;38(05):37–40.
[10] Finland M. Exploration of production process of anhydrous magnesium chloride. Salt Science and Chemical Industry 2017;46(08):20–2.
[11] Jinsheng L, Jie L, Li L, et al. New research on the preparation of boric acid and recovery of magnesium sulfate from boron rich slag. Mineral Comprehensive Utilization 2011(03):29–32.
[12] Zhiqiang N, Yuchun Z, Di Z, et al. Research on the preparation of magnesium sulfate heptahydrate using boron mud. Light Metals 2007(07):61–3.
[13] Liangqing Z, Lixiang P, Lifang S, et al. Research on the process of producing magnesium sulfate from by-products of magnesium based flue gas desulfurization. Environmental Engineering 2014;32(02):91–4.
[14] Ji T. Research on the by-product sodium chloride in the production of magnesium sulfate from white sodium magnesium aluminate and magnesium chloride. Salt Science and Chemical Industry 2018;47(09):38–9.
[15] Rui G, Xuefeng S, Xianyun Z, et al. The influence of different phosphates on the hydration and hardening properties of magnesium phosphate cement. Journal of Ceramics 2014;33(02):346–50.
[16] Jierui L, Kai Z, Fanghai L, et al. A review of methods for preparing potassium dihydrogen phosphate from wet process phosphoric acid. Modern Salt Chemical Industry 2021;48(01):3–4.
[17] Wei W, Yongqi W, Ming L, et al. Research on the process of preparing high-purity potassium dihydrogen phosphate by multi-step method. Shanghai Chemical Industry 2023;48(04):25–8.
[18] Zifu W, Hongtao C, Tianjian C, et al. The role and mechanism of modifiers in magnesite cementitious materials. New Building Materials 2011;38(06). 18-20+43.
[19] Y. Zhengbin. Research on improving the early strength and engineering performance of low temperature magnesium oxychloride cement. Changzhou University, 2022.
[20] Liu J.. The effect of retarder on the properties of magnesium oxychloride inorganic plywood. Zhejiang Agriculture and Forestry University, 2023.
[21] Yaowen C, Shigui T. Research on water resistance of magnesium magnesium cement products. New Building Materials 1991(6):4.

[22] Jiaohua S, Weiqiang L. Study on flexible acrylate lotion adhesive modified magnesium oxychloride cement. Bonding 2008(09). 38+53.
[23] Zhou W, Jianfang S, Haiyuan X, et al. Research progress on toughening modification of magnesium phosphate cement. Materials Review 2016;30(S1):454–7.
[24] Xiaojia Z, Xiaotao Z, Yujie Z, et al. Experimental study on improving the performance of magnesium oxychloride cement with new additives. Building Materials Technology and Application 2018(02):6–11.
[25] Z. Qinyu. Performance and interface study of modified magnesium phosphate cement. Nanchang University, 2023.
[26] X. Ming. Research on fast hardening high strength soil consolidating agent and engineering application technology. Chengdu University of Technology, 2012.
[27] Linlin Z, Yu Q, Xiuwen Q, et al. The effect of permeation crystallization reaction on the porosity of cementitious materials. Journal of Shihezi University (Natural Science Edition) 2018;36(02):201–6.
[28] Xiaojia Z, Yujie Z, Qianqian Z, et al. Experimental study on the influence of water-resistant and waterproof agents on the properties of magnesium oxychloride cement. Jiangsu Building Materials 2018(04):40–4.
[29] Wen X, Guanlong Q, Zilong H. Experimental study on chemical preparation of foam concrete. Concrete and Cement Products 2011(12):1–4.
[30] Hailong S, Chunlei G, Dong W. Research progress on inorganic chemical foaming cement technology. Commercial Concrete 2012(07):29–32.
[31] L. Qijin. Study on preparation and performance of low-density foam concrete thermal insulation material. Jinan University, 2014.
[32] X. Ye. Study on the adaptability of chemical foaming agent in alkali slag cementitious material system. Chongqing University, 2021.
[33] WONGKEO W, THONGSANITGARN P, PIMRAKSA K, et al. Compressive strength, flexural strength and thermal conductivity of autoclaved concrete block made using bottom ash as cement replacement materials. Materials & Design 2012;35(434-9).
[34] LACH M, KORNIEJENKO K, MIKULA J. Thermal insulation and thermally resistant materials made of geopolymer foams. Procedia Engineering 2016;151(410-6).
[35] ZHOU JIAQI, LIU PANPAN, WU CHENGYOU, et al. Properties of foam concrete prepared from magnesium oxychloride cement. Ceramics-Silik áty 2020;64(2):200–14.
[36] Weixin Z, Chenggong C, Xueying X, et al. Study on the suitability of foaming agent for magnesium oxychloride cement material. Salt Lake Research 2019;27(04):62–9.
[37] Dianzhong L, Shuming H, Ying L. Discussion on foam concrete foaming agent. Wall Material Innovation and Building Energy Conservation 2013(08):39–41.
[38] X. Lingyun. Application of foaming agent in lightweight grouting materials. Central South University, 2011.
[39] W. Zhijie. Research on Inorganic Composite Thermal Insulation Materials for Exterior Walls. Dalian Maritime University, 2011.
[40] Y. Xin. Research on foaming agent for low-density foam concrete. Northeastern University, 2015.
[41] Z. Wanfa. Experimental study on preparation of foam concrete thermal insulation material with rosin gum foaming agent. Hubei University of Technology, 2012.
[42] Zhisi C, Chang Y, Huan C, et al. Design and research of new energy-saving foam wall materials. Guangdong Building Materials 2011;27(11):77–9.
[43] Zhizhen X. Research on concrete foam agent. Jiangxi Building Materials 2000(03):5–8.
[44] Chengcan G, Yuansheng W, Hongwei Y, et al. Research progress of foam and defoamer. Material Development and Application 2010;25(06):81–5.

[45] Botao L. A brief discussion on green design in China—research on bamboo culture design. Art and Design (Theory) 2012;2(12):130–1.
[46] X. Yunchuan. Research on alkaline color distressed printing of polyester/cotton fabric. Suzhou University, 2010.
[47] Haimei Z. Application of polypropylene fiber in crack resistance of composite floor slabs. Building Structures 2009;39(S1):899–902.
[48] Promotion of new building materials. Architecture, 2003; (07):77.
[49] Q. Xiangpeng, R. Zhijin. Alternative materials for wooden packaging of machine tools//Beijing Society of Mechanical Engineering, Chongqing Society of Mechanical Engineering, Gansu Society of Mechanical Engineering, Heilongjiang Society of Mechanical Engineering, Jilin Society of Mechanical Engineering. In: Proceedings of the 5th science and technology forum of mechanical engineering societies in thirteen provinces, regions and cities Qizhong CNC Equipment Co., Ltd; 2009:286-288.
[50] Yuning W, Qiang P. Research progress on fiber-reinforced cement-based materials from agricultural waste. Journal of Guangdong University of Petrochemical Technology 2023;33(04). 70-73+81.
[51] Feng C, Hongxia Q, Shuangying L, et al. Mechanical properties and mechanism of action of biomass silicon modified magnesium oxychloride cement composite materials. Journal of Composite Materials 2023;40(10):5859–70.
[52] Bingbing Z, Yan X, Fei Y. Application of polyphenyl foam core formwork in medium and small span continuous beam slab bridge. Jiangsu Water Conservancy 2017(03). 11-13+20.
[53] Shuhua L, Peiyu Y. The mechanism of limestone powder in the hydration of composite cementitious materials. Cement Engineering 2008(06):6–8.
[54] Yixiong C. When economic leverage focuses on environmental protection. Modern Township 2008(07):24–9.
[55] W. Yaning. Research on the design of indoor public space modular intelligent garbage sorting bin. North China Electric Power University, 2022.
[56] Pingtao T. Misconceptions and analysis of production of magnesium oxychloride products—composition and main materials. Wall Material Innovation and Building Energy Efficiency 2005(12). 40-41+3-4.
[57] Qifeng L, Ying X. Optimization of pretreatment method for lead adsorption by perlite filter aid. Food Safety Guide 2022(25):79–81.
[58] Hong X. Application types and prospects of building insulation materials. China New Technology and New Products 2010(17):182.
[59] Zhifu L. Development of hydrophobic expanded perlite. Popular Standardization 2003(12):48–9.
[60] S. Guoqiang. The mountain is strong Study on the influencing factors of the properties of microporous lightweight mullite synthetic materials. Henan University of Science and Technology, 2010.
[61] L. Jianfang. Research on new high temperature resistant multilayer thermal insulation structure. Harbin Institute of Technology, 2007.
[62] Kistler SS. Coherent expanded-aerogels. The Journal of Physical Chemistry 1932;36(1):52–64.

CHAPTER 2

Theoretical ratio

This chapter primarily focuses on the theoretical composition ratios of magnesium-based cementitious materials, including chloromagnesium cementitious materials, sulfomagnesium cementitious materials, sulfur-chlorine composite cementitious materials, and magnesium phosphate cementitious materials. The transformation from various raw materials into a functional inorganic nonmetallic cementitious material involves complex chemical reactions. Throughout this process, continuous chemical reactions contribute to the formation of the cementitious matrix. However, if the initial raw material ratio exceeds the reasonable range, it can significantly adversely affect the final cementitious body, particularly in terms of mechanical strength, which is a key consideration in industrial applications.

To minimize the occurrence of the issues mentioned above, the appropriate range of feeding ratios for these four cementitious materials was determined through the research and analysis of a series of experimental data. Specifically, the theoretical molar ratio (based on experimental data and empirical ratios) was identified as follows:

1. $5Mg(OH)_2 \cdot MgCl_2 \cdot 8H_2O$ (5·1·8 phase) is the main strength phase of the chloromagnesium oxide cementitious material. As the molar ratio of MgO to $MgCl_2$ increases, this strength phase gradually forms. The research found that within the reference range of 6 to 10 for the theoretical molar ratio, the 5·1·8 phase was the main structural phase. Moreover, by controlling the molar ratio of $H_2O/MgCl_2$, the 5·1·8 phase could also be induced to form. At the same time, it should be noted that when the molar ratio of $H_2O/MgCl_2$ is too low, a layer of salt scale will form on the surface of the gel material.
2. $5Mg(OH)_2 \cdot MgSO_4 \cdot 7H_2O$ (5·1·7phase) is the main strength phase of the sulfur-magnesium gel material. As the molar ratio of $MgO/MgCl_2$ increases, the strength phase gradually forms. The research found that within the theoretical molar ratio range of 5 to 8, the 5·1·7 phase was the main structural phase. When the molar ratio of $H_2O/MgSO_4$ is

Magnesium-Based Cementitious Materials
DOI: https://doi.org/10.1016/B978-0-443-54075-2.00001-1

between 21 and 23, the 5·1·7 phase is the most stable, and the strength of the sulfur-magnesium cement is higher.

3. The sulfur-chlorine composite cementitious material is a four-component system. In fact, it still stems from the inherent strength phases of chloromagnesium cement and sulfomagnesium cement, which are produced individually. By combining the two in production, the aim is to reduce the free chloride content in the pure chloromagnesium cementitious system and enhance the stability of the product. When the raw material ratio is controlled within an appropriate range, the microstructure of the hydration products of sulfur-chloride composite magnesium cement becomes dense, thereby imparting high strength and good water resistance.
4. $Mg(M)PO_4 \cdot 6H_2O$ is the main strength phase of the phosphate magnesium cementitious material. By compounding with the commonly used potassium dihydrogen phosphate, the appropriate mass ratio to be referred to is calcined magnesium oxide: potassium dihydrogen phosphate: water: retarder=4:1:0.9:0.04. In fact, while ensuring the working performance, reducing the water-cement ratio can bring higher strength.

2.1 Chloromagnesium oxide cementitious material

Chloromagnesium cement was discovered by Sorel [1]. It is an air-hardening cement, which is formed by mixing light-burned magnesium oxide (usually calcined at 700°C–900°C) with a magnesium chloride ($MgCl_2$) solution. As one of the main representatives of mineral cement, chloromagnesium cement has attracted widespread attention due to its superior performance compared to ordinary Portland cement in many aspects, such as fast setting, low density, high mechanical strength, wear resistance, fire resistance, and resistance to the penetration of oils and paints [2]. Chloromagnesium cement exhibits excellent bonding capacity with a wide range of various fillers, such as gravel, sandstone, marble powder, asbestos, wood particles, and expanded clay. It also features a fast-setting time and high early strength [3]. Furthermore, the calcination temperature of MgO is lower than that of traditional Portland cement, and chloromagnesium cement products can absorb and sequester carbon dioxide from the atmosphere, making magnesium oxychloride cement (MOC) an environmentally friendly “carbon-neutral” cement.

Common chloromagnesium cement products include residential building floor materials, insulation and soundproofing materials, fire-resistant bricks, polished stone, artificial ivory, etc. However, due to the poor water resistance of chloromagnesium cement and the extremely high chloride ion concentration in the system, it is not suitable to use untreated reinforcing bars for reinforcement, which limits the application of chloromagnesium cement materials in various fields.

2.1.1 Reaction mechanism

Over the past few decades, extensive research has been conducted on the microstructure, reaction mechanism, and strength development of the MgO-$MgCl_2$-H_2O ternary system composed of the three basic raw materials. In the MgO-$MgCl_2$-H_2O ternary system, the hydrated phase is usually represented as x Mg $(OH)_2 \cdot yMgCl_2 \cdot zH_2O$. After the chloromagnesium cement solidifies, there are generally four main reaction crystal phases in the system, namely 2 phase (2Mg $(OH)_2 \cdot MgCl_2 \cdot 4H_2O$), 3 phase (3Mg $(OH)_2 \cdot MgCl_2 \cdot 8H_2O$), 5 phase (5Mg $(OH)_2 \cdot MgCl_2 \cdot 8H_2O$), and 9 phase (9Mg $(OH)_2 \cdot MgCl_2 \cdot 4H_2O$) [4]. The 3 phase and 5 phase remain stable at temperatures lower than 100°C, while the 2 phase and 9 phase only remain stable when the solidification temperature exceeds 100°C [5]. The 3 phase and 5 phase exhibit well-crystallized needle-like whiskers, which are also known as scroll-tubular whiskers [6] (Fig. 2.1). Due to the formation of interlocked and dense microstructures during the formation of these whiskers, this is also considered to be the main reaction product that causes MOC to harden and contribute to its strength. The 5 phase appears shortly after the chloromagnesium slurry is mixed (approximately 2 hours after mixing), has good spatial filling performance, a dense microstructure, and the lowest porosity among the four crystal phases. Therefore, in the design of the MOC system, it is more preferred to promote the formation of the 5 phase. According to the writing convention, the 3Mg $(OH)_2 \cdot MgCl_2 \cdot 8H_2O$ phase is generally abbreviated as 3·1·8 phase, and the 5Mg $(OH)_2 \cdot MgCl_2 \cdot 8H_2O$ phase is abbreviated as 5·1·8 phase.

$Mg(OH)_2$ is a poorly soluble substance, with a very low solubility in water. Even in a 30°Bé $MgCl_2$ solution, the solubility of $Mg(OH)_2$ is only 0.1903 g/L [7]. $Mg(OH)_2$ is difficult to ionize into Mg^{2+} and OH^- in brine, so no 5·1·8 phase and 3·1·8 phase will spontaneously form in brine. Research has found that by using X-ray diffraction (XRD) phase analysis technology to track the hydration reaction of chloromagnesium cement, it

Figure 2.1 Scanning electron microscopy image of chloromagnesium.

was not observed that the $Mg(OH)_2$ phase was formed in the early stage of the chloromagnesium cement reaction, that is, before the 5·1·8 phase appeared. Therefore, the "unified chemical theory" is questioned, and the view of basic salts is proposed. It is believed that the magnesium cement hydrate is formed by the combination of water-hydrogenated magnesium ions, Cl^-, OH^-, and H_2O. The hydration process and mechanism of chloromagnesium cement are explained [5,8].

The formation process of the hydrated phase in the MOC gel can be summarized as follows: $MgCl_2$ crystals dissolve in water, MgO powder is neutralized and dissolved, Mg^{2+} hydrolyzes and bridges to form $[Mg_x\ (OH)_y\ (H_2O)_z\ ]^{(2x-y)+}$, forming the initial state of the MOC gel, and then the amorphous gel compound transforms into a crystalline hydrated phase. The overall reaction equation is as follows:

$$3MgO + MgCl_2 + 11H_2O \rightarrow 3Mg(OH)_2 \cdot MgCl_2 \cdot 8H_2O \quad (2.1)$$

$$5MgO + MgCl_2 + 13H_2O \rightarrow 5Mg(OH)_2 \cdot MgCl_2 \cdot 8H_2O \quad (2.2)$$

2.1.1.1 Dissociation of $MgCl_2$ crystals in water

When $MgCl_2{\cdot}6H_2O$ is dissolved in water, the $MgCl_2$ molecules can be electrolyzed to generate chloride ions (Cl^-) and water-magnesium ions ($[Mg(H_2O)_6]^{2+}$). Due to the polarization effect of Mg^{2+}, some $[Mg(H_2O)_6]^{2+}$ can further hydrolyze, generating mononuclear water-hydroxylated magnesium ions and hydrogen ions, causing the solution to exhibit weak acidity. The following equation is presented:

$$MgCl_2{\cdot}6H_2O \rightarrow [Mg(H_2O)_6]^{2+}(aq) + 2Cl^-(aq) \tag{2.3}$$

$$[Mg(H_2O)_6]^{2+}(aq) \rightleftharpoons [Mg(OH)(H_2O)_5]^{+}(aq) + H^{+}(aq) \tag{2.4}$$

2.1.1.2 Reaction of MgO in $MgCl_2$ solution

When MgO is mixed with the $MgCl_2$ solution, the H^+ ions formed by the hydrolysis in the $MgCl_2$ solution undergo a neutralization reaction with MgO, causing the MgO powder to dissolve and the pH value of the system to increase. The possible reactions are as follows:

$$MgO + 2H^{+}(aq) + 5H_2O \Rightarrow [Mg(H_2O)_6]^{2+}(aq) \tag{2.5}$$

$$MgO + H^{+}(aq) + 5H_2O \Rightarrow [Mg(OH)(H_2O)_5]^{+}(aq) \tag{2.6}$$

2.1.1.3 Hydrolytic bridging reaction of mononuclear complexes

The dissolution of MgO in $MgCl_2$ solution and the increase in pH value not only led to the formation of mononuclear complexes but also induced these mononuclear complexes to connect with each other through shared ligands (OH^- or H_2O), resulting in the formation of some complexes with uncertain components between the two Mg^{2+}, as shown in the following equation:

$$\begin{aligned}&[(H_2O)_5Mg(OH)]^{+} + [H_2O - Mg(OH)(H_2O)_4]^{+} \\ &\Leftrightarrow [(H_2O)_5Mg - OH - Mg(OH)(H_2O)_4]^{2+} + H_2O\end{aligned} \tag{2.7}$$

$$\begin{aligned}&[(H_2O)_4Mg{-}OH{-}H_2O]^{+} + [H_2O{-}H_2O{-}Mg(H_2O)_4]^{2+} \\ &\Leftrightarrow [(H_2O)_4Mg{-}OH{-}H_2O{-}Mg(H_2O)_4]^{3+} + 2H_2O\end{aligned} \tag{2.8}$$

The bridging reaction produces bidentate, tridentate and polydentate complexes with higher charges. These complexes are further hydrolyzed to reduce the charge and release H^+ as follows:

$$\begin{aligned}&\left[(H_2O)_4Mg{-}OH{-}H_2O{-}Mg(H_2O)_4\right]^{3+}\\&\Leftrightarrow [(H_2O)_4Mg{-}OH{-}OH{-}Mg(H_2O)_4]^{2+} + H^+\end{aligned} \tag{2.9}$$

These hydrolysis and bridging reactions occur alternately and increase continuously as MgO dissolves in the $MgCl_2$ solution. This leads to the formation of multinuclear water-hydroxylated magnesium ions with an uncertain composition. This process involves the following order of hydrolysis-bridging reactions:

$$\begin{aligned}&xMg^{2+}(aq) + (z + y)H_2O\\&\rightleftharpoons \left[Mg_x(OH)_y(H_2O)_z\right]^{(2x-y)+} + yH^+\end{aligned} \tag{2.10}$$

2.1.1.4 Crystallization of the hydrate phase

The free H^+ generated from the hydrolysis bridging reaction is exactly what is needed for the hydrolysis and neutralization of MgO. The hydrolysis products of MgO further hydrolyze and bridge to form polynuclear complexes. In the early age, the content of MgO in the MOC gel bodies decreases, and the number of unidentifiable hydrolysis products increases.

The formation of less soluble hydrolysis products means a reduction in the water content of the MOC gel materials, and amorphous gel phases rapidly form within a few hours. Then, the amorphous gel bodies convert to crystalline phases after several days or weeks. The formation of these crystals is as follows:

$$\begin{aligned}&\left[Mg_x(OH)_y(H_2O)_z\right]^{(2x-y)+} + Cl^- + H_2O\\&\rightarrow \left[Mg_3(OH)_5(H_2O)_m\right]^+ Cl^-(4-m)H_2O\end{aligned} \tag{2.11}$$

$$\begin{aligned}&\left[Mg_x(OH)_y(H_2O)_z\right]^{(2x-y)+} + Cl^- + H_2O\\&\rightarrow \left[Mg_2(OH)_3(H_2O)_3\right]^+ Cl^- H_2O\end{aligned} \tag{2.12}$$

The theory regarding the formation of the gel phase in MOC is merely a relatively reasonable hypothesis. The formation mechanism of the hydrates in chloromagnesium cement still needs further exploration. This is extremely necessary for understanding the essence of chloromagnesium cement and for the research on the modification of chloromagnesium cement [9].

2.1.2 Molar ratio design

The hydration products in chloromagnesium oxide cement are rather complex. At room temperature, the main hydration products include $5Mg(OH)_2{\cdot}MgCl_2{\cdot}8H_2O$ (referred to as the 5·1·8 phase), $3Mg(OH)_2{\cdot}MgCl_2{\cdot}8H_2O$ (referred to as the 3·1·8 phase), and the $Mg(OH)_2$ gel [10] (Fig. 2.2).

In the chloromagnesium cement system, the composition, stability, and physical properties of the hydrate phase are related to many factors. During the setting and hardening process of chloromagnesium cement, the formation of the strength phase is a process controlled by ionic reaction kinetics. The ratio of the solid-liquid system and the temperature change during the reaction will affect the composition form of the final product and also determine the stability of chloromagnesium cement products under natural conditions.

2.1.2.1 The crystal structure of the chloromagnesium cement phase

In the chloromagnesium system, the structure of the hydrate is similar to that of the water-magnesite ($Mg(OH)_2$). The crystal structure of the chloromagnesium cement hydrate can be regarded as the product formed

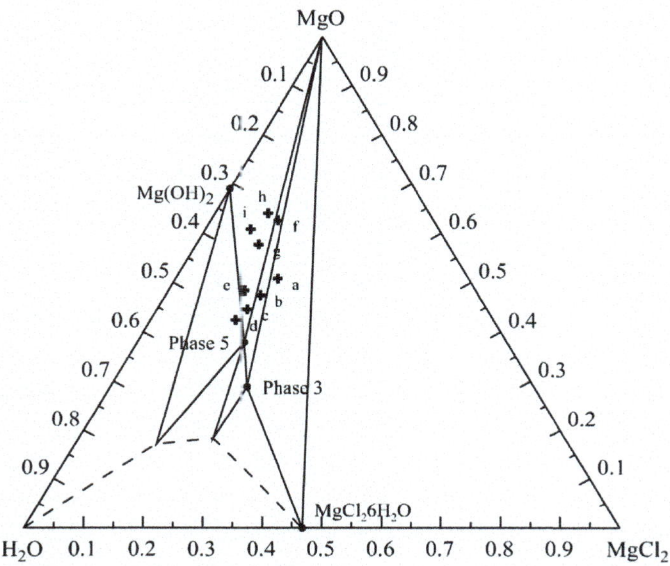

Figure 2.2 The ternary system of MgO-$MgCl_2$-H_2O at room temperature (the dotted line represents the approximate limit for the formation of uniform gel).

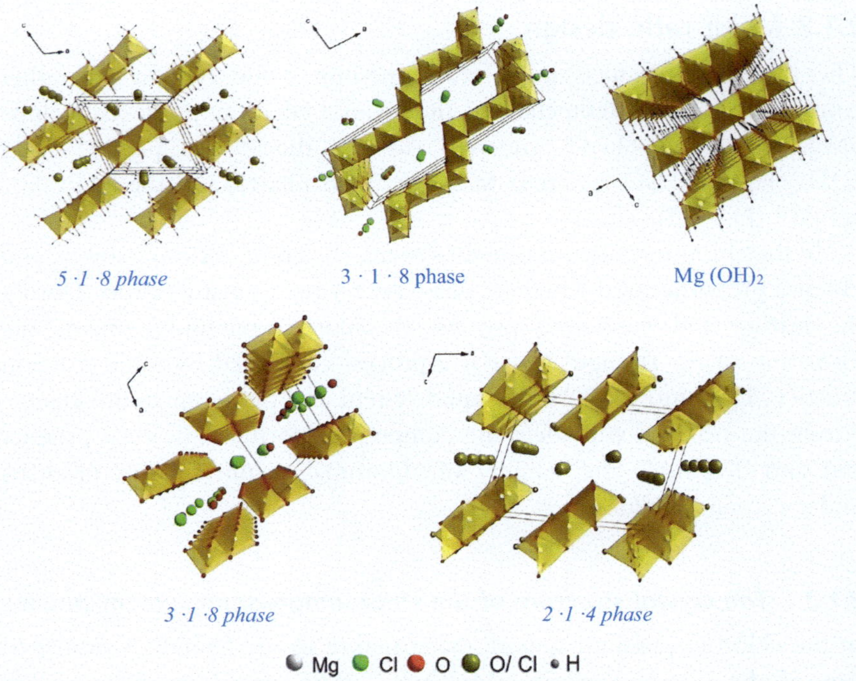

Figure 2.3 The crystal structure of the phase present in the chloromagnesium cement system.

when water-magnesite is eroded by chloride ions and water molecules. The crystal structure of $Mg(OH)_2$ is an octahedron formed by Mg^{2+} surrounded by oxygen atoms in the form of hydroxide ligands. Many octahedrons share the edges and connect with each other to form a layered structure (Fig. 2.3). From the perspective of chemical reactions, water-magnesite can be generated by the reaction of MgO with water. In the presence of a certain concentration of Cl^-, Cl^- and water can invade the closed octahedral layer sequence of water-magnesite and form a new octahedral structure in the form of chloromagnesium hydrate at a certain temperature.

The hydroxide ions in these octahedral layered structures combine with water molecules, and the chloride ions and water molecules between the layers form stable hydrogen bonds. The detailed crystal structure analysis is as follows [11].

The monoclinic 9·1·4 phase is composed of corrugated layers of distorted MgO_6 octahedra, separated by disordered Cl^- and H_2O molecules in "zigzag" interlayer gaps. Half of the atomic positions are occupied. Each MgO_6 octahedron has a coordination structure of only hydroxide ions, which are connected to eight adjacent octahedra through a common angle and to six octahedra through a common edge.

The monoclinic 5·1·8 phase has a crystal structure composed of an infinite-extending MgO_6 octahedron trichain framework and disordered chloride ions and water molecules inserted between them. The magnesium ions are fully coordinated with hydroxide and water molecules, similar to the structure of water mica and the 9·1·4 phase.

The triclinic 3·1·8 phase has a crystal structure composed of double chains of distorted $Mg(OH)_4(H_2O)_2$ octahedra. These double-chain structures form a layered structure similar to hydrotalcite with ordered chain-like structures composed of chloride ions and water molecules.

The main components of the monoclinic 2·1·4 crystal structure are infinitely extended three-chain edge-connected distorted octahedra, in which magnesium ions are not only coordinated with hydroxide ions and water molecules, but also with chloride ions. When the concentration of $MgCl_2$ solution is high, the 2·1·4 phase can crystallize at high temperatures. The priority of coordination of hydroxide and water molecules with the magnesium ions at the center of the unit cell is not high.

2.1.2.2 Molar ratio of magnesium oxychloride cement raw materials

Early studies have shown that the stability of 5 phase in water is better than that of 3 phase, and the spatial filling capacity of 5 phase is better. The microstructure is more compact, which significantly reduces the porosity of the product and accordingly improves the strength of the product [12]. When producing chloromagnesium cement products, efforts should be made to promote the formation of 5 phase to obtain stronger magnesium-based products.

The reaction of the chloromagnesium system is not complete. When designing the ratio, it is not possible to produce the corresponding crystal phase products simply by following the ratio of 5 phase or 3 phase. In fact, to ensure the large-scale formation of 5 phase, an excessive amount of magnesium oxide (MgO) needs to be added, that is, the molar ratio of $MgO/MgCl_2$ should be greater than 5.

Chloromagnesium cement is an MgO-$MgCl_2$-H_2O ternary system, and there exists an optimal ratio for the molar composition of these three substances. Under the optimal ratio, chloromagnesium cement exhibits excellent application performance. The molar ratio of the ternary system is generally divided into two parts: the molar ratio of $MgO/MgCl_2$ and the molar ratio of $H_2O/MgCl_2$. Here, the molar quantity of MgO is measured by the molar ratio of active MgO, while the molar ratio of $H_2O/MgCl_2$ determines the concentration and dosage of brine, which is subject to the workability and processability of the chloromagnesium cement paste [13].

Studies have shown that when the molar ratio of $MgO/MgCl_2$ is constant, as the molar ratio of $H_2O/MgCl_2$ decreases, the strength of the product increases; when the molar ratio of $H_2O/MgCl_2$ is fixed, the higher the molar ratio of $MgO/MgCl_2$, the higher the strength of the product. Assuming an active magnesium oxide content of 60%, when the molar ratio of $MgO/MgCl_2$ is lower than 6, the solidification time of the ternary system is longer, and the proportion of the 3 phases generated is higher; when the molar ratio of $MgO/MgCl_2$ is higher than 12, the content of $Mg(OH)_2$ in the chloromagnesium slurry significantly increases, and the strength of the product drops significantly [14].

Studies have shown that when the molar ratio of $H_2O/MgCl_2$ is fixed at 12, for different $MgO/MgCl_2$ molar ratios:

(1) When the $MgO/MgCl_2$ molar ratio is 4, more 3-phase structures are formed, and a small amount of 5-phase structures are formed;

(2) When the $MgO/MgCl_2$ molar ratio is 6, fewer 3-phase structures are formed, while more 5-phase structures are formed;

(3) When the $MgO/MgCl_2$ molar ratio further increases to 8–10, only 5-phase structures are formed, and no 3-phase structures are observed.

At lower $MgO/MgCl_2$ molar ratios, there is not enough MgO to form 5-phase structures, and the system tends to form 3-phase structures. As the $MgO/MgCl_2$ molar ratio increases, the available MgO increases, and the proportion of 5-phase structures also increases [15].

Further research indicates that the molar ratio of $MgO/MgCl_2$ is not the only variable that determines whether the crystal phase of the system is dominated by 3 phases or 5 phases. The molar ratio of $H_2O/MgCl_2$ also has a significant impact on the composition of the crystals within the system. When the molar ratio of $MgO/MgCl_2$ is fixed at 3 and the molar ratio of $H_2O/MgCl_2$ is 11, the chloromagnesium oxide system is more likely to form 3 phases; while when the molar ratio of $H_2O/MgCl_2$ is 13, 5 phases can be observed to dominate in the ternary system. This means that even in the case of a relatively small molar ratio of $MgO/MgCl_2$, by controlling the molar ratio of $H_2O/MgCl_2$, the formation of 5 phases can be guided.

It is worth noting that when the fixed molar ratio of $MgO/MgCl_2$ remains unchanged, as the molar ratio of $H_2O/MgCl_2$ continuously increases, the content of $Mg(OH)_2$ in the system will also gradually increase. This is because as the molar ratio of $H_2O/MgCl_2$ increases, the concentration of $MgCl_2$ in the mixed system decreases, and MgO slowly hydrates to form $Mg(OH)_2$. When the molar ratio of $H_2O/MgCl_2$ is too high, a large amount of MgO tends to form insoluble, amorphous $Mg(OH)_2$, and no longer participates in the formation of strong-phase structures such as 3-phase or 5-phase, instead causing the proportion of the strong-phase to decrease, and thus the strength of the product also significantly decreases. Based on the actual production conditions, usually under the condition of higher summer temperatures, it is appropriate to reduce the brix of the brine and increase the water-to-cement ratio to adapt to the production operation. However, the brix of the brine generally will not be lower than 20°Bé; otherwise, it will lead to poor strength of the product after processing and solidification and reduce the functionality of use. By simply calculating with 20°Bé brine and 60% active calcined magnesium oxide as raw materials, the reference material usage is 1 kg of magnesium oxide and 1.2 kg of brine, and the calculated molar ratio of $H_2O/MgCl_2$ is 26, which is far higher than the appropriate theoretical ratio. The water usage is too large, so it will cause the moisture content in the solidification process of the product to be too high, affecting the formation of the 5·1·8 strength phase, and thus affecting the strength of the product.

When the molar ratio of $H_2O/MgCl_2$ is too low, that is, when the brix degree of the brine is too high, a large amount of $MgCl_2$ will not participate in the chemical reaction and instead will exist in the pores between the crystals. Under certain conditions, it will reconvert into hydrated $MgCl_2$ salt. Excessive $MgCl_2$ is prone to absorb water in a humid environment, and water droplets tend to appear on the surface of the product. When these water droplets evaporate, the hydrated $MgCl_2$ salt in the interior of the product will dissolve and, under capillary action, accumulate along the pores between the crystals to form salt frost on the surface of the product. According to production experience, when the temperature is low in winter, and during the production of chlorine magnesium cement products, it is advisable to appropriately increase the brix degree of the brine to adapt to the production operation. However, usually the brix degree of the brine will not exceed 28°Bé; otherwise, an obvious salt frost reliquefaction phenomenon will occur on the surface of the product. With a 28°Bé brine and 60% active magnesium oxide content in the calcined magnesium oxide as raw materials, a simple calculation shows that the molar ratio of $MgO/MgCl_2$ is 3.8, which is far below the appropriate theoretical ratio. The amount of magnesium chloride is too large, resulting in a large amount of magnesium chloride not participating in the reaction, thus causing the above-mentioned reliquefaction and salt frost problem.

The influence of moisture on chloromagnesium oxide products is throughout their entire life cycle. Under the action of moisture, the crystalline phases in the products will continuously develop and transform. The various strength phases in chloromagnesium cement are not water-resistant by themselves. Whether it is the excess water in the slurry or the later infiltrating moisture, it will cause the crystal phases in the products to decompose and accumulate $MgCl_2$ with higher solubility, leaving the insoluble $Mg(OH)_2$. The erosion speed of chloromagnesium materials by water depends on factors such as the degree of moisture invasion, time, and speed. In addition, chloromagnesium cement products will also be eroded by carbon dioxide and undergo chemical reactions to be carbonized, generating chlorocarbonate magnesium salts, and then continuously transforming into hydrated magnesite, water-magnesite, and magnesite, etc. The presence of moisture often accelerates the carbonization process of chloromagnesium cement. The carbonization process may involve the following reactions:

$$5MgO \cdot MgCl_2 \cdot 8H_2O + CO_2 + H_2O \rightarrow 2MgCO_3 \cdot Mg(OH)_2 \cdot MgCl_2 \cdot 6H_2O \quad (2.13)$$

$$3MgO \cdot MgCl_2 \cdot 8H_2O + CO_2 + H_2O \rightarrow 2MgCO_3 \cdot Mg(OH)_2 \cdot MgCl_2 \cdot 6H_2O \quad (2.14)$$

$$2MgCO_3 \cdot Mg(OH)_2 \cdot MgCl_2 \cdot 6H_2O + CO_2 + H_2O \rightarrow 4MgCO_3 \cdot Mg(OH)_2 \cdot MgCl_2 \cdot 4H_2O + MgCO_3 \cdot Mg(OH)_2 \cdot 3H_2O + MgCO_3 \quad (2.15)$$

2.1.2.3 The influence of temperature on the phases of magnesium oxychloride cement

The solid-liquid equilibrium in the MgO-$MgCl_2$-H_2O ternary system is related to temperature, and the solidification temperature of the system also affects the types of hydrated products formed within the system. The solidification of chloromagnesium cement is a exothermic reaction. For every gram of magnesium oxide participating in the reaction, the released heat is approximately 1000–1350 J, while the hydration heat of ordinary silicate cement is only 300–400 J/g. The chloromagnesium system has a fast solidification rate and concentrated heat release. The internal temperature of large-volume chloromagnesium cement products can reach above 140°C, and unstable crystal phases are generated. At higher solidification temperatures, chloromagnesium cement products often deform and crack, and these structural defects will greatly accelerate the softening and brittleness of the products.

Studies have shown that the different solidification temperatures of the chloromagnesium system result in different hydration products. At lower temperatures (5°C–10°C), the chloromagnesium cement system is more likely to generate 3-phase products; under 20°C–50°C conditions, the 5-phase dominates the system; when the environmental temperature during solidification reaches above 65°C, 9-phase begins to appear in the system, indicating that the actual internal temperature of the sample block has already reached above 80°C, and the chemical reaction process is as follows:

$$3MgO + \left\{MgCl_2 + 11H_2O\right\}_{mixingliquid} \xrightarrow{T\geq 80^{\circ}C,\ t} \frac{1}{3}\left[9Mg(OH)_2 \cdot MgCl_2 \cdot 4H_2O\right] + \left\{\frac{2}{3}MgCl_2 + \frac{20}{3}H_2O\right\}_{poresolution} \quad (2.16)$$

Through XRD analysis, the composition of the chloromagnesium oxide strength phase was studied under different curing temperatures. In the sample blocks cured in an environment of 20°C–35°C, no phase 9 was observed at all. However, in the sample blocks cured at 50°C, a small amount of phase 9 had already appeared. This indicates that when the curing temperature of the chloromagnesium cement products reaches above 50°C, the final strength of the products has begun to decline. Further research revealed that in the sample blocks cured at 50°C and 65°C, the content of unreacted MgO decreased, which also corroborated with the formation of phase 9—in cases where the molar ratio of MgO to $MgCl_2$ is high, the amount of MgO consumed by phase 9 is greater [15]. In production practice, this is usually manifested as an excessive reaction, a weakened ability of MgO to combine with $MgCl_2$, resulting in excess $MgCl_2$ and subsequently the occurrence of the back-saltation phenomenon.

In summary, during the curing process of chloromagnesium cement, the formation of various strength phases is a process controlled by kinetics. The equilibrium of the strength phase precipitation–dissolution reaction is a function of temperature. By controlling a certain material ratio, the types of strength phases can be artificially controlled. In the production of chloromagnesium cement products, the molar ratio of $MgO/MgCl_2$ and the molar ratio of $H_2O/MgCl_2$ affect the development of the internal microstructure of the products, and the curing environmental temperature selectively promotes the formation of certain crystal phases. These three indicators play a decisive role in the final performance of the products. The specific numerical values of the molar ratios need to be determined according to the performance requirements of the products, the production process and other conditions. The material ratio should also be dynamically adjusted according to changes in environmental temperature. In the appendix of this book, there are basic formulas for common types of chloromagnesium products for colleagues to exchange and use.

2.2 Magnesium sulfate cementitious material

2.2.1 Reaction mechanism of magnesium sulfate cementitious materials

Magnesium oxide sulfur (MOS) cement, like chlorinated magnesium cement, is also an air-hardening cementitious material. It was first invented and reported by Belgian scholar Thaisa Demediuk in 1957. MOS cement is produced by mixing a certain amount of calcined magnesium oxide with a certain concentration of magnesium sulfate solution. Compared with traditional silicate cement, MOS cement has the characteristics of fast hardening, good fire resistance, low thermal conductivity, and low density. It also has excellent reinforcing properties, wear resistance, and chemical resistance. MOS cement is widely used and can be used as an adhesive for lightweight panels or as a fireproof material or insulating material. Due to the lower strength of MOS cement products compared to chlorinated magnesium cement products, MOS cement is limited in its application in products with certain load requirements [16].

Depending on the temperature and formation conditions, it is known that in the $MgO\text{-}MgSO_4\text{-}H_2O$ system, within the temperature range of 30°C–120°C, there may exist four primary phases, and four types of sulfate oxygen complexes can be formed. The main hydrates of MOS clay include $5Mg(OH)_2{\cdot}MgSO_4{\cdot}3H_2O$ (5·1·3 phase), $3Mg(OH)_2{\cdot}MgSO_4{\cdot}8H_2O$

(3·1·8 phase), $Mg(OH)_2 \cdot MgSO_4 \cdot 5H_2O$ (1·1·5 phase), and $Mg(OH)_2 \cdot 2MgSO_4 \cdot 3H_2O$ (1·2·3 phase). The main mineral phases in the hardened body of sulfur-magnesium cement formed at room temperature are the 3·1·8 phase, as well as the metastable 1·1·5 phase, along with a large amount of unreacted $MgSO_4 \cdot nH_2O$ (n=7, 6, 4, 1) and MgO. The 5·1·3 phase can only be stable under steam curing conditions [17].

Through continuous experimental research, the researchers found that in the modified magnesium oxide-sulfur cement, a new hydration product, $5Mg(OH)_2 \cdot MgSO_4 \cdot 7H_2O$ (5·1·7 phase), was successfully extracted [18]. Currently, the mechanism of inducing the formation of the 5·1·7 phase by chemical admixtures is still controversial. The influence laws of the material composition of the ternary system on the types, formation process, and production quantity of the hydration product phases also needs to be further studied.

Among the various crystal phases in the magnesium oxide-sulfur cement system, $Mg(OH)_2$ is loose and porous, which has a negative impact on strength. The needle-like 5·1·7 phase has a higher strength and is basically insoluble in water, and its performance is superior to the 3·1·8 phase and the 5·1·3 phase. After modification, the crystallinity of the 3·1·8 phase or other phases in the magnesium oxide-sulfur cement significantly decreases, and the 5·1·7 phase exists in the form of needle-like crystals or whiskers, which is the reason why the magnesium oxide-sulfur cement has good water resistance [19–21].

$$3MgO + MgSO_4 + 11H_2O \rightarrow 3Mg(OH)_2 \cdot MgSO_4 \cdot 8H_2O \quad (2.17)$$

3·1·8 phase formation mechanism: MgO dissolves in the $MgSO_4$ solution and forms $[Mg(H_2O)_xOH]^+$ and OH^-. As the alkalinity of the system increases, the surface of MgO adsorbs SO_4^{2-} and Mg^{2+}. After the pH value continues to rise, $[Mg(H_2O)_xOH]^+$ transforms into $Mg(OH)_2$, generating a phase with lower crystallinity, 3·1·8, resulting in a decrease in the pH value. Subsequently, the 3·1·8 phase is continuously generated and deposited. The reaction equation is as follows (2.17).

$$5MgO + MgSO_4 + 12H_2O — 5Mg(OH)_2 \cdot MgSO_4 \cdot 7H_2O \quad (2.18)$$

Under the action of the modifier, phase 5.1.7 is continuously formed. The reaction equation is as shown in Eq. (2.18). The specific hydration process is divided into five stages [22]:

2.2.1.1 The initial induction stage (not exceeding 3 hours)

MgO dissolves in the $MgSO_4$ solution. The chemical reaction occurs rapidly, generating a large amount of OH^-. This corresponds to the time point when the system releases a lot of heat. The formation of $[Mg(H_2O)_xOH]^+$ leads to a rapid increase in the pH value of the solution. The reaction process is as follows:

$$MgO(s) + (x + 1)H_2O \rightarrow [Mg(H_2O)_xOH]^+(sf) + OH^-(aq) \quad (2.19)$$

2.2.1.2 Induction period (3–5.4 hours)

$[Mg(H_2O)_xOH]^+$ is coated on the surface of MgO particles, hindering their further hydration reaction. Under alkaline conditions, $Mg(OH)_2$ is formed in the system, resulting in a decrease in hydration heat. According to coordination theory, magnesium ions are more inclined to form magnesium hydroxide rather than magnesium bisulfate hydroxide. The process is as follows:

$$[Mg(H_2O)_xOH]^+(sf) + OH^-(aq) \rightarrow Mg(OH)_2 + xH_2O \quad (2.20)$$

Without the effect of additives, the unstable $[Mg(H_2O)_xOH]^+$ hydration layer will rapidly react to form $Mg(OH)_2$. However, under the influence of additives, the $[Mg(H_2O)_xOH]^+$ hydration layer in the cementing system can interact with the additives to form an organic-magnesium complex layer $[CA^{n-} \rightarrow Mg(OH)(H_2O)_{x-1}]$, reducing the positive charge on the hydration layer of magnesium oxide surface and lowering the surface energy of $[Mg(H_2O)_xOH]^+$, thereby inhibiting the rate at which it reacts with water to form $Mg(OH)_2$ [23]. The reaction process is as follows:

$$\begin{aligned} &CA^{n-} + [Mg(H_2O)_xOH]^+ \\ &\rightarrow [CA^{n-} \rightarrow Mg(OH)(H_2O)_{x-1}] + H_2O \end{aligned} \quad (2.21)$$

2.2.1.3 Acceleration period (5.4 ~ 14.7 hours)

As MgO continues to hydrate, the rate of water release and heat increases, and the organic-magnesium complex layer continuously adsorbs SO_4^{2-} and Mg^{2+}. When OH^- reaches a certain concentration, the organic-magnesium complex layer generates a low-crystalline phase nucleus $xMg(OH)_2{\cdot}yMgSO_4{\cdot}zH_2O$ in the SO_4^{2-} solution. The formation of the crystalline phase continuously destroys the $[Mg(H_2O)_xOH]^+$ layer, causing the exposure of the active surface of the MgO that was previously covered, allowing the MgO to further hydrate. The acceleration of the reaction rate in this stage is mainly attributed to the formation of the crystal phase and the hydration of the new MgO. The reaction process is as follows:

$$\begin{aligned} &\{SO_4^{2-} \rightarrow [CA^{n-} \rightarrow Mg(OH)(H_2O)_{x-1}]_4 \rightarrow Mg^{2+}\}_{(surface)} \\ &+6OH^- \rightarrow 5Mg(OH)_2\cdot MgSO_4\cdot 7H_2O_{(nucles)} \\ &\qquad + (4x-13)H_2O + CA^{n-} \end{aligned} \tag{2.22}$$

$$\begin{aligned} &5Mg(OH)_2\cdot MgSO_4\cdot 7H_2O_{(nucles)} \\ &\rightarrow 5Mg(OH)_2\cdot MgSO_4\cdot 7H_2O_{(s)} \end{aligned} \tag{2.23}$$

2.2.1.4 Deceleration period (14.7–58 hours)

As the crystal phases continue to form, the ion concentration in the solution gradually decreases, and the slurry begins to solidify and harden [22]. Therefore, the diffusion rate of ions in the aqueous phase towards the MgO surface decreases, resulting in a reduction in the pH value of the system and the heat release rate during hydration.

2.2.1.5 Stable period (more than 58 hours)

The heat release rate during hydration remains at a relatively low level. The decrease in the ion concentration in the liquid phase and the reduction in the contact area between MgO and water molecules lead to a very slow formation of crystal phases. During this stage, the heat release rate remains at a low level, and the formation of phase 5.1.7 is very slow [24].

2.2.2 Molar ratio

2.2.2.1 Sulfur-magnesium crystal structure

The lattice of the 5·1·7 phase crystal is an infinite three-chain structure, mainly formed by MgO_6 coordination octahedra through self-assembly processes. Water molecules and sulfate tetrahedra are embedded into the crystal lattice through lattice forces (Fig. 2.4).

2.2.2.2 Molar ratio

Research indicates that the phase transformation and stability of hydrates in MOS cement slurry are mainly related to the molar ratio of $MgO/MgSO_4$ and $H_2O/MgSO_4$ (Fig. 2.5). The value of the molar ratio has a significant impact on the compressive strength of MOS cement [25].

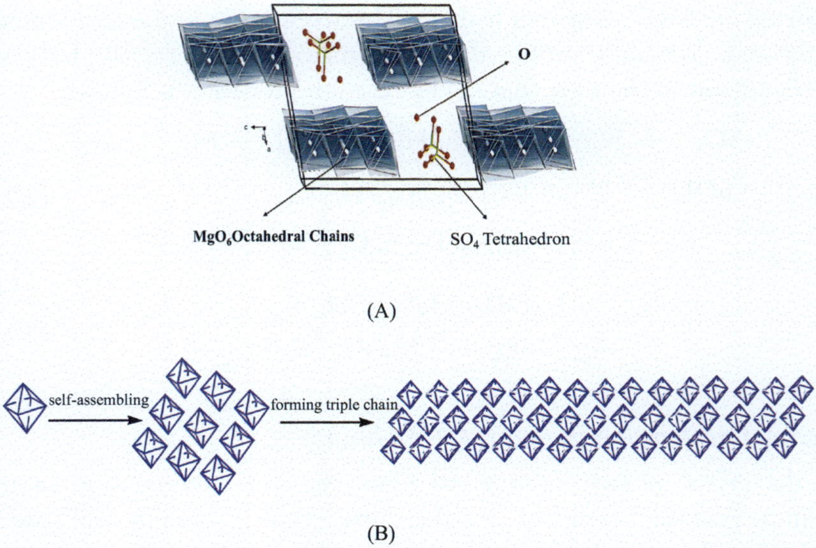

Figure 2.4 Structure of magnesium sulfate crystal. (A) Unit cell structure of the 5·1·7 phase. (B) Self-assembly process of the 5·1·7 phase.

Studies have shown that, when the molar ratio of $H_2O/MgSO_4$ is fixed, as the molar ratio of $MgO/MgSO_4$ increases (not exceeding 11), the compressive strength of the sulfur-magnesium cement paste increases. When the molar ratio of $MgO/MgSO_4$ is fixed, the lower the molar ratio of $H_2O/MgSO_4$ (that is, the higher the concentration of magnesium sulfate), the higher the strength of the sulfur-magnesium cement system. Under different ratios, the final strength of the test blocks can reach 20–100 MPa.

When the molar ratio of $MgO/MgSO_4$ is fixed at 9, compared with the system with a $H_2O/MgSO_4$ molar ratio of 28, the sulfur-magnesium system with a $H_2O/MgSO_4$ molar ratio of 20 can generate more 5·1·7 phases and has a lower production of $Mg(OH)_2$. When the molar ratio of $H_2O/MgSO_4$ is fixed at 20, under the condition of $MgO/MgSO_4$ molar ratio of 9, more 5·1·7 phases are generated compared to the condition of $MgO/MgSO_4$ molar ratio of 7, and the production of $Mg(OH)_2$ also decreases accordingly.

Furthermore, when the molar ratio of $H_2O/MgSO_4$ remains unchanged, as the molar ratio of $MgO/MgSO_4$ increases, the content of residual MgO in the system slightly increases. This indicates that MgO has not fully reacted. An excessively large molar ratio of $MgO/MgSO_4$ will result in more residual MgO. The residual magnesium oxide can play a role

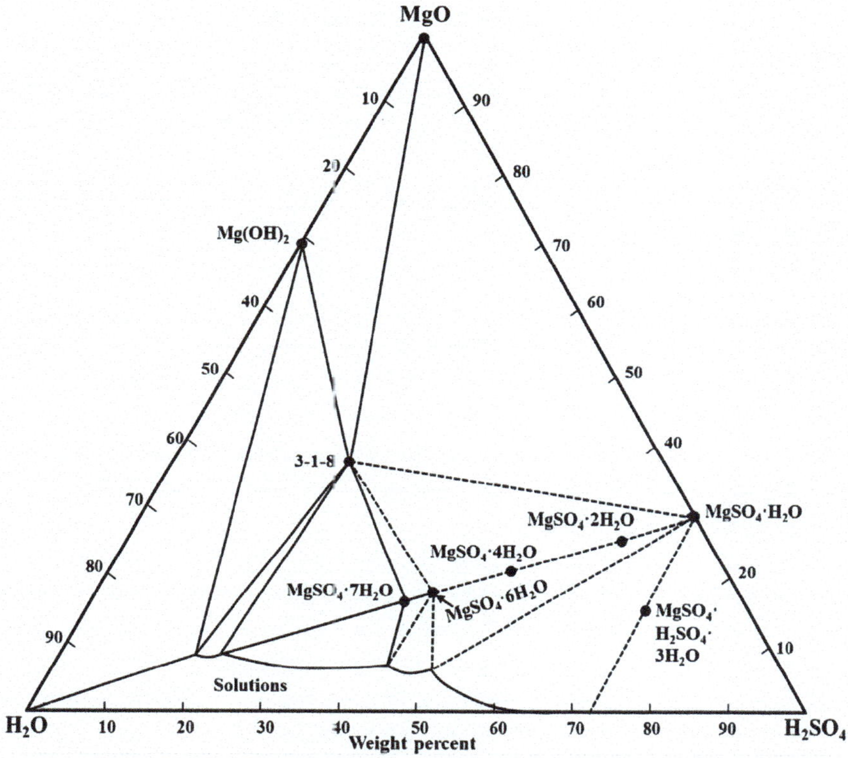

Figure 2.5 The ternary system of MgO-$MgSO_4$-H_2O at room temperature (the dotted line represents the approximate limit for the formation of uniform gel).

as a filler in the hardened MOS cement paste, which is beneficial for improving the strength of the system. Generally, a lower molar ratio of $H_2O/MgSO_4$ (i.e., a higher concentration of magnesium sulfate solution) and a higher molar ratio of $MgO/MgSO_4$ are conducive to the formation of phase 5·1·7, and the higher the molar ratio of $H_2O/MgSO_4$, the greater the amount of $Mg(OH)_2$ produced.

Further research indicates that the composition of the hydration products in the magnesium oxide-sulfur cement system is relatively sensitive to the amounts of MgO and H_2O. The proportion of the 5·1·7 phase increases with the increase in the amount of MgO, while the 3·1·8 phase gradually disappears.

The 3·1·8 phase exists only when the water usage is low, and the 5·1·7 phase gradually appears as the amount of H_2O increases. When the molar ratio of H_2O to $MgSO_4$ is sufficiently large, only $Mg(OH)_2$ is present, and the strength and water resistance of the magnesium oxide-sulfur cement slurry are greatly reduced.

At a molar ratio of $H_2O/MgSO_4$ of 18, the internal composition of the magnesium oxide-sulfur system changes as the $MgO/MgSO_4$ molar ratio increases as follows:

When the $MgO/MgSO_4$ molar ratio is lower than 3, the hydration products are mainly the 3·1·8 phase.

When the molar ratio of $MgO/MgSO_4$ is 3, the 5·1·7 phase begins to appear in the sulfur-magnesium system. As the molar ratio of $MgO/MgSO_4$ increases, the 5·1·7 phase gradually emerges and coexists with the 3·1·8 phase. As the molar ratio of $MgO/MgSO_4$ increases from 3 to 5, the compressive strength of the sulfur-magnesium cement products increases, which may be attributed to the accumulation of the 3·1·8 phase and the 5·1·7 phase, indicating that in the case of a high MgO content, the 5·1·7 phase is more likely to form.

When the molar ratio of $MgO/MgSO_4$ is greater than 5, as the 3·1·8 phase disappears, the 5·1·7 phase becomes the dominant product. Due to the lack of sufficient $MgSO_4$ solution for the reaction, the amount of MgO is excessive, and the amount of $Mg(OH)_2$ also increases.

When the molar ratio of $MgO/MgSO_4$ increases continuously from 5 to 13, the compressive strength of the sulfur-magnesium cement decreases, and the amount of $Mg(OH)_2$ increases.

When the molar ratio of $MgO/MgSO_4$ is fixed at 8, the internal composition of the sulfur-magnesium system changes with the increase of the $H_2O/MgSO_4$ molar ratio as follows:

When the molar ratio of $H_2O/MgSO_4$ is relatively small, only phase 3·1·8 exists in the gel.

When the molar ratio of $H_2O/MgSO_4$ is around 15, the stage 5·1·7 begins to appear in the 3.3 Sulfur-Chlorine Composite Clinker Materials system, and phase 3·1·8 gradually decomposes.

When the molar ratio of $H_2O/MgSO_4$ reaches 23, $Mg(OH)_2$ begins to appear, and the proportion of phase 5·1·7 starts to decrease at this time.

When the molar ratio of $H_2O/MgSO_4$ approaches 32, the phase 5·1·7 completely disappears, and the sulfur-magnesium cement forms a loose gel structure. The further increase of the molar ratio of $H_2O/MgSO_4$ leads to the accumulation of $Mg(OH)_2$ and gradually causes the collapse of the gel structure, resulting in a significantly lower water resistance of the sulfur-magnesium cement. When the molar ratio of $H_2O/MgSO_4$ is 27, the XRD results show that the peak intensity of phase 5·1·7 is stronger than that at a molar ratio of 33 for $H_2O/MgSO_4$, and the peak intensity of Mg $(OH)_2$ is weaker than that at a molar ratio of 33 for $H_2O/MgSO_4$, indicating that phase 5·1·7 has decomposed and generated more $Mg(OH)_2$.

When the molar ratio of $H_2O/MgSO_4$ is between 21 and 23, phase 5·1·7 is the most stable, and the strength of the sulfur-magnesium cement is higher. Therefore, in the practical application of sulfur-magnesium cement, the molar ratio of $H_2O/MgSO_4$ should be between 21 and 23, at which time the mechanical strength and water resistance of the sulfur-magnesium cement system are higher.

2.3 Sulfur-chlorine composite clinker materials

Magnesium-based cementitious materials include chloromagnesium cement, phosphomagnesium cement, and sulfomagnesium cement. These are all ternary cementitious system materials, each with its own advantages and disadvantages. Chloromagnesium cement has the advantages of light weight, high strength, heat insulation, and noncombustibility, but it has poor water resistance and corrodes steel bars. Sulfomagnesium cement has good water resistance but low mechanical properties. Phosphomagnesium cement is fast-setting and high-strength, but it has high cost and difficult process operation. To further modify magnesium-based cementitious materials, a magnesium-based cementitious material composite system was adopted. The magnesium-based cementitious material composite system refers to a four-component cementitious material formed by blending two magnesium cement modifiers and reacting with magnesium oxide. The currently studied four-component composite magnesium cement materials mainly include sulfur-chlorine composite magnesium cement [26]. Currently, related research work has gradually started, but there are not many relevant materials. However, by exploring the hydration phase composition of the system, a scheme to improve the water resistance of magnesium-based basic salt cement can be obtained [27]. In the sulfur-chlorine composite magnesium cement four-component system, by adding cementitious liquid seawater, aggregate coral reef, and modifying materials citric acid and dihydrogen phosphate ammonium, an excellent mechanical performance magnesium-based basic salt coral concrete material is obtained. This provides a reference for the construction of islands in our country and broadens the application fields of magnesium-based cementitious materials [28]. Since the new composite magnesium cement was only studied in recent years, there are not many related literature materials. For this reason, the author analyzed based on practical experience and test data.

This section studied the influence of different raw material ratios on the early hydration behavior, strength development, and water resistance of sulfur-chlorine composite magnesium cementitious materials.

2.3.1 The influence of raw material ratio on setting time

The initial setting time of pure chloromagnesium cement is less than or equal to 6 hours, and the initial setting time of pure soda magnesium sulfate cement is less than or equal to 9 hours. The initial setting time of sulfur-chlorine composite magnesium cement is within 8 hours, and the final setting time is less than or equal to 9 hours. This meets the technical indicators for ordinary silicate cement products, where the initial setting time is greater than or equal to 45 minutes and the final setting time is less than or equal to 10 hours [26].

There is some controversy regarding the influence of raw material ratio on the setting time of sulfur-chlorine composite magnesium cement. Some scholars believe that when the brine dosage is low, the concentration of magnesium sulfate is high, which promotes the formation of hydration products, so the setting time is shorter; while other scholars believe that the concentration of magnesium chloride is low, and the consistency of the paste is low, resulting in a longer setting time of the mixed system than pure magnesium oxide-sulfur cement. As the mass fraction of magnesium chloride increases, the concentration of the solution gradually decreases, and there is an interference between SO_4^{2-} and Cl^- in the paste, which slows down the generation rate and crystallization rate of hydration products, prolonging the setting time of the paste. When the mass fraction of magnesium chloride is greater than 70%, the concentration of SO_4^{2-} in the paste decreases, and the concentration of Cl^- increases, promoting the formation of phase 5.1.8, and the setting time of the paste is shortened. Therefore, as the mass fraction of magnesium chloride increases, both the initial and final setting times of the paste show a trend of first prolonging and then shortening [29]. Some scholars have proposed that the initial setting time and final setting time of sulfur-chlorine composite magnesium cement are higher than those of pure chlorine-oxygen magnesium cement but lower than those of pure sodium bicarbonate magnesium sulfate cement. As the mass fraction of magnesium chloride changes from 0%–30% to 50%–100%, the interval between the initial and final setting times of the paste gradually shortens. The reduction of the sulfur-chlorine ratio in the raw materials of magnesium-based composite cement has a promoting effect on the setting of the magnesium-based composite cement. In magnesium-based composite cement, the reduction of the sulfur-chlorine ratio has a stronger promoting effect on setting than the retardation effect of citric acid, because the reduction of the sulfur-chlorine ratio in the raw materials leads to the faster hydration process of the hydration products formed by magnesium chloride and magnesium oxide, which shortens the setting time. The shorter the setting time, the more and larger the pores left after the hydration of the cement paste may be, and the faster the hydration rate will cause cracks to appear in the paste during the curing process. All of these are likely to have adverse effects on the later performance of the material [26].

The Shandong Meijiatu New Materials Technology Co., Ltd.(MJT) experimental study has found that the setting time of chloromagnesium cement is shorter than that of sulfomagnesium cement. As the mass fraction of magnesium chloride in the sulfur-chlorine composite solution increases, the setting time of the paste also gradually shortens. When the concentration of magnesium sulfate in the composite system is relatively high, there is more SO_4^{2-} in the paste, and the reaction rate is slow, resulting in a longer setting time. With the increase in the mass fraction of magnesium chloride, the concentration of SO_4^{2-} in the paste decreases, while the concentration of Cl^- increases, promoting the formation of phase 5·1·8, and shortening the setting time of the paste.

2.3.2 The influence of raw material ratio on mechanical properties

Strength is one of the indicators reflecting the performance of cementitious materials. The compressive and flexural strengths of the sulfur-chlorine composite magnesium cement vary with the change in magnesium chloride mass fraction. Firstly, the change in compressive strength is analyzed.

Through experiments, it was determined that the strength of the sulfur-chlorine composite cementitious material always remains between that of pure sulfur-MOC and pure chlorine-MOC. The molar ratio of magnesium oxide, magnesium sulfate, and magnesium chloride all affect the compressive strength of the magnesium-based alkali-based salt cement paste specimens. When the molar ratio of magnesium oxide and magnesium chloride remains unchanged, as the content of magnesium sulfate increases, the compressive strength of the magnesium-based alkali-based salt cement paste specimens decreases. This is because the compressive strength of sulfur-MOC is relatively low, and at the same time, as the molar ratio of magnesium oxide and magnesium sulfate decreases, the compressive strength of the cement also decreases. When the molar ratio of magnesium oxide and magnesium sulfate remains unchanged, as the content of magnesium chloride increases, the compressive strength of the magnesium-based alkali-based salt cement first increases and then decreases. This is because as the molar ratio of magnesium oxide and magnesium chloride decreases, the compressive strength of the cement first increases and then decreases. That is, both chlorine-MOC and sulfur-MOC require an appropriate molar ratio [27]. With the extension of the curing age, the compressive strength of chlorine-MOC, alkali-based sulfur-MOC, and sulfur-chlorine composite magnesium cement all increase with the extension of the age. When the curing age is the same, the compressive strength of magnesium-based composite cementitious materials decreases as the sulfur-chlorine ratio decreases. However, due to the influence of the sulfur-chlorine ratio in the raw materials, the strength at different ages varies. At 1 day of curing, the strength of the magnesium-based composite cementitious material is much lower than that of pure chlorine-MOC. When the curing time reaches 3 days and 7 days, the materials with sulfur-chlorine molar ratios of 7:3 and 5:5 have a faster strength increase. At 28 days of curing, the compressive strengths of the three materials with sulfur-chlorine molar ratios of 7:3, 5:5, and 3:7 are not much different [26].

The molar ratio of magnesium oxide to magnesium chloride will affect the flexural strength of the magnesium-based alkali-activated salt cement paste specimens. When the contents of magnesium oxide and magnesium sulfate remain unchanged, as the content of magnesium chloride increases, the flexural strength of the magnesium-based alkali-activated salt cement paste specimens first increases and then decreases. When the contents of magnesium oxide and magnesium chloride remain unchanged, as the content of magnesium sulfate increases, the flexural strength of the magnesium-based alkali-activated salt cement paste specimens decreases, indicating that the sulfur-magnesium cement exhibits lower flexural strength. At ages of 3 days and 28 days, as the mass fraction of magnesium chloride increases, the flexural strength of the cement specimens first increases, then decreases, and then increases again. The mass fraction of 50% and 70% magnesium chloride is the two inflection points of the change in the flexural strength of the specimens [29].

The results show that during the hydration stage of sulfur-chlorine composite magnesium cement, the reaction speed of MgO with $MgCl_2$ solution to form phase 5.1.8 is faster than that of MgO with $MgSO_4$ solution to form phase 5.1.7. To some extent, phase 5.1.8 and phase 5.1.7 exist in a competitive relationship, and at the same time, it hinders the formation of $Mg(OH)_2$. Therefore, with the increase of the content of magnesium chloride in sulfur-chlorine composite magnesium cement, the content of $Mg(OH)_2$ first decreases and then increases. The strength of sulfur-chlorine composite magnesium cement changes with age, and whether the mass fraction of magnesium chloride (i.e., the sulfur-chlorine molar ratio) or the molar ratio of magnesium oxide to the admixture affects the material strength, the specific products need to adjust the raw material ratio according to their own requirements. However, in general, compared with the alkaline magnesium-sulfate magnesium cement, the compressive and flexural strengths of the composite magnesium cement have significantly increased. This indicates that adding an appropriate amount of $MgCl_2$ to the $MgSO_4$ solution can obtain a magnesium-based cementitious material with excellent mechanical properties, which provides experience for the preparation of actual magnesium-based composite cementitious materials.

2.3.3 Phase analysis and microstructure of composite curing materials

Through XRD spectrum analysis, it was found that the hydration products of chloromagnesium cement are mainly the 5·1·8 phase, while the hydration products of sulfur-magnesium cement are mainly the 5·1·7 phase. By tracking the crystalline phase changes of the products, it was discovered that the ratio of raw materials, especially the molar ratio of magnesium oxide and the regulator, that is, the molar ratio of magnesium oxide to calcium chloride, and the molar ratio of magnesium oxide to magnesium sulfate, all affect the hydration process of magnesium cement and the types and quantities of hydration products. Therefore, products with different ratios have different microstructures after hardening. This part has been explained in the previous pure chloromagnesium and pure alkali-type sulfur-magnesium cement mechanisms. Among them, the microstructure of 5-phase cement stone becomes denser, while the structure of 3-phase cement stone is loose.

The hydration products of sulfur-chlorine composite magnesium cement mainly consist of needle-shaped 5·1·8 phase crystals and needle-fork-shaped

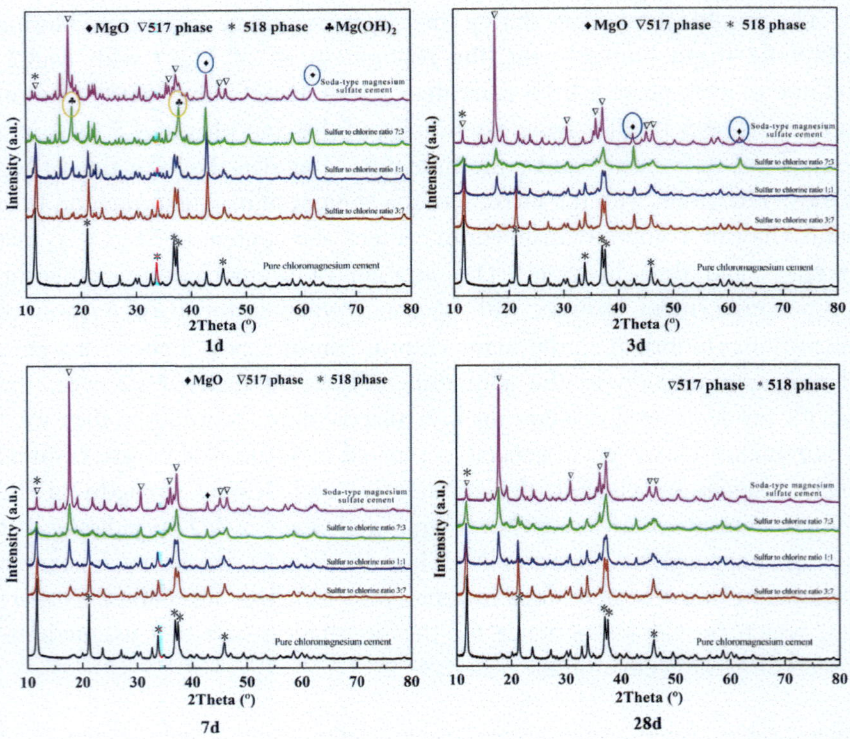

Figure 2.6 XRD spectra of magnesium-based composite cementitious materials after being cured in air for 1 day, 3 days, 7 days, and 28 days.

5·1·7 phase crystals. However, the specific content of each type depends on the sulfur-chlorine ratio in the raw material ratio. Additionally, as the curing time increases, the hydration of sulfur-chlorine composite magnesium cement becomes more complete, resulting in more and more complete crystal growth. The two phases grow in an interlaced manner and are closely connected, changing the micro-pore structure and reducing the porosity of the material. In sulfur-chlorine composite magnesium cement, the free ions tend to aggregate in the gel phase, forming aggregates, improving the mechanical properties and water resistance of sulfur-chlorine composite magnesium cement [26] (Figs. 2.6 and 2.7).

Through energy-dispersive X-ray spectroscopy (EDS) energy spectrum analysis, the morphology of the needle-shaped crystals 5·1·8 on the surface of the mixed system matrix shows significant differences from that of the pure chloromagnesium oxide system 5·1·8. When the raw material ratio is controlled within an optimal range, the microstructure of the hydration

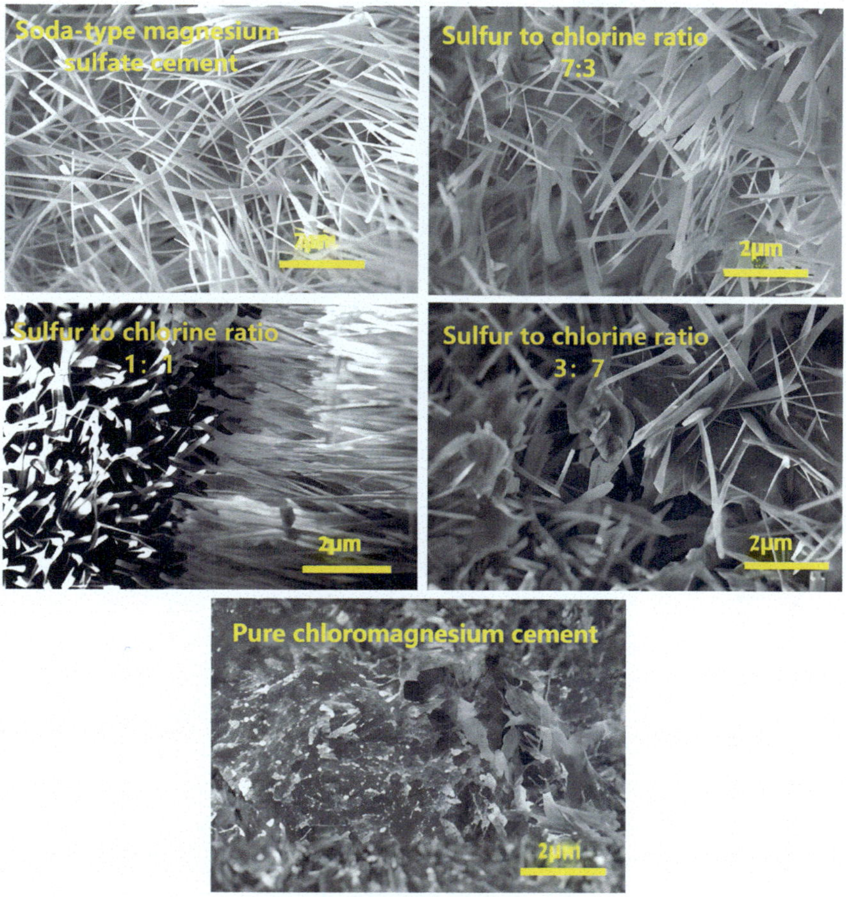

Figure 2.7 Scanning electron microscopy image of the cementitious material after 28 days of curing in air.

products of the sulfur-chlorine composite magnesium cement is dense, with uniform crystal distribution, sufficient quantity, closely arranged, interconnected, and interlaced in all directions, and it also has high strength and good water resistance [27] (Fig. 2.8).

2.3.4 Pore size distribution of composite curing materials

Through comprehensive analysis using XRD/scanning electron microscopy /EDS, the hydration products of sulfur-chlorine composite magnesium cement still consist of pure sodium-based sulfur-magnesium hydrate

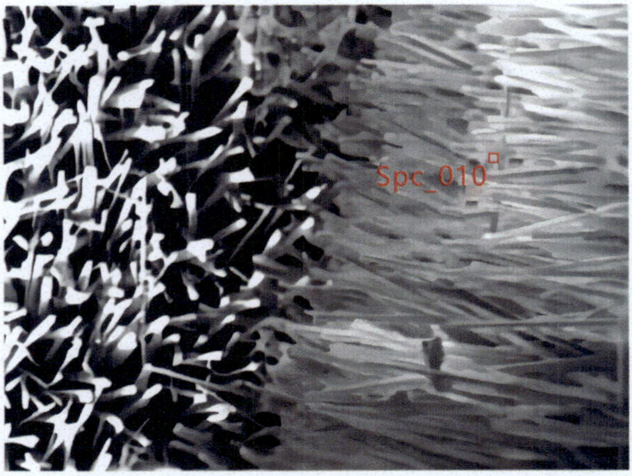

Figure 2.8 EDS image of the substrate surface after 28 days of sulfur-chlorine 1:1 maintenance.

and pure chlorine-based magnesium hydrate. No new crystal phases were formed. However, compared with pure sodium-based sulfur-magnesium cement and pure chlorine-based magnesium cement, the total porosity and pore diameter of sulfur-chlorine composite magnesium cement were significantly reduced. The 5.1.7 phase of the sodium-based sulfur-magnesium hydration product is needle-like crystals, while the 5.1.8 phase of the chlorine-based magnesium hydration product is needle-rod-like crystals. The space occupied by the 5.1.8 fibers is larger than that of the 5.1.7 fibers. Affected by the sulfur-chlorine molar ratio of the admixture, when the sulfur-chlorine molar ratio is greater than 1 (i.e., more magnesium sulfate), the pore volume distribution pattern of sulfur-chlorine composite magnesium cement leans towards that of pure sodium-based sulfur-magnesium cement, and the material pores are mainly distributed in the gel pores (< 5 nm). When the sulfur-chlorine molar ratio is less than 1 (i.e., more magnesium chloride), the pore volume distribution pattern of sulfur-chlorine composite magnesium cement leans towards that of pure chlorine-based magnesium cement, and the material pores are mainly distributed in the capillary pores (10–100 nm). When the sulfur-chlorine ratio is 1, that is, the amount of magnesium sulfate is the same as that of magnesium chloride, the pore volume distribution pattern of sulfur-chlorine composite magnesium cement combines the characteristics of both, but the volume of pores distributed in the gel pores (< 5 nm) is slightly larger than that distributed in the capillary pores (10–100 nm).

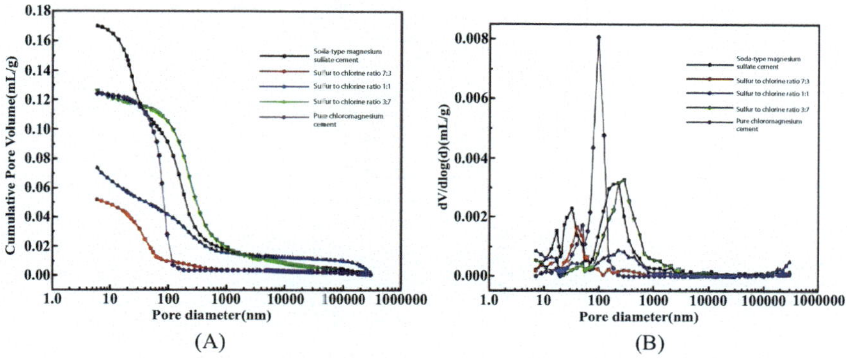

Figure 2.9 Porosity distribution curve of the cementitious material after 28 days of curing. (A) Cumulative pore volume distribution curve; (B) Differential pore size distribution curve.

A comprehensive comparison shows that the pore distribution of the composite material makes its mechanical properties and water resistance properties fall between those of pure sodium-based sulfur-magnesium cement and pure chlorine-based magnesium cement, as shown in Fig. 2.9 and Table 2.1.

2.3.5 The hydration reaction mechanism of composite cementitious materials

By analyzing the XRD spectra of the hydration products of sulfur-chlorine composite magnesium cement at 2–72 hours of hydration reaction, it is inferred that the hydration reaction mechanism of sulfur-chlorine composite magnesium cement is as follows: At the beginning of the hydration reaction, the 5·1·8 phase of chloro-magnesium cement predominates and inhibits the formation of the 5·1·7 phase of sulfur-chlorine composite magnesium cement. After 48 hours of hydration, a large number of 5·1·8 phase crystals are formed, while the 5·1·7 phase begins to grow, but during the early hydration process, the 5·1·8 phase still dominates [26] (Fig. 2.10).

2.3.6 The influence of raw material ratio on water resistance

The composition and structure of the hydration products are the fundamental factors influencing the water resistance of magnesia-based cementitious materials. As the mass fraction of magnesium chloride increases, the

Table 2.1 Analysis of pore structure of cementitious materials after 28 days of curing in air.

Sample	Total porosity/%	Porosity diameter/nm	Total infiltration volume/(mL/g)	Porosity volume distribution/%			
				< 0.05 μm	0.05 ~ 0.1 μm	1.1 ~ 1 μm	> 1 μm
Sodium bicarbonate magnesium sulfate cement	25.79	43.34	0.17	37.52	14.38	37.75	10.35
Sulfur-chlorine ratio 7:3	8.73	27.66	0.05	64.33	17.35	10.93	7.39
Sulfur-chlorine ratio 1:1	11.96	28.49	0.07	36.75	11.91	30.83	20.51
Sulfur-chlorine ratio 3:7	18.93	72. 14	0. 12	11.01	10.96	62.9	15.13

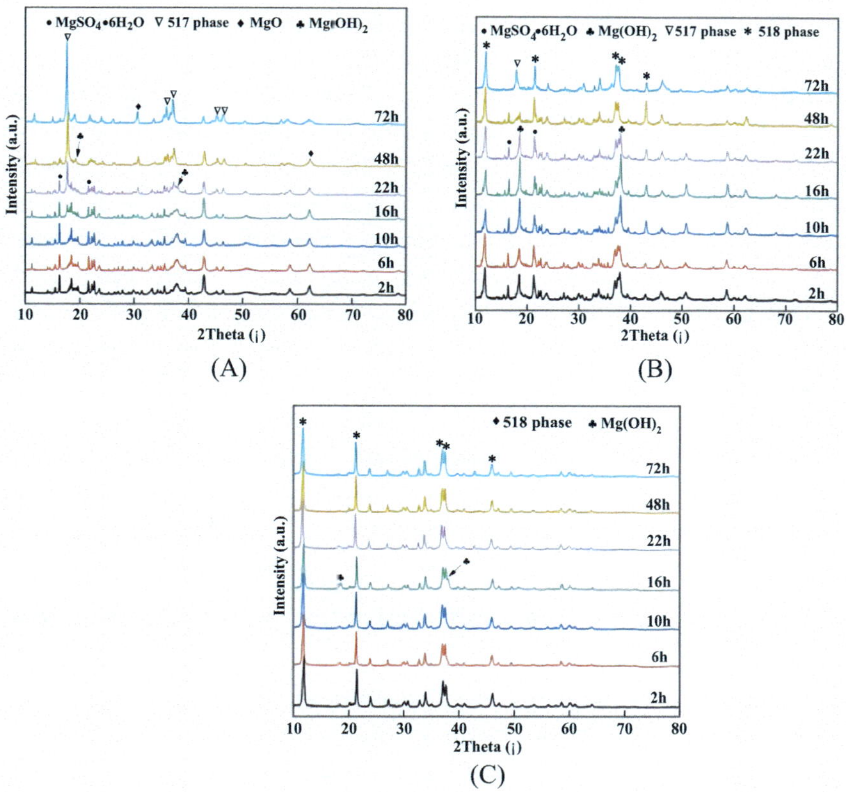

Figure 2.10 XRD patterns of the products at the early stage of hydration. (A) Pure magnesium sulfate cement. (B) Sulfur-to-chlorine ratio of 5:5. (C) Pure chlorine magnesium cement.

softening coefficients of the compressive strength and flexural strength of the sulfur-chlorine composite magnesia cement specimens show a trend of first increasing and then decreasing. Compared with pure soda-type sulfur-oxygen magnesia cement and pure chlorine-oxygen magnesia cement, the mixed system exhibits excellent water resistance. Through tracking the water immersion experiments of sulfur-chlorine composite magnesia cement with different magnesium chloride mass fractions, it was found that after 7 days of immersion, as the magnesium chloride mass fraction increased, the compressive and flexural strengths gradually decreased, and the water resistance deteriorated.

By observing the XRD spectra, after the chloromagnesium cement specimens were soaked in water for 28 days, the main strength crystal of chloromagnesium cement, namely the 5.1.8 crystal phase, exhibited a significant weakening in its characteristic diffraction peaks. At the same time, the characteristic peak of $Mg(OH)_2$ appeared. These changes in the diffraction peaks proved that chloromagnesium cement underwent further hydration during the soaking process [26]. The new crystal phase $Mg(OH)_2$ formed during the hydration process would gradually grow. Due to the internal stress, many microcracks would be generated within the hardened body, and when the hardened body was immersed in water, water would penetrate into the body through the pores and cracks, weakening the bonding force between the hydration product particles, causing the strength of the specimen to decrease in water [29]. However, the strength phase 5.1.7 of sulfur-magnesium cement in water has certain stability, so the strength loss is not significant. Thus it can be concluded that compared with chloromagnesium cement and sulfur-magnesium cement, the sulfur-chlorine composite magnesium cement system exhibits excellent water resistance.

2.3.7 The influence of raw material ratios on free chlorine ions

The reoxidation of chloromagnesium cement products can easily cause steel corrosion, resulting in a rapid decline in product strength and a shortened service life. A large amount of experimental data indicates that magnesium cement products with a free chlorine ion content of less than 3% basically do not reoxidize. The content of free chlorine ions in sulfur-chlorine composite magnesium cement after adequate and reasonable curing is tested (Fig. 2.11). From the figure, it can be seen that the free chlorine content of sulfur-chlorine composite magnesium cement with a magnesium oxide content of no more than 90% can be controlled below 3%, which can effectively reduce the probability of products absorbing moisture and reoxidation.

2.3.8 Production implementation case

The chloromagnesium core board of the chloride-containing rusted steel fireproof door panel has the problem of reoxidation and corrosion of the steel plate. To prevent its development, a study was conducted to replace the chloromagnesium core board with a sulfur-oxygen-magnesium core board. However, in regions such as Sichuan, Chongqing, and Guizhou, agricultural-grade magnesium sulfate is mostly used, which contains aminosulfonate components. Using it to produce sulfur-oxygen-magnesium core boards will

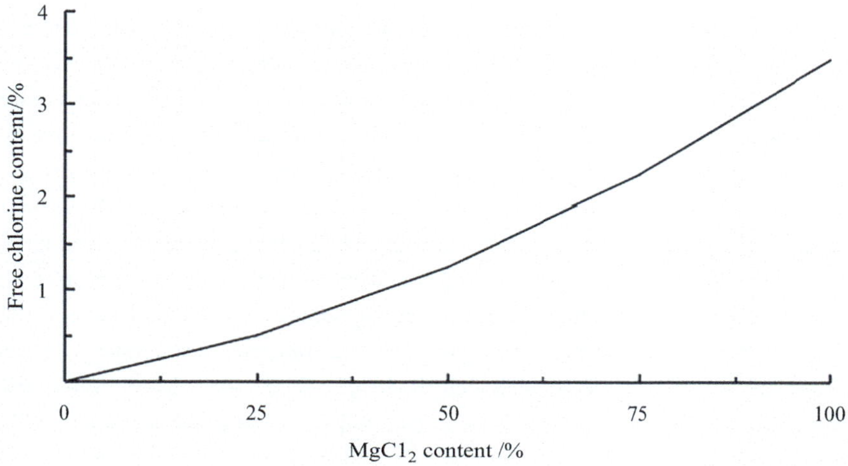

Figure 2.11 The influence of magnesium chloride content on the free chlorine content of sulfur-chlorine composite magnesium cement.

release ammonia gas, causing adverse effects on workers and the environment. Moreover, the core boards produced with it may experience collapse problems. The purchase of high-quality industrial-grade magnesium sulfate from other places is excessively expensive, limiting the production and promotion of sulfur-oxygen-magnesium core boards in these regions. Based on the above situation, MJT mixed local agricultural-grade magnesium sulfate with anhydrous magnesium chloride to formulate a binder and developed a production technology for sulfur-chlorine composite core boards, reducing the free chlorine content and solving the problem of moisture absorption, reoxidation, and rusting of the steel plate.

Currently, there are two preparation methods for the sulfur-chlorine composite solution: one involves mixing anhydrous magnesium chloride and agricultural-grade magnesium sulfate at a mass ratio of 2:8 or 3:7, which is used to produce steel-based fireproof door core boards. After adequate curing, the free chlorine content can be controlled below 3%; the other method involves mixing anhydrous magnesium chloride and agricultural-grade magnesium sulfate at a mass ratio of 9:1, which is used for the production of wooden fireproof door molds.

When preparing the composite solution with a mass ratio of anhydrous magnesium chloride to agricultural-grade magnesium sulfate of 2:8, first add the anhydrous magnesium chloride to the water and stir rapidly to form a brine with a temperature of 18–20°Bé. Then add the agricultural-grade magnesium sulfate and fully dissolve it. After that, measure and adjust the temperature and specific gravity. During the preparation of the sulfur-chlorine composite solution, the anhydrous magnesium chloride dissolves and releases heat, while the magnesium sulfate dissolves and absorbs heat. The temperature of the solution is controlled by adjusting the ratio of anhydrous magnesium chloride and magnesium sulfate. In actual production, if the environmental temperature is too high, the amount of anhydrous magnesium chloride should be appropriately reduced; conversely, if the environmental temperature is too low, the amount of anhydrous magnesium chloride should be appropriately increased to raise the temperature of the composite solution. In summer, the temperature of the composite solution is controlled below 30°C, 27°Bé; in winter, when the temperature is low, the temperature of the composite solution is controlled at around 38°C, 28–29°Bé.

In the actual production process of the core board, if the flat mold production method is adopted, the mass ratio of light-burned magnesium oxide (85 powder, with an activity of over 63%) to the composite solution is 1: (1.1–1.3); if the vertical mold production method is adopted, the mass ratio of light-burned magnesium oxide (85 powder, with an activity of over 63%) to the composite solution is 1: (1.2–1.5). The modifier is a sulfur-magnesium oxide modifier (GX-15#) or a neutral sulfur-magnesium oxide modifier (GX-15–2#).

2.4 Magnesium phosphate cementitious material

During the exploration and mining of phosphate ore, it was discovered that in the habitats of seabirds and bats on some low-latitude islands, there was a large accumulation of bird droppings. Under the action of bacteria, this gradually formed the bird droppings stone ($Mg(NH_4)PO_4{\cdot}H_2O$) structure, which was the rudiment of the phosphate cement structure.

Phosphate cement, as a ternary cementing system composed of magnesium oxide, phosphate salts, and water, has the characteristics of fast setting, small shrinkage upon drying, high strength, corrosion resistance, and resistance to high and low temperatures.

2.4.1 Mechanism of magnesium phosphate cement hydration

Unlike chloromagnesium oxide and sulfomagnesium oxide, the mechanism of hydration of magnesium phosphate cement is quite simple. The essence of its hydration process is a neutralization reaction. Specifically, the hydrogen ions of the acidic phosphate salts combine with the hydroxide

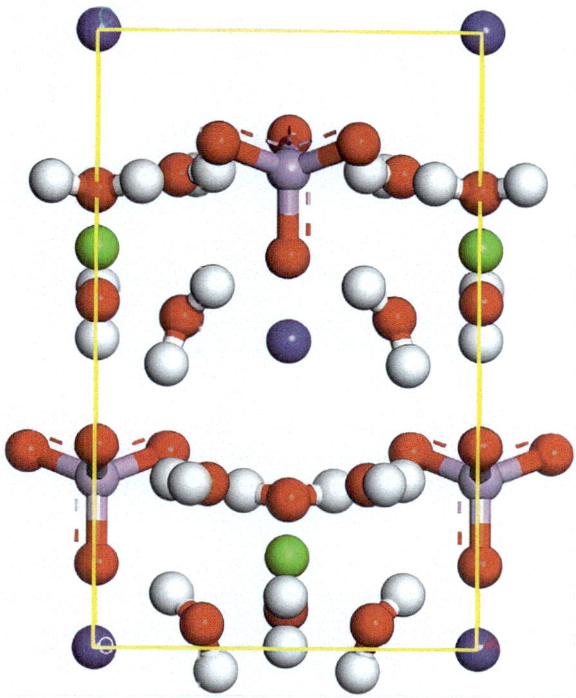

Figure 2.12 Molecular structure of K⁻ bird's-eye quartz crystal.

ions produced by the hydration of heavy-burned magnesium oxide in water to form water. Subsequently, phosphate ions, magnesium ions, and ammonium ions (or potassium ions, sodium ions) combine in the aqueous phase to form the hydrated structure of magnesium phosphate ammonium, and then a process of generating struvite precipitate structure occurs.

The K^-chlorite structure is composed of phosphorus-oxygen tetrahedra, hydrated magnesium ions, potassium ions, and water molecules. Different structures are connected through covalent bonds and hydrogen bonds. The hydrogen bonds between water molecules and the phosphorus-oxygen tetrahedra have a significant effect on the development of the strength of K^-chlorite (Fig. 2.12).

The hydration process of magnesium phosphate cement can be divided into four stages. Taking the phosphate dihydrogen potassium–calcined magnesia–water ternary system as an example for a brief introduction:

(1) The soluble phosphate ions release H_2PO^{4-}, HPO_4^{2-}, PO_4^{3-}, H^+, and K^+; the reaction system becomes acidic;

(2) In the acidic system, calcined magnesia hydrates, generating Mg^{2+}, Mg^{2+} forms hydrated magnesium ions, and at the same time, the pH value of the system rises, and a large amount of heat is released;
(3) The hydrated magnesium ions combine with the H_2PO^{4-}, HPO_4^{2-}, PO_4^{3-}, and K^+ ions released by the dissociation of phosphate, generating hydrated potassium magnesium phosphate colloid;
(4) The hydrated potassium magnesium phosphate colloid forms crystals, and these crystals interlock with each other to form a strong network structure.

In some reports, the solid-state magic-angle spinning nuclear magnetic resonance technique was used to determine the structure of the phosphate magnesium cement products. It was found that in addition to a large amount of monohydrate calcium phosphate structure, there were also a considerable amount of amorphous orthophosphate magnesium ($Mg_3(PO_4)_2$) and a small amount of magnesium hydrogen phosphate trihydrate [30]. Magnesium hydrogen phosphate trihydrate is an easily soluble salt, which has a negative impact on the water resistance of the products. Therefore, the ratio design of phosphate magnesium cement requires certain considerations to inhibit the formation of magnesium hydrogen phosphate trihydrate.

Other literature shows that when phosphate magnesium cement is solidified, if the reaction is too rapid, there will be a large amount of unreacted magnesium oxide remaining, promoting the formation of amorphous one-water product ($MgKPO_4 \cdot H_2O$) [31].

2.4.2 Design of magnesium phosphate cement ratio

2.4.2.1 Selection of phosphate salts

The phosphate section of this book briefly introduces the phosphates suitable for the production of magnesium phosphate cement, mainly including potassium dihydrogen phosphate and ammonium dihydrogen phosphate. Using other phosphates and phosphates to produce magnesium phosphate cement will result in serious structural defects in the products. In actual production, different phosphates are used to prepare magnesium phosphate cement products, each having its own unique advantages. The solubility of ammonium dihydrogen phosphate is relatively high (37.4 g/100 g water), so when it is applied in the magnesium phosphate cement system, it is more likely to react with magnesium oxide and release a large amount of heat, resulting in a faster curing speed and faster strength development. However, during the reaction, ammonium dihydrogen phosphate is prone to release unpleasant ammonia gas, which affects the health of the production personnel. Moreover, the rapid reaction also makes it difficult to remove the bubbles produced in the system. The solubility of potassium dihydrogen phosphate is relatively low, and its reaction with magnesium oxide is slower, resulting in slightly slower strength development of the magnesium phosphate cement products. However, it does not release harmful gases.

In fact, whether using ammonium dihydrogen phosphate or potassium dihydrogen phosphate, the main phase formed is the carnallite structure. However, in the carnallite structure of the magnesium phosphate cement products prepared using potassium dihydrogen phosphate, the ammonium ions are replaced by potassium ions. During production, according to one's own needs, potassium dihydrogen phosphate and ammonium dihydrogen phosphate can be mixed for use, and certain trade-offs can be made in terms of reaction speed, early strength, and ammonia gas release.

Although the structure of the products prepared by using sodium dihydrogen phosphate is defective and the strength is low when making magnesium phosphate cement, some studies have shown that when using potassium dihydrogen phosphate to prepare magnesium phosphate cement and replacing potassium dihydrogen phosphate with a certain amount of sodium dihydrogen phosphate, the hydration heat rate of the system can be increased, the system can be solidified more quickly, and the flexural and compressive strengths of the products can be significantly improved. Under the action of a certain amount of sodium dihydrogen phosphate, the K-eggshellite crystals in the potassium magnesium phosphate system change from block-like and rod-like to finer needle-like Na, K-eggshellite structure, while the production volume of the eggshellite structure increases and the porosity significantly decreases. This may be the reason for the improvement in the strength of the composite phosphate salt system blocks. When the amount of sodium dihydrogen phosphate is too high, the eggshellite structure in the system decreases, the crystals transform into an amorphous structure, and the strength of the products significantly decreases. The amount of sodium dihydrogen phosphate should be below 60%.

In actual production, it is necessary to combine with one's own needs and conduct small-scale tests based on production conditions to select the required types of phosphate salts. Overall, the comprehensive performance of the system in terms of setting speed and strength is better when different types of phosphate salts are compounded and used than when using a single type of phosphate salt.

2.4.2.2 Magnesium phosphorus ratio design

The main strength phase of magnesium phosphate cement is Mg(M) $PO_4 \cdot 6H_2O$, where M can be NH_4^+, K^+, Na^+, etc. In addition, there are amorphous phases such as $MgHPO_4$ and $Mg_3(PO_4)_2$ in magnesium phosphate cement. To obtain high-quality magnesium phosphate cement products, it is necessary to generate the $Mg(M)PO_4 \cdot 6H_2O$ phase as much as possible. That is, the formula for designing the recipe should be based on the molar ratio of Mg/PO_4^{3-}.

The molar ratio and the amounts of phosphate salts and calcined magnesium oxide can be expressed by the following formula:

$$K = \frac{\frac{X\eta}{40.3}}{\frac{Y\delta}{Z}}$$

In the formula: K represents the molar ratio of calcined magnesia to phosphate; X represents the actual amount of calcined magnesia in grams; η represents the purity of calcined magnesia, expressed as a percentage; 40.3 represents the molar quantity of magnesia; Y represents the actual amount of phosphate in grams; δ represents the purity of phosphate, expressed as a percentage; and Z represents the molar mass of phosphate.

Generally speaking, under the condition that the phosphate is completely reacted, when the ratio K of Mg/PO_4^{3-} in magnesium phosphate cement increases, the setting time of magnesium phosphate cement shortens, and the proportion of the calcite structure formed in the magnesium phosphate system also decreases. When K is too large, the proportion of calcined magnesia participating in the reaction is low, and most of the calcined magnesia exists in the form of a filler. Although the reactivity of calcined magnesia is low, in the long run, it still causes the products to crack. If K is too small, there is a surplus of phosphate, and the product is prone to absorb water and expand, which also easily leads to the cracking of the products. Generally, K is between 9 and 16, and magnesium phosphate cement performs well in mechanical properties and durability.

2.4.2.3 Water-cement ratio design

In the magnesium phosphate cement system, under a reasonable K design, the amount of calcined magnesium oxide is obviously not completely utilized. The water consumption only needs to ensure that all phosphate salts are consumed and $Mg(M)PO_4 \cdot 6H_2O$ is generated, so as to meet the fluidity of the system. When the water consumption is too low, the amount of monohydrate and tetrahydrate structures in the system is less, and the soluble tetrahydrate and monohydrate increase, resulting in a lower strength of the system. In the later stage, these amorphous structures can further develop and age in a humid environment, undergoing the following reactions, as shown in the following formula, which leads to an increase in the strength of the product [32].

$$MgO + 2NH_4H_2PO_4 + 3H_2O \rightarrow Mg(NH_4)_2(HPO_4)_2 \cdot 4H_2O \quad (2.24)$$

$$Mg(NH_4)_2(HPO_4)_2 \cdot 4H_2O + MgO + 7H_2O \rightarrow Mg(NH_4)PO_4 \cdot 6H_2O \quad (2.25)$$

$$Mg(NH_4)_2(HPO_4)_2 \cdot 4H_2O + 2H_2O \rightarrow Mg(NH_4)PO_4 \cdot 6H_2O + NH_4^+ + H_2PO_4^- \quad (2.26)$$

When the water consumption is excessive, after the system solidifies, the excess water gradually evaporates, leaving a large number of micropores. This not only affects the strength of the product but also makes the product more susceptible to water intrusion, further reducing the water resistance of the product.

2.4.2.4 Modifier

Due to the essence of the reaction of magnesium phosphate cement, which is an acid-base neutralization reaction, the reaction is intense, and the ion concentration in the solution is high, causing the slurry to solidify too quickly, and there is no operation time. Therefore, a certain amount of modifier needs to be added to control the reaction rate. Currently, the most effective retardation series products are magnesium phosphate retardants. After adding magnesium phosphate retardants to the magnesium phosphate cement system, the magnesium phosphate retardants dissolve and absorb heat, reducing the system temperature and pH value. The dissolved magnesium phosphate retardants generate $B_4O_7^{2-}$, which combines with Mg^{2+}, adsorbing on the surface of the calcined magnesium oxide particles, inhibiting the hydration of magnesium oxide in an acidic environment, thereby slowing down the reaction rate of the system. Considering the working performance and mechanical strength of magnesium phosphate cement, there is an optimal dosage of magnesium phosphate retardants. If the dosage of magnesium phosphate retardants is too low, the magnesium phosphate cement develops too quickly, the crystal connections are not tight enough, the product strength is too low. If the dosage is too high, the magnesium phosphate system develops slowly, thereby affecting the final strength. Generally, the dosage of magnesium phosphate retardants is around 4% of the calcined magnesium oxide mass.

In the magnesium phosphate cement system, the magnesium ions originate from the hydration of calcined magnesia and inhibit the hydration of calcined magnesia, which is one of the research directions of the retardant for magnesium phosphate cement. Besides the magnesium phosphate retardant, some weak acid salts or alkaline substances added to the magnesium phosphate system can also adjust the pH value of the system and reduce the hydration rate of calcined magnesia. Moreover, using complexes to coordinate the magnesium ions generated during hydration can also reduce the reaction rate of the system.

The rate reduction effect of a single substance on the magnesium phosphate cement system is limited. Using the compound products of these several retardants has a lower cost-effectiveness ratio.

If pure calcined magnesia with a purity of 95% and pure potassium dihydrogen phosphate with a purity of 99% are used to produce magnesium phosphate cement, with a magnesium-phosphorus ratio of 13 (i.e., K=13) and a design ratio of the components, as much monocalcium phosphate rock salt structure as possible is generated. Then, the ratio of calcined magnesia: potassium dihydrogen phosphate: water: magnesium phosphate retardant=4:1:0.9:0.2 is required. In practice, under the condition of ensuring the working performance, reducing the water-cement ratio can lead to higher strength. In some reports, when K is greater than 10, a sample block with excellent strength can be produced at a water-cement ratio of 0.11 [33].

References

[1] Sorel S. On a new magnesium cement. Compt Rend, 1867;65:102–4.
[2] Altiner M, Yildirim M. Influence of filler on the properties of magnesium oxychloride cement prepared from dolomite. Emerging Materials Research 2017;6(2):417–21.
[3] Gu K, Chen B, Bi W, et al. Recycling of waste gypsum in preparation of magnesium oxychloride cement (MOC). Journal of Cleaner Production 2021;313:127952–8.
[4] Sorrell CA, Armstrong CR. Reactions and equilibria in magnesium oxychloride cements. Journal of the American Ceramic Society 1976;59(1):51–4.
[5] Dehua D. Theoretical and applied research on improving the performance of magnesium basic salt cement. Doctoral Dissertation" Central South University; 2005.
[6] Chau CK, Li Z. Microstructures of magnesium oxychloride Sorel cement. Advances in Cement Research 2008;20(2):85–92.

[7] Sorll CA, Armstrong C. Reaction and equilibria in magnesium oxychloride cements. American Ceramic Society. 1976;59(1-2):51–4.
[8] Dehua D., Chuanmei Z. The formation mechanism of the hydrate phases in magnesium oxychloride cement. Cement and Concrete Research. 1999;29(9):1365–71.
[9] Gó chez R, Wambaugh J, Rochner B, et al. Kinetic study of the magnesium oxychloride cement cure reaction. Journal of Materials Science. 2017;52(13):7637–46.
[10] Cole WF, Demediuk T. X-rar, thermal, and dehydration studies on magnesium oxychlorides. Australian Journal of Chemistry 1955;8(2):234–51.
[11] Pö llmann, Herbert. Cementitious materials (composition, properties, application). Berlin, Boston: De Gruyter; 2017.
[12] Guo Y, Zhang Y, Soe K, et al. Recent development in magnesium oxychloride cement. Structural Concrete 2018;19(5):1290–300.
[13] Li Z, Chau CK. Influence of molar ratios on properties of magnesium oxychloride cement. Cement and Concrete Research 2007;37:866–70.
[14] A. TA, Mark R, Daniel M, et al. Effect of molar ratios and curing conditions on the moisture resistance of magnesium oxychloride cement. Journal of Materials in Civil Engineering 2022;34(2).
[15] Liu Z, Balonis M, Huang J, et al. The influence of composition and temperature on hydrated phase assemblages in magnesium oxychloride cements. Journal of the American Ceramic Society 2017;100(7):3246–61.
[16] Beaudoin JJ, Ramachandran VS. Strength development in magnesium oxysulfate cement. Cement and Concrete Research 1978;8(1):103–12.
[17] Demediuk T, Cole WF. A study of mangesium oxysulphates. Australian Journal of Chemistry 1967;10(3):287.
[18] Runcˇevski T, Wu CY, Yu HF, Yang B, Dinnebier RE. Structural characterization of a new magnesium oxysulfate hydrate cement phase and its surface reactions with atmospheric carbon dioxide. Journal of the American Ceramic Society 2013;96:3609–16.
[19] Urwongse L, Sorrell CA. Phase relations in magnesium oxysulfate cements. Journal of the American Ceramic Society 2010;63(9-10):523–6.
[20] Shengwen T, Junhui Y, Rongjin C. In situ monitoring of hydration of magnesium oxysulfate cement paste: effect of MgO/MgSO4 ratio—science direct. Construction and Building Materials 2024:251.
[21] Jiasheng H, Wenwei L, Desheng H, et al. Fractal analysis on pore structure and hydration of magnesium oxysulfate cements by first principle, thermodynamic and microstructure-based methods. Fractal and Fractional 2021;5(4). 164-164.
[22] Huihui D, Jiajie L, Wen N, et al. The hydration mechanism of magnesium oxysulfate cement prepared by magnesium desulfurization byproducts. Journal of Materials Research and Technology 2022:171211–20.
[23] Wu C, Xing S, Zhang W, et al. Research on the hydration law of basic magnesium sulfate cement. Functional Materials 2016;47(11):11120–4. +11130.
[24] Tao G, Hefang W, Hongjian Y, et al. The mechanical properties of magnesium oxysulfate cement enhanced with 517 phase magnesium oxysulfate whiskers. Construction and Building Materials 2017;150:844–50.
[25] Guo ZL, Sheng ZJ. Effect of molar ratios on compressive strength of modified magnesium oxysulfate cement. International Journal of Hybrid Information Technology 2015;8(6):87–94.
[26] Yujie L. Hydration mechanism and properties of magnesium based composite cementitious materials. Shanxi University; 2023.
[27] Zhao C. Study on the microstructure and self toughening mechanism of magnesium based alkaline salt cementitious materials. Harbin Institute of Technology; 2017.

[28] Jiajun X. Experimental study on quaternary magnesium based cementitious materials and their self toughening properties. Harbin Institute of Technology; 2019.
[29] Shengxia A, Xueying X, Ying L. etc. Solidification and hardening properties and microstructure of sulfur and magnesium oxychloride mixed cementitious system. Silicates Bulletin 2017;36(08):2607–13. +2624.
[30] Scrimgeour SN, Chudek JA, Lloyd CH. The determination of phosphorus containing compounds in dental casting investment products by 31P solid-state MAS-NMR spectroscopy. Dental Materials 2007;23(4):415–24.
[31] Pengxiao L, Liangbo D, Dongxu L. Preparation and performance study of new early strength magnesium phosphate cement. Portland Bulletin 2008(01):20–5.
[32] Kato K, Shiba M, Aono H, et al. The study on phosphate bonded investment (II) On effect to physical properties by reaction during kneadling(author's transl). Iyo Kizai Kenkyujo Hokoku. Reports of the Institute for Medical and Dental Engineering, Tokyo Medical and Dental University 1979;13:1–12.
[33] Chau C.K., Qiao F., Li Z. Microstructure of magnesium potassium phosphate cement. Construction and Building Materials. 2011;25(6):2911–17.

CHAPTER 3

Process and related equipment

Producing high-quality magnesium materials requires the coordination of multiple processes and cannot be completed in a single step. Even with a fixed formula and using raw materials of the same specifications for blending, the quality of magnesium products can vary significantly among different manufacturers. For the same magnesium product, while the production processes of different manufacturers may differ considerably, the parameter control at each step varies from one manufacturer to another.

Even for a simple mixing process, the appropriate equipment needs to be selected based on the different products, and different parameters need to be set, such as mixing time and speed, to ensure the materials are mixed evenly and always maintained in the best condition. The foaming process is generally divided into physical foaming and chemical foaming. The selection between these two processes is related not only to the product type but also to other processes on the production line. Typically, the physical foaming process is chosen, and the mixing process usually employs a horizontal mixer to better incorporate the foam produced by the foaming machine with the slurry. For the chemical foaming process, there are specific requirements for the mixing speed to ensure bubble stability. In addition, a relatively strict curing process is usually required to ensure the stability of the bubbles.

With the continuous advancement of technology, the molding process of magnesium products has evolved from manual and semimanual methods to mechanized and automated ones. Mold design has also been continuously optimized, leading to increasingly stringent requirements for the demolding process. To ensure a certain level of production efficiency, industry professionals have meticulously studied the curing process and constructed curing rooms, making full use of various methods such as residual heat, steam, and microwaves to control temperature and humidity.

To meet the market's demand for the aesthetic appearance of products, the industry has also begun to further process magnesium products. From the most basic cutting, sanding, and traditional object-based processing methods

Magnesium-Based Cementitious Materials
DOI: https://doi.org/10.1016/B978-0-443-54075-2.00006-0

to decorative processes such as coating and spraying for the consumer market, the added value of magnesium products is consistently rising.

3.1 Mixing process

3.1.1 Mixing

Mixing involves the uniform blending of two or more substances with different states, properties, or temperatures. It is a key production process in the preparation of magnesium-based cementitious materials, significantly affecting their performance. Correct and scientific mixing techniques help to better utilize the many advantages of magnesium-based cementitious materials and are an important part of the process modification technology system for these materials.

3.1.2 Classification of mixers

The mixer is the primary tool for the mixing process. Based on the axial orientation of the mixer, it can be classified into horizontal mixers (Fig. 3.1) and vertical mixers. Generally, for mixing magnesium-based cementitious materials, horizontal mixers are more commonly used. This section exclusively introduces the horizontal mixing method.

3.1.3 How to select a blender

1. Select the mixer based on the product's volume, output, efficiency, and so forth.

   For the product's volume, select a mixer that is slightly larger than the product's volume. Generally, the product's volume should not exceed two-thirds of the mixer's capacity.

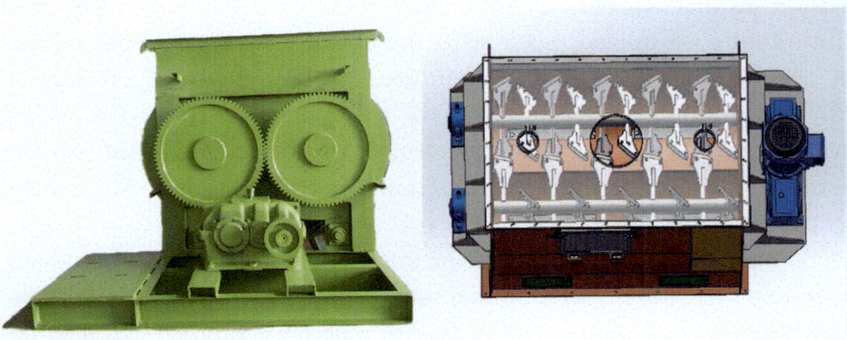

Figure 3.1 Double-axis horizontal mixer.

According to the production output requirements, a mixer with a larger capacity should be chosen if the output is high.

According to production efficiency, large volume and relatively high rotational speed are usually chosen for the mixer.

2. Since magnesium-based cementitious materials are air-hardening materials, the slurry has a higher viscosity and hardens quickly. Unlike ordinary Portland cement, when using a standard cement mixer to prepare magnesium-based cementitious materials, certain common issues may arise: (1) The traditional mixing drum wall and blades are prone to adhere to the slurry, forming mixing dead zones, which affects the uniformity of the slurry; (2) The conventional mixer has a low rotational speed, which easily causes uneven mixing and uneven product quality. Therefore when mixing magnesium-based cementitious materials, a mixer with a high rotational speed should be reasonably selected. The rotational speed of the mixer should be 50–150 r/minute, and it is recommended to use a rotational speed of 60–120 r/minute, which is 2–6 times the rotational speed of conventional mortar or concrete mixing. For the mixing of semidry magnesium-based cementitious materials, the rotational speed can be reduced to 50–80 r/minute to protect the mixer's rotating shaft.
3. Technical parameters

The technical parameters of the double-axis horizontal mixer are shown in Table 3.1.

3.1.4 The modification principle of the high-speed opposite-direction reversing mixer

The high-speed bidirectional reversing mixer can meet the mixing requirements of magnesium-based cementitious materials and produce a highly uniform slurry. It successfully enables the mixing and modification of magnesium-based cementitious materials. The principle of its mixing modification is as follows:

Table 3.1 Technical parameters of double-axis horizontal mixer.

Volume (m^3)	Actual volume (m^3)	Power (kW)
1.5–4	1.3–3.5	15–30

3.1.4.1 High-speed stirring effect

The effects of high-speed stirring on magnesium-based cementitious materials are as follows:

1. The high-speed rotation of the blades has a crushing effect, which can effectively crush the magnesium oxide lumps and some fillers.
2. Due to the fast-stirring speed, the mixing frequency of each material is high, and it is easy to mix evenly. The more uniform the mixing of magnesium oxide and the admixture solution is, the more complete the reaction will be, and the better the various properties will be.
3. It strengthens the shear mixing force of the slurry. Shear mixing is the most efficient mixing method. The shear force is proportional to the speed; the higher the speed, the stronger the shear force, and the higher the mixing efficiency, making it easier to mix the materials evenly.
4. High-speed stirring generates an impact mixing force. Under high-speed rotation, the mixer blades create an impact force. Driven by the blades, the slurry generates centrifugal force that impacts the inner wall of the mixing cylinder. Reflected and dispersed by this centrifugal impact, the slurry enhances mixing uniformity.

3.1.4.2 Dual-axis counter-rotating mixing modification effect

Due to the high viscosity and resistance of magnesium-based cementitious materials, their resistance to breakage under high-speed stirring is low, and the relative resistance increases. The higher the stirring speed, the lower the resistance to breakage, indicating that stirring speed is inversely proportional to the resistance to breakage. By using counter-rotating dual-spiral stirring, the blades of the spiral stirrers are narrow and inclined push blades, significantly reducing resistance. The dual-axis counter-rotation can bring slurry from both sides into the mixing zone between the two axes simultaneously. The dual-axis blades forcibly shear and mix the slurry, making it easier to achieve uniform mixing. After high-speed counter-rotation mixing, the magnesium-based cementitious material slurry becomes fine and smooth, with a uniform and viscous consistency and a shiny appearance, exhibiting good dispersion properties. This provides a foundation for the complete reaction of magnesium oxide and additives. The mixing modification alters the crystal structure of the hydration products of the magnesium-based materials, leading to a predominance of large blade-shaped structures and a reduction in the proportion of needle-like structures.

3.2 Foaming process

3.2.1 Overview of foaming

3.2.1.1 Foaming

Foaming can generally be divided into two types: physical foaming and chemical foaming. Among these, physical foaming involves mixing a foaming agent with a foaming solution and utilizing the compressor in the foaming device to pressurize the mixture, thereby forming foam with a certain tension. This foam is then stirred together with the magnesium-based cementitious material slurry. During the stirring process, the volume of the magnesium-based cementitious material slurry continuously increases. When the expansion reaches the predetermined density, a foamed magnesium-based product material is formed. Chemical foaming involves adding the foaming agent (liquid or powder) to the magnesium-based cementitious material slurry and stirring it to induce a chemical reaction that releases gas, causing the volume of the magnesium cement to continuously expand. When the expansion reaches the preset formulation state, a magnesium-based foamed cement material is formed.

3.2.1.2 Characteristics

Magnesium-based foaming products have the characteristics of lightweight, heat insulation, sound absorption, and low cost.

3.2.1.3 Applications

The foaming process is widely used in building materials fields such as fireproof door core boards, exterior wall insulation boards, partition boards, decorative boards, roofs, floors, and floor heating.

3.2.2 Foaming method

3.2.2.1 Physical foaming

3.2.2.1.1 Common types, characteristics, and applications of physical foaming agents

1. Polymer foaming agents

   This foaming agent exhibits slightly lower stability compared with protein and rosin foaming agents. It features lower water content, a high foaming multiple with a dilution range of 80–200 times, good fluidity, and low production cost. It is suitable for producing products such as flowing-edge framed door core boards, homogeneous boards, and exterior wall insulation boards. It can be used with magnesium cement as well as silicate cement.

2. Protein foaming agents
   This foaming agent exhibits good stability, a higher water content, a dilution factor of approximately 40 times, poor fluidity, and simpler operation compared with rosin-based foaming agents. It can be used for manufacturing flowing-edge flat-mold door core boards that demand higher production requirements and lack a frame. Protein foaming agents exhibit strong adaptability and can be used with both magnesium cement and silicate cement.
3. Rosin foaming agents

This foaming agent has poor stability, a higher water content, a low foaming multiple, a dilution multiple of 30–40 times, and is more complex to operate. The two components, A and B, are mixed in proportion, then mixed with water to form a solution, which is subsequently foamed using a foaming machine. It is generally used for high-density foamed concrete with a density of 600 kg/m^3.

3.2.2.1.2 How to judge the quality of physical foaming foam

The performance of a foaming machine is measured by the quality of the foam it produces. The foam must meet three specific criteria: fineness, uniformity, and the water content of the foaming agent.

Foam fineness refers to the size of the bubbles. Smaller bubble diameters and finer foam lead to better foam stability, resulting in higher strength and improved insulation performance of the final product.

Foam uniformity refers to the uniformity of the bubble diameter size. The more uniform the bubble diameter size and the narrower the distribution range, the more uniform the stress on the product, and the better the usage effect.

Foam foaming agent water content refers to the amount of foaming agent water solution produced after the foam breaks. Generally speaking, the lower the water content of the foam, the lower the foaming agent water content, indicating better foaming performance of the foaming machine.

To meet the quality requirements for foam, the foaming device independently developed and designed by Shandong Meijiatu New Materials Technology Co., Ltd.(MJT), featuring its own intellectual property rights (Application Number: CN202310769538.8) (Fig. 3.2), is installed between the foaming tank and the output pipe that carries the foam from the tank. There is a transition cavity between them.

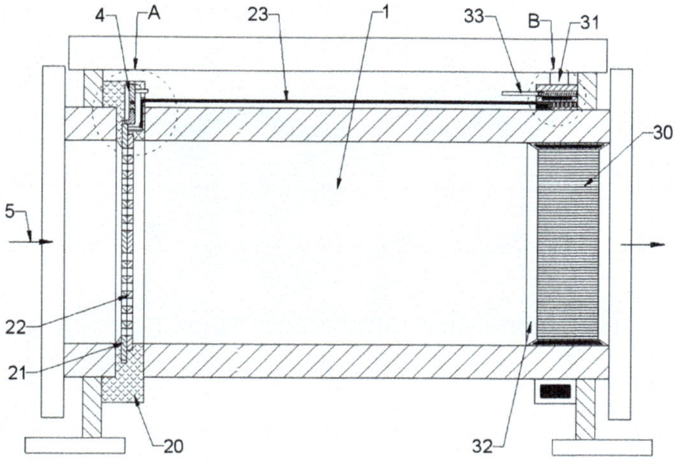

Figure 3.2 Foaming device. 1—Transition cavity; static filter assembly: 2—static frame, 3—main filter plate; 4—secondary filter plate; 5—pulling rope; active filter assembly: 6—filter plate; 7—wire block; 8—scraper; 9—trigger component; 10—limiting component; 11—fcam flow direction.

The transition cavity contains a static filter component and a dynamic filter component arranged along the direction of foam movement. The filter holes on the static filter component are larger than those on the dynamic filter component, and the static filter component opens and closes its filter holes based on the movement of the dynamic filter component. When the dynamic filter component moves near the static filter component, the filter holes on the static filter component gradually close. When the moving filter component approaches the stationary filter component, the filter holes on the stationary filter component close. The moving filter component and the stationary filter component can compress the foam filtered by the moving filter component, causing these foams to rupture and form foams capable of passing through the filter holes of the moving filter component. The foam size can be effectively controlled within a certain range, producing sponge-like foam with a high water content, low density, and a uniform, fine structure.

3.2.2.1.3 Working principle of the foaming machine

There are two common physical foaming processes.

1. Production of foam through the utilization of high-speed rotating blades. The foaming agent solution and raw materials are introduced into a high-speed mixer simultaneously. Foam is generated by the rapid

Figure 3.3 MC-900 foaming machine.

rotation and agitation of the mixer blades. Nonetheless, the foam produced by this technique exhibits uneven bubble diameters, and the cured products demonstrate variations in volume density, leading to the products not meeting the necessary mechanical properties.

2. Foam is generated by pressure-induced air injection. Using the MC-900 foaming machine as an example (Fig. 3.3), it employs the pressure-induced foaming principle. An air compressor pressurizes the air and injects it into the storage tank. The foaming agent aqueous solution is drawn into the plunger pump via the inlet pipe. Both components are thoroughly mixed in the foaming generator, and then, by utilizing the pressure difference between the interior and exterior, the formed foam is expelled from the foaming tube. The equipment structure is depicted in Fig. 3.4, and the specific parameters are provided in Table 3.2.

Comparing the two physical foaming processes, the pressure-induced air foaming process is more complex, but it has higher foaming efficiency and uniform bubble diameters. The foams are directly mixed with the well-mixed magnesium-based cement slurry, and the slurry density is precisely controlled.

3.2.2.1.4 Physical foaming process flow

The preparation of foamed magnesium cement slurry is divided into two stages: the first stage is the preparation of magnesium cement slurry; the second stage is the preparation of foamed magnesium cement slurry. In simple terms, the foaming process of foamed magnesium cement is to add the prepared foams to the magnesium cement slurry through a mixer (Fig. 3.5).

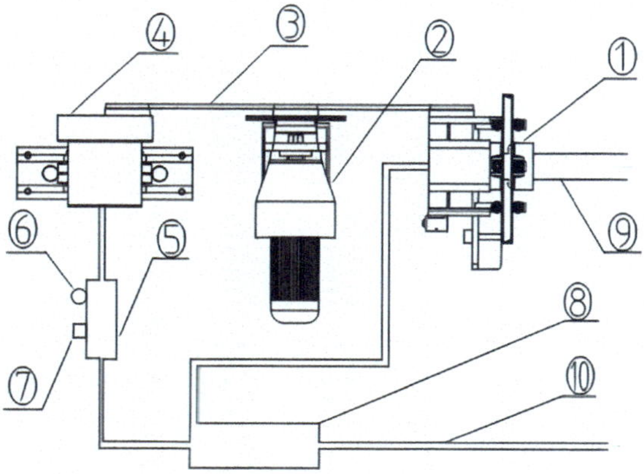

1 - Plunger pump; 2 - Motor; 3 - Belt; 4 - Air compressor; 5 - Air storage tank; 6 - Pressure gauge;7 - Safety valve; 8 - Foam generator; 9 - Inlet pipe; 10 - Foaming pipe.

Figure 3.4 Structural schematic diagram of MC-90. 1—Plunger pump; 2—motor; 3—belt; 4—air compressor; 5—air storage tank; 6—pressure gauge;7—safety valve; 8—foam generator; 9—inlet pipe; 10—foaming pipe.

Table 3.2 Parameters of MC-900 foaming machine.

Technical specifications	**Model**	
	MC-600	**MC-900**
Foaming volume (m^3/min)	0.6	0.8
Installed power (kW)	4	7.5
Machine weight (kg)	98	118
Overall dimensions (mm)	1100 × 700 × 750	1100 × 700 × 750
Power supply voltage (V)	380	380
Foam diameter (mm)	0.1–2	0.1–2

1. Preparation of magnesium cement slurry

Add the admixture to the mixer, start the mixer to stir, add the magnesium oxide powder while stirring, stir for 3 minutes, slowly add the modifier, and stir evenly to obtain the magnesium cement slurry.

Foamed magnesium cement uses magnesium-based cementitious materials as the base material. Sometimes a small amount of fly ash is added. For example, when preparing lightweight partition panels with a volume density of about 600 kg/m^3, 25% sawdust will be added.

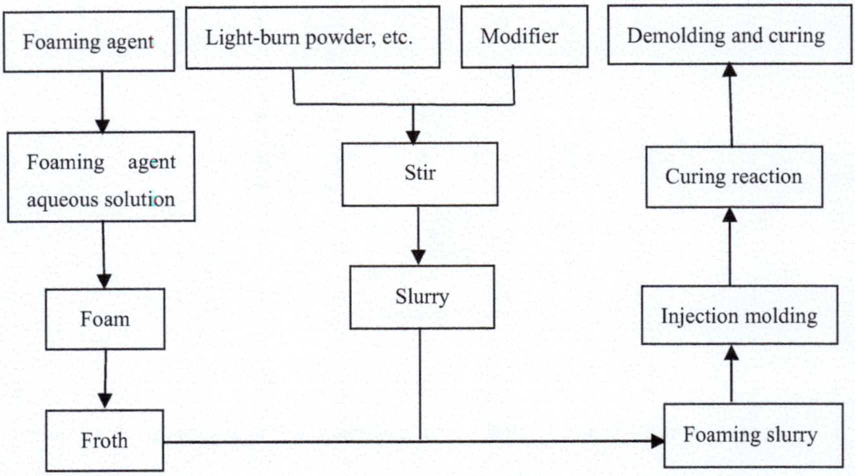

Figure 3.5 Preparation process of foamed magnesium cement slurry.

Products with special requirements, such as exterior wall insulation boards, will use some lightweight aggregates, such as expanded perlite and polystyrene particles.

2. Preparation of foamed magnesium cement slurry

After the magnesium cement slurry is thoroughly mixed, fresh foam is added to the mixer while it is still mixing. The stirring time should be about 1 minute, which can be adjusted according to the situation. When the foam is evenly mixed and no layer of floating foam is visible on the slurry surface, the foamed magnesium cement slurry is prepared and ready for discharge and pouring.

3.2.2.2 Chemical foaming (hydrogen peroxide)

3.2.2.2.1 Characteristics and applications of chemical foaming agents

The prominent feature of chemical foaming is that it does not require any foaming equipment. By releasing a large number of bubbles through a chemical reaction, it can achieve the foaming effect. All that is needed is to fully mix the chemical foaming agent with the magnesium cement slurry and allow it to react. The foaming process lasts from 5 to 15 minutes. After foaming, the pores typically form a honeycomb-like closed-cell structure.

Products produced using chemical foaming agents exhibit higher strength than those produced by physical foaming (at the same density). The main reasons are that the pores are larger and evenly distributed, the

Table 3.3 Performance indicators of chemical foaming magnesium-based exterior wall insulation board.

Performance indicators		Parameter
Dry apparent density (kg/m^3)		≤250
Compressive strength (MPa)		≥0.40
Flexural strength (MPa)		≥0.10
Moisture content (%)		≤5.0
Volume water absorption rate (%)		≤10
Thermal conductivity (W/(m·K))		≤0.069
Dry shrinkage value (rapid method) (mm/m)		≤2.0
Freezing resistance (15 times)	Quality loss (%)	≤5.0
	Strength loss (%)	≤25
Softening coefficient		≥0.75
Fire resistance rating		A1

cement slurry on the bubble walls is concentrated, and thus the strength is high. Moreover, the bubbles are mostly closed pores, resulting in a low water absorption rate. Specific parameters are shown in Table 3.3. The disadvantage of chemical foaming is that the process is difficult to control. The size, quantity, durability, and distribution of the bubbles in the slurry are difficult to precisely control. Large-scale production requires a constant temperature workshop or curing room. Additionally, the foaming agent dosage is large, and the foaming cost is high.

3.2.2.2.2 Principles and precautions for selecting chemical foaming agents

The selection of chemical foaming agents should follow the following four principles:

1. The decomposition temperature should be compatible with the curing temperature of the magnesium-based foaming slurry.
2. The foaming capacity should be large.
3. It should be relatively safe.
4. The price should be low.

The heat release peak during the curing of magnesium cement should be controlled within 60°C. If the decomposition temperature of the chemical foaming agent needs to be 100°C or higher, it will interfere with the curing of magnesium cement. Therefore chemical foaming agents that can undergo decomposition reactions at room temperature can be used for

Table 3.4 Reference formula for chemical foamed magnesium cement (unit: kg).

Light calcined magnesia	Brine (with water temperature ranging from 30°C to 35°C)	Polypropylene fiber	Hydrogen peroxide	Zinc stearate
100	90	0.5	10–11	2.3–2.5

the production of foamed magnesium cement. The phrase "the reaction products are relatively safe" contains two meanings: first, the substances produced by decomposition or reaction have no adverse effects on the curing process of magnesium cement; second, the chemical foaming agent itself or the gas produced during the reaction will not cause harm to the environment when it escapes.

3.2.2.2.3 Chemical foaming mechanism

The chemical foaming agents currently used for the production of foamed magnesium cement via chemical methods are peroxides, such as hydrogen peroxide (H_2O_2). The reference formula for foamed magnesite cement with a density of 300 kg/m^3 is shown in Table 3.4.

Hydrogen peroxide is relatively stable in acidic and neutral media, but it decomposes easily in alkaline media. Its decomposition rate increases under conditions of light and heat. The foaming principle of hydrogen peroxide is as follows:

$$2H_2O_2 \xrightarrow{\text{catalyst}} 2H_2O + O_2\uparrow$$

Theoretically, the maximum gas production capacity of 1 kg of 30% hydrogen peroxide is 98 L. In actual production, considering that some gas will escape and the volume will be lost, the volume loss rate should be considered. Therefore the hydrogen peroxide dosage should be slightly larger than the theoretical calculation value. The dosage of hydrogen peroxide should be adapted to the density of magnesium-based foaming cement. To achieve the expected gas production volume and obtain a uniformly pored chemical foaming product, the following adjustments should be made.

1. Control the production temperature and brine temperature, and take measures to adjust the solidification rate of magnesium cement, or add appropriate catalysts, such as manganese dioxide and copper oxide, to

regulate the gas production rate of hydrogen peroxide decomposition, so that the decomposition rate of hydrogen peroxide matches the solidification and curing rate of magnesium-based cement slurry.

2. Stabilizer. Generally, the hydrogen peroxide used contains a certain amount of stabilizer to reduce the decomposition of hydrogen peroxide during storage. Common stabilizers include sodium stannate, sodium pyrophosphate, and organic phosphorates.
3. Foaming stabilizer. The function of the foaming stabilizer is to reduce the surface tension of the liquid phase of the slurry to stabilize the bubbles and maintain the stability of the bubble liquid membrane structure. Common foaming stabilizers include soluble oil and surfactants.

In the reference formula, 1%–3% zinc stearate is used as a foaming stabilizer, which is based on the thickening effect of zinc stearate on the magnesite slurry. The foaming principle is to increase the resistance when bubbles escape, making the pore structure of the solidified slurry fine and uniform; at the same time, zinc stearate is a strong hydrophobic substance, and after curing, it forms a layer of hydrophobic film on the gas pores, reducing the water absorption rate of the foaming products in a humid environment.

3.2.2.2.4 Chemical foaming process (hydrogen peroxide)

1. Raw material preparation: The raw materials for foaming molding primarily include foaming agents, substrates and additives (such as stabilizers). Before use, they need to be mixed in a certain proportion.
2. Foaming: The mixed raw materials are placed in a chemical foaming device and undergo chemical reactions through heating and pressurization, thereby generating gas and causing the raw materials to expand into a foamed state.
3. Forming: After foaming, the raw materials are poured into molds and formed through heating and pressure. The shapes and sizes of the molds vary and can be designed and manufactured according to product requirements.
4. Processing: The processed products need to undergo subsequent processing, such as cutting, grinding, and painting, to meet the customer's requirements.
5. Inspection: The final step is to inspect the finished products to ensure that their quality meets the standards, including appearance inspection, size measurement, and physical performance testing.

3.3 Molding process

3.3.1 Definition and development

Molding is a crucial step in the production process. By applying pressure, the well-mixed slurry is poured into various forming devices (or molds) to achieve the desired shape and solidify the product.

The molding processes for different products vary, and even for the same product, different molding processes are employed at different production stages. The molding process evolves from manual to semi-mechanized, mechanized, and then automated production, eventually progressing towards intelligence. Production efficiency continuously increases, while the demand for manual labor decreases, and production becomes more refined, with product precision also improving.

3.3.2 Classification and characteristics of molding processes

This book, based on the actual production of magnesium-based cementitious materials, only introduces several representative molding processes, including mold molding, flat plate rolling, vertical mold pouring, atmospheric extrusion, and hot press extrusion.

3.3.2.1 Mold forming

Mold forming is a process that achieves the shaping of an object's appearance by altering the physical state of the material being molded. Depending on the shape of the desired product, different molds are selected for processing and manufacturing. The following example is given using a magnesium flower pot.

Pour the mixed slurry into the mold and use a vibration table or vibrator to remove entrapped air and ensure that the slurry completely fills the mold. Usually, to obtain a smoother outer surface, a defoamer (GX-5#) can be added to the slurry, combined with vibration or vacuum adsorption methods, to produce high-quality flower pot products (Fig. 3.6).

3.3.2.2 Flat plate rolling

Flat plate rolling is primarily used for large-area, flat-shaped products of varying thicknesses, typically relatively thin. It shifts the previously labor-intensive, low-efficiency production model to a semimechanized mode, where machinery is the main force and manual labor supports it, significantly enhancing production efficiency. It is commonly employed

Figure 3.6 Mold forming.

in manufacturing fireproof boards, door core boards, purification boards, and other products.

The pouring process of the flat mold production line can be further divided into two types: framed pouring and unframed roller pressing. The framed pouring process uses the original molds of manual flat mold pouring, changing the original material transportation and pouring method to a storage material trough that remains stationary, with the belt-driven mold running to pour. After pouring, it is stacked by a lifting platform or a palletizer, and then deformed and flattened or sanded. This process reduces the manual workload, improves production efficiency to some extent, and reduces the loss of slurry during transportation by concentrating the material supply. However, the mold investment is relatively large, and the number of processes has not been reduced. On the other hand, the unframed roller pressing process requires a higher state of the slurry, with the slurry having a certain viscosity. It is discharged from the storage trough to the mold, and after being rolled to a fixed size, it takes on a tofu-like shape, not flowing or collapsing. It can basically achieve automated production, and the production efficiency is greatly improved. Ordinary high-molecular foaming agents cannot meet its usage requirements. Generally, animal protein or rosin two-component foaming agents are used. These types of foaming agents have high stability, but the dilution ratio is slightly lower (generally 20–50 times diluted with water). The stacking and curing process after the unframed roller pressing remains unchanged. Due to the absence of a frame, the size and specification of the plates after demolding are not standard. In addition to the conventional flattening, they also need to undergo four-sided cutting and shaping, which will increase additional processes and bring a large amount of edge and corner waste.

Although both flat mold production line forming processes can improve production efficiency, each has its own advantages and disadvantages. They are suitable for producing 500−800 door core boards per day.

3.3.2.3 Mold laying and pouring

Mold laying and pouring is a highly mechanized and automated production process primarily used for manufacturing large-sized foamed products of varying thicknesses. This method boasts high efficiency and yield, generally producing aesthetically pleasing product shapes without requiring further processing. It is commonly employed in the production of partition wall panels and door core panels. The following example illustrates the process using fireproof door core panels.

Molding and pouring (Fig. 3.7). The forming process of the core board of fireproof doors is divided into two types: steel fireproof core material pouring and wooden fireproof core material pouring. The steel fireproof core material pouring molding process was born and developed along with the sulfur-magnesium fireproof core board. The properties of the sulfur-magnesium fireproof core board material are different from those of the chlorine-magnesium fireproof core board material. The traditional stacking-style curing method is difficult to reach the required temperature for the reaction, resulting in slow solidification, poor strength, and low mold turnover efficiency. The molding process of the standing mold pouring makes the core materials separated only by aluminum alloy partitions, allowing for rapid heat accumulation and slow heat dissipation. A small increase in environmental temperature can achieve the temperature requirement for triggering the reaction of the slurry. The equipment for the standing mold pouring of the core board can reach 40–50 sheets per group, with a small floor space and high production efficiency. The formed boards do not need to be flattened or sanded. The author has summarized through practice that after the core material is poured for 8–12 hours, the temperature of the board can reach 50°C–60°C. At this time, the strength of the board has met the requirements for demolding, and after demolding,

Figure 3.7 Mold making and pouring.

the board heats up, and when the temperature drops, the board can be taken out. When the core material temperature exceeds 60°C, if the demolding is not timely, the temperature will continue to rise, and the final temperature can exceed 70°C–80°C. At this time, there is a risk of burning the board. After demolding, the burned board has no strength and a loose structure. Wooden fireproof doors generally adopt the direct pouring process. The wooden door frame is prefabricated and the pouring holes are left, and then it is fixed to the pouring frame after the composite fireproof board and the upper and lower uncoated panels. The movable end screw expands to prevent the uncoated panel from loosening and bulging or sinking during the pouring process, affecting the flatness. Wooden fireproof doors mostly adopt the standing mold pouring process, so after the foaming slurry is well-mixed, it is mostly pumped to the pouring frame through a flexible hose (or screw pump) for pouring. It should be noted that during the pumping process of the flexible hose (or screw pump), about 15% of the foam has a damage rate, resulting in a decrease in the fineness of the slurry. When formulating the production formula, this issue should be considered to prevent foam damage from causing the core board to collapse.

3.3.2.4 Cold press extrusion

When producing antique brick-like products, it is essential that the appearance of the products conforms to classical design specifications, possessing high strength and excellent durability. Additionally, it is

Figure 3.8 Cold-pressed brick equipment.

important to minimize labor intensity during production. To meet these requirements, cold pressing brick equipment (Fig. 3.8) is predominantly used. This equipment mixes the slurry based on a rational formula and pours it into the designed mold. Typically, under a pressure of 6000−8000 kN, multiple brick-like products can be manufactured simultaneously. The operation is straightforward, and the production efficiency is high.

3.3.2.5 Hot pressing extrusion

According to the requirements of the scientific formula, magnesium-based cementing materials and various auxiliary materials are uniformly mixed into a plastic state. This mixture is then transported to the high-pressure extrusion thermosetting machine device. Under conditions of heating and pressure, it is extruded for a specific period of time. This method produces boards with high density, good structural strength, rapid curing and formation, and high production efficiency. (Refer to Section 5.15. No further elaboration is provided here.)

3.4 Demolding process

3.4.1 Preparation work

To facilitate the demolding of the final product, a layer of demolding agent should be evenly sprayed on the template before the normal production of the product. The demolding agent must match the product and the template; otherwise, it will not only cause the defoaming phenomenon of the foamed product but also fail to achieve the demolding effect. Moreover, it will make the product stick to the template, which is even more unfavorable for demolding.

3.4.2 Pneumatic demolding

The pneumatic mold needs to be connected to an air pump, which is an air compressor. Press the pneumatic switch to demold. The thickness of the finished product is fixed and cannot be adjusted. Its advantages are easy demolding, fast demolding speed, and the finished product has a beautiful plate shape. It is suitable for mass production, but the cost of the mold is relatively high.

3.4.3 Selection of templates

Polyvinyl chloride (PVC) is a polymer formed by the free radical polymerization reaction mechanism of vinyl chloride monomer in the presence of initiators such as peroxides and azo compounds, or under the action of light and heat. Vinyl chloride homopolymers and vinyl chloride copolymers are collectively referred to as PVC resin.

PVC is a white powder with an amorphous structure, having a relative density of approximately 1.4. Its glass transition temperature ranges from 77°C to 90°C, and it begins to decompose at around 170°C. It has poor stability to light and heat. It will decompose and produce hydrogen chloride at temperatures above 100°C or after prolonged exposure to sunlight, and further undergoes autocatalytic decomposition, causing discoloration. Its physical and mechanical properties also rapidly decline. In practical applications, stabilizers must be added to enhance its stability to heat and light.

PVC board is also known as Schüco board or Flon board. It is a type of board formed by extrusion using PVC as the main material. This board has a smooth and flat surface, and its cross-section has a honeycomb texture. It has high strength and good weather resistance. PVC has a high flame-retardant performance, with a flame-retardant value of over 40. When burning under a flame, it extinguishes immediately and produces a yellow flame, with a green lower end and white smoke. PVC exhibits good chemical resistance, being able to resist the corrosion of concentrated hydrochloric acid, concentrated sulfuric acid (with a concentration of 90%), concentrated nitric acid (with a concentration of 60%), and concentrated sodium hydroxide (with a concentration of 20%).

Table 3.5 Relevant parameters of acrylonitrile, butadiene, and styrene extruded sheets.

Test items	Technical requirements	Measured results	Individual judgment
Appearance	The surface should be smooth and flat. There should be no waves, shortage marks, scratches, black spots, impurities, bubbles or cracks that affect its usability.	Complies with the inspection requirements	Up to standard
Longitudinal tensile yield stress (MPa)	≥32.0	44.0	Up to standard
Transverse tensile yield stress (MPa)	≥32.0	41.6	Up to standard
Longitudinal dimension change rate (%)	−20.0 to +5.00	−4.8	Up to standard
Transverse dimension change rate (%)	−20.0 to +5.00	−4.2	Up to standard
Vishay softening Temperature (°C)	≥80.0	84.6	Up to standard

PVC board has a high mechanical strength at room temperature, with a higher wear resistance than vulcanized rubber. In addition, its hardness and rigidity are also superior to polyethylene. At the same time, it has good tolerance to nonpolar solvents such as petroleum and mineral oil.

The ABS plastic template is a terpolymer composed of acrylonitrile (A), butadiene (B), and styrene (S), characterized by its white color. It integrates the properties of its three constituent components. Acrylonitrile provides the template with high hardness, strength, heat resistance, and corrosion resistance; butadiene contributes impact resistance and toughness; styrene imparts high gloss, ease of coloring, and convenient processing properties. This makes it suitable for manufacturing glass fiber fireproof boards, lightweight fiber fireproof boards, and construction templates. The specific parameters of ABS extrusion board are shown in Table 3.5.

PVC and ABS boards are commonly used molds in the production of magnesium-based cement materials. However, PVC boards have a lower glass transition temperature. When the product's reaction temperature rises excessively, they are prone to warping deformation, affecting product flatness. ABS boards have a relatively higher glass transition temperature and exhibit good heat resistance. During high-temperature curing, the deformation rate of ABS boards is significantly lower than that of PVC boards.

As the duration of use increases, an oil film will develop on the surface of the template sheet. This oil film can negatively impact the film-forming effectiveness of newly sprayed release agents, and in severe cases, it may also affect the flatness of the finished products. Typically, each production facility arranges for workers to clean the templates periodically. Common cleaning methods involve using a putty knife to scrape off the film, or applying acidic substances. However, regardless of the method employed, it will influence the smoothness of the template surface. Over time, the service life of the sheet gradually declines. Additionally, the cost of purchasing templates is relatively high, so each factory often contacts the template manufacturer to recycle and reprocess used templates to alleviate financial pressure. The standard-thickness templates produced by manufacturers are mostly sold based on quality. To maximize profits, manufacturers frequently incorporate certain amounts of talcum powder, calcium carbonate, or similar additives into the PVC or ABS raw materials during production. Adding an appropriate amount of filler can enhance the hardness of the sheet. However, excessive filler may lead to increased brittleness, making the sheet prone to breakage during use.

3.5 Maintenance procedures

Maintenance is an essential step in product manufacturing. The duration of maintenance, temperature levels, and moisture content significantly impact product quality. While ensuring product quality, accelerating turnover rate and improving production efficiency are also concerns for production units [1].

3.5.1 Factors and principles to consider in maintenance

3.5.1.1 Temperature

Taking chloromagnesium oxide products as an example, the mechanical properties and durability of these products primarily derive from phase 5·1·8. This crystalline phase is generally formed when the internal temperature of the product does not exceed 60°C. If the internal temperature of the product during the curing process reaches 80°C or higher, then phase 5·1·8 will no longer be formed and instead, phase $9Mg(OH)_2·MgCl_2·4H_2O$ (9·1·4 phase) and a large amount of $Mg(OH)_2$ will be generated. The excessive $MgCl_2$ remains in a free state, causing the product to have severe moisture absorption and reoxidation and the overall mechanical properties and durability to deteriorate. Practice has proved that the optimal peak internal temperature range of the product is 60°C–80°C (Tables 3.6 and 3.7); if it is lower than 40°C, it indicates that the hydration reaction of the product is incomplete and the degree of bonding reaction is poor, and there is too much free $MgCl_2$ in the product. Such products are prone to moisture absorption and reoxidation and usually start to soften after 3 months, with strength gradually decreasing. If it is higher than 60°C, the hydration reaction is excessive, the types of crystalline phases change, and the heat release and solidification promote each other. The thermal expansion stress and crystalline expansion stress inside the product overlap, causing damage to the internal structure of the product and a sharp decline in quality.

Therefore during the production process, when the medium temperature is low, measures such as insulation or heating should be taken; when the medium temperature is high, timely heat dissipation or adjustment of the ratio should be implemented to control hydration heat and stabilize product quality.

3.5.1.2 Moisture

Water is one of the important components of the solid phase formed by magnesium-based cementitious materials, and it also serves as the medium for the curing reaction of magnesium-based cementitious materials. In actual production, an excessive amount of water is used to facilitate operation and ensure the complete progression of the reaction. This water will be retained or lost in three forms: The first is bound water, which will be consumed by the formation of 5·1·8 phase, 3·1·8 phase or $Mg(OH)_2$; the second is that water changes into water vapor and escapes during the reaction process; the last is that a part of the water remains in the product in the form of free water, that is, the moisture content.

Practical experience has proved that when MgO and $MgCl_2$ are mixed together and left for a long time, they will not react to produce strength. When the water content is sufficient, during the curing process of the product, attention should be paid to the rate of water loss, as if the water

Table 3.6 Correlation between the hardened phase of magnesium cement and temperature.

Maintenance conditions	Hardened crystalline phase and content (%)					
	5·1·8 phase	3·1·8 phase	Mg $(OH)_2$	MgO	9·1·4 phase	Pending crystalline phase
Indoor temperature: 15°C–25°C, moisture-preserving curing for 28 days	66	0	5	8	0	0
Cure at 40°C for 24 hours, then move to indoor moisture-preserving curing for 28 days	53	0	5	9	0	0
Cure at 60°C for 3 days, then move to indoor moisture-preserving curing for 7 days	15	0	7	12	0	A large amount
Cure at 80°C for 24 h, then move to indoor moisture-preserving curing for 28 days	0	0	23	2	0	A large amount
Cure at 100°C for 3 days, then move to indoor moisture-preserving curing for 28 days	0	0	21	2	0	A large amount
Form at (110 ± 5)°C, with a pressure of 5 kg/cm^2	Thimbleful	0	Principal quantity	—	—	—

Table 3.7 Correlation between internal peak temperature of products and solidification quality.

Curing medium temperature (°C)	Test specimen peak temperature (°C)	Water immersion 1 month water absorption rate (%)	28d strength (MPa)		Water immersion 1 month intensity (MPa)		Soaking in water - 1 month softening coefficient		Volume stability after being immersed in water for half a month	Moisture absorption and reverse condensation situation
			Flexural	compressive	Flexural	compressive	Flexural	compressive		
Indoor temperature	28	5.9	9.5	55.6	7.7	46.5	0.84	0.85	Good	10 days later: Absorbs moisture and condenses
40	54	6.2	8.6	56.3	7.4	41.8	0.85	0.75	Good	Always dry
60	81	8.3	8.6	33.8	2.1	7.1	0.26	0.21	Severe cracks	10 h: Surface condensation
80	108	9.5	7.8	33.2	2	10.5	0.25	0.31	General crack disintegration	12 h: Surface condensation
100	110	10.4	7.4	26.2	2.4	10.5	0.33	0.40	General crack disintegration	12 h: Surface condensation

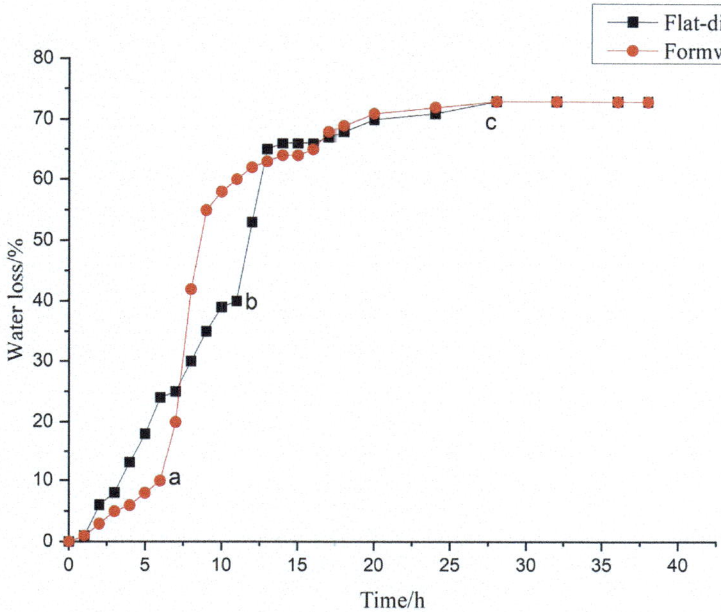

Figure 3.9 Relationship between door core board time and water loss under different production modes.

loss is too fast, the water required for crystal development is insufficient, resulting in insufficient product strength; if the water loss is too slow, the chemical reaction rate will slow down, and the production efficiency will decrease. Moreover, the presence of excessive water will cause the arrangement of crystals to be less dense, the structure to be loose, and the strength to be lost.

A comparative test was conducted between the production of magnesium oxide core board using flat molds and vertical molds. As shown in Figs. 3.9 and 3.10, with the passage of time, the temperatures of both methods were increasing, and at the same time, moisture was being lost. As time extended, the moisture content of the boards produced by the two methods gradually stabilized and approached each other.

Under the flat mold production mode, the contact area between the board and the air is large, and as the temperature rises, water vapor can volatilize normally. When the temperature reaches point e in Fig. 3.10, it reaches its peak. At this time, the mold needs to be removed promptly. After removing the core board, its lower surface is exposed to air, significantly accelerating moisture and temperature loss, leading to the

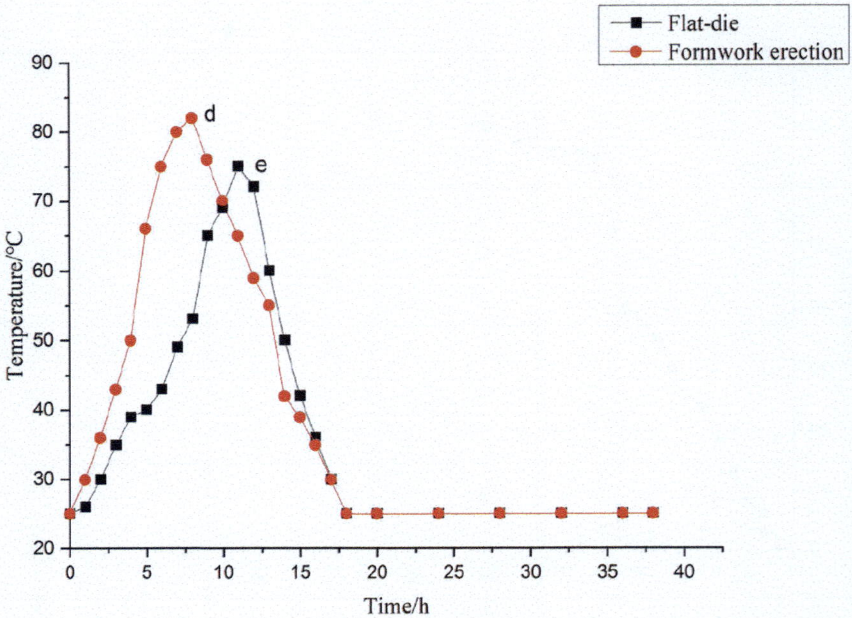

Figure 3.10 Relationship between door core board time and temperature under different production modes.

emergence of point b in Fig. 3.9. The temperature gradually declines, accompanied by substantial moisture evaporation. Under the vertical mold production mode, the temperature rises rapidly, and the highest temperature point d (Fig. 3.10) appears swiftly. Due to the lack of ventilation on both sides, the area for moisture dispersion and heat dissipation is small, causing water vapor to be unable to escape until the vertical core board is removed at point a in Fig. 3.9. At this time, the temperature and moisture are significantly different from the natural environment outside, and there will be intense water loss and heat dissipation, entering a period of rapid water dispersion loss. With the passage of time, the rate of water loss for both production modes tends to stabilize, and eventually, it does not change after point c (Fig. 3.9).

During summer production, the issue known as "plate burning" is prone to occur. The primary cause is that heat and moisture are not dissipated promptly. Whether using flat or vertical mold production, in dry and high-temperature conditions, the surface of the plate often reacts and forms a hard shell. This shell acts like a "blanket" covering the core board, obstructing the normal escape of water from the slurry into vapor. As a result, internal looseness and softness occur, leading to another type of "plate burning" phenomenon. Typically, a retarder is added to the reaction system to mitigate crust formation, ensuring uniform slurry reaction and timely dissipation of heat and moisture to avoid this problem.

In conclusion, water is the prerequisite for the hardening of magnesia-based cementitious materials. Apart from ensuring the formation of the 5·1·8 phase and the evaporated water, the amount of excess free water should be as little as possible. During the early stage of curing, moisture control and temperature regulation must be implemented.

3.5.1.3 Demolding time

The curing of magnesia-based cementitious materials is a process where chemical and physical reactions occur simultaneously. These reactions vary in speed depending on environmental temperature and changes in hydration heat. To ensure good product quality, it is necessary to master an appropriate curing time. The initial reaction of the slurry requires an activation temperature. If the activation temperature is too low, the reaction speed is too slow, and the demolding time will be too long, resulting in poor product strength; if the activation temperature is too high, combined with the heat release from the hydration reaction of the slurry, the two temperatures will superimpose, and the reaction speed will increase. If the demolding time is not well controlled, the "burning board" phenomenon will occur, and the product will have no strength and be directly scrapped.

As shown in Fig. 3.11, as time progresses, the temperature begins to rise and then starts to decrease. As shown in Curve 1 in Fig. 3.11, when the environmental temperature is above 30°C, the temperature rises rapidly within a short period of time. If the demolding time is not controlled properly, the temperature will exceed 80°C, and even if the demolding is delayed by more than 100°C, the "burning board" phenomenon will occur, and the product will have no strength and be directly scrapped. As shown in Fig. 3.10, when the temperature is 20°C at normal temperature, generally, the temperature begins to rise normally after 2 hours. When the temperature rises for 6–8 hours, it reaches 60°C. At this time, timely demolding and cooling to reduce moisture, and then normal temperature (generally 20°C−25°C) secondary curing will result in the mechanical properties of the product approaching the ideal state. As shown in Curve 3 in Fig. 3.11, when the environmental temperature is below 10°C, without heating intervention, the slurry slowly heats up. Sometimes it takes more

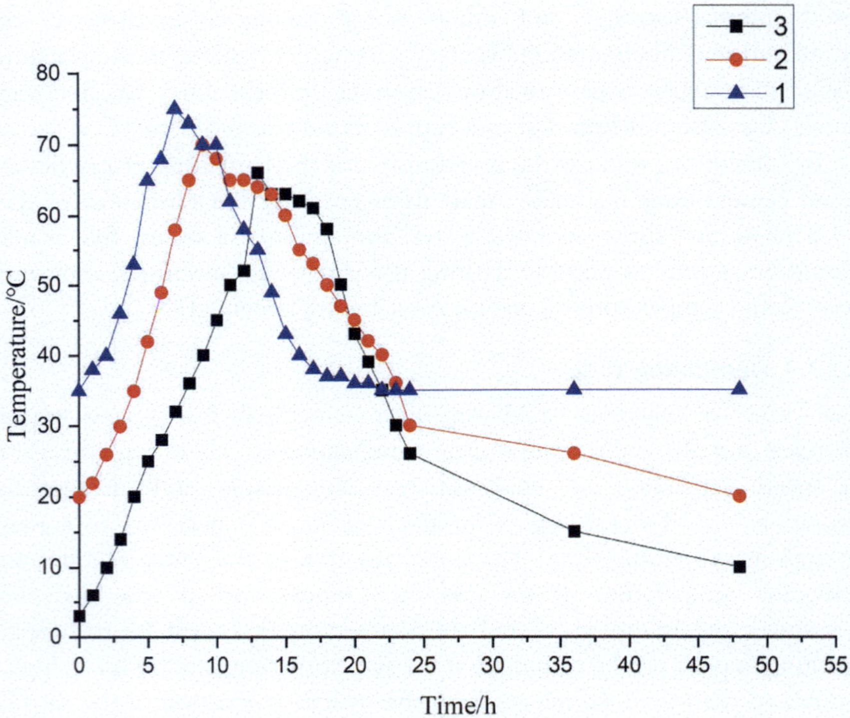

Figure 3.11 Relationship between time and temperature.

than 12 hours to reach the activation temperature. The demolding time even exceeds 24 hours, affecting the production cycle and efficiency.

The same requirements apply to the products made from magnesium thiosulfate and chlorine thiosulfate, as well as magnesium phosphate gelling materials. It is crucial to meticulously consider the three elements of time, temperature, and moisture and to precisely control their "degree," as this is of utmost importance.

3.5.2 Maintenance methods

With the deepening understanding of magnesium-based cementitious materials and the continuous development of production technologies, the methods of curing have become increasingly diverse. For different products, the curing methods also vary. This book, based on actual production, introduces several common curing methods.

Figure 3.12 Heating curing room.

3.5.2.1 Heating curing room (electric heating, residual heat, and diesel)

The heating curing room (Fig. 3.12) is a curing method adopted based on the characteristics of magnesium-based cementitious material products, the conditions of the production unit, and the surrounding environmental resources.

Some production units are located near power plants. In winter, the residual heat from the power plants can be utilized to heat the curing rooms, which is a cost-saving method. For smaller or less-equipped production units, curing rooms can be constructed using oil sheets (or thick polyurethane films). Heating methods such as electric heating or stove heating can be employed inside. However, these types of curing rooms do not distribute heat evenly. Products near the heat source react and cure quickly, while those farther away experience slower heating and curing rates. For production units requiring strict control of product moisture content, a separate curing room can be constructed when temperatures are low. For example, foam magnesium-based cementitious materials like door core boards, after demolding, can be placed in the curing room for a period to remove excess internal moisture, significantly reducing moisture content while enhancing strength.

Figure 3.13 Sunshade.

3.5.2.2 Sunshade

The sunshade (Fig. 3.13) is a curing method used in the production of magnesium permeable insulation boards (or silica insulation boards) when the temperature is low and rapid curing and packaging are required.

3.5.2.3 Microwave heating

Microwave heating curing is a modern curing method. It utilizes the principle of microwave heating, significantly reducing the initial triggering temperature time, allowing the MgO–$MgCl_2$–H_2O ternary phase to fully and rapidly react, forming crystalline phases, and solidifying and molding before normal curing (this microwave heating device refers to the patent number: CN202310378332.2; patent name: A microwave heating system for rapid curing and molding of foamed sheet materials) [2] (Fig. 3.14). Taking the foamed core board products as an example, this paper introduces the microwave heating curing process. The working principle of microwave heating is based on the principle of microwaves and molecular vibration. Microwaves are a type of electromagnetic wave, with their frequency typically around 2.45 GHz. These electromagnetic waves are generated by the magnetron in the microwave device and propagated through the metal waveguide within the microwave device. When these electromagnetic waves encounter the foamed core board sheet materials, they are absorbed by the water molecules in the foamed core board sheet materials. When the microwave radiation is absorbed by the foamed core board sheet materials, the water molecules start to vibrate. Due to the friction between these molecules, this vibration causes the

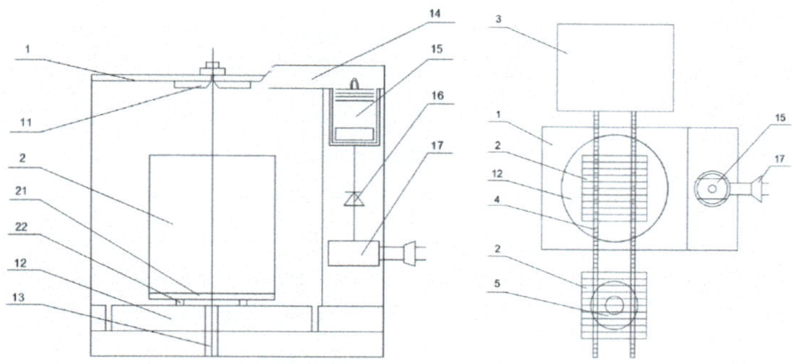

1 - Microwave heating device for core board (foamed board); 11 - Stirrer (made of metal); 12 - Rotating worktable;13 - Rotating shaft; 14 - Filter tube; 15 - Magnetron; 16 - High-voltage rectifier; 17 - Transformer; 2 - Core board mold kit set; 21 - Core board mold kit set; 22 - Pulley; 3 - Curing room; 4 - Guide rail; 5 - Slurry storage bin.

Figure 3.14 Main view and top view schematic diagram of microwave heating device. 1—Microwave heating device for core board (foamed board); 11—stirrer (made of metal); 12—rotating worktable; 13—rotating shaft; 14—filter tube; 15—magnetron; 16—high-voltage rectifier; 17—transformer; 2—core board mold kit set; 21—core board mold kit set; 22—pulley; 3—curing room; 4—guide rail; 5—slurry storage bin.

temperature of the foamed core board sheet materials to rise rapidly. The microwave radiation is only absorbed by the water molecules in the foamed core board sheet materials, so only foamed core board sheet materials containing water molecules can be heated by microwave radiation. When the microwave heating device is powered on, the alternating current power supply passes through a transformer and a high-voltage rectifier, converting the voltage into a high-voltage direct current power supply. This high-voltage direct current power supply is used to drive the magnetron. The magnetron in the microwave heating device is used to generate microwave radiation. The magnetron contains a cathode and an anode. When electrons pass through the magnetic field, they form a ray under the action of the magnetic field, which propagates along the waveguide and generates microwave radiation. The microwave heating device also includes a reflector (mixer), which functions to reflect the microwave back into the microwave heating device to ensure that the microwave can be evenly distributed throughout the foamed core board sheet materials. The microwave heating device also includes a turntable, which is used to uniformly expose the foamed core board sheet materials to the microwave radiation, so that the foamed core board sheet materials can be uniformly heated.

The microwave heating device for the core panel (foam board) is basically a microwave heating space made of metal with front and rear doors. This device uses metal as the microwave reflection wall. When the front and rear doors are closed and the sealing effect is good, the microwave will not leak out. There is a rotating worktable in the center of the device. The middle of the rotating worktable has a rotating shaft. The rotating shaft can be controlled by a switch to provide power and rotate the rotating platform. There are guide rails on the top of the rotating platform. Although the guide rails of the microwave heating device and the maintenance chamber are disconnected, when the rotating worktable is at the specified position, the gap at the connection point of the three sections of guide rails is very small, which does not affect the continuous rolling of the pulley. On the right side of the device, a transformer is installed. The top of the transformer is connected to a high-voltage rectifier, and above the high-voltage rectifier is a magnetron. When the transformer is powered on, the magnetron will release microwaves, which will be reflected by the rotating mixer and hit the core panel (foam board) of the mold kit for the inorganic cementitious material that is to be cured. The inorganic cementitious material core panel is generally made of cement or magnesium-based cementitious materials and is produced through foaming, modification, and so forth. Some core panels may also contain short fibers, waterproofing agents, and so forth.

This device enables the rapid formation of the internal lattice before the foaming of the foam breaks, solving the technical problem that the physical foaming plates collapse due to the premature rupture of the foaming foam over a long period. Through this device, more foaming foams can also be added to the slurry to achieve foaming boards with a lower density, breaking through the previous lower density limit.

The curing methods introduced in this book include both traditional and innovative ones, but the curing methods are not limited to these few. With the continuous development of technology and the integration of innovative thinking, new technologies will emerge.

3.6 Postprocessing techniques

3.6.1 Cutting

Magnesium plate products were initially processed using reciprocating four-sided cutting machines. The stacked plates of a certain height were placed on a fixed platform and then pushed into position by pneumatic

Figure 3.15 L-shaped cutting machine.

cylinders to be arranged neatly. Afterward, they were driven by a transmission motor into the cutting area. First, a fixed longitudinal motor trimmed the edges, and then the limit switch was triggered. The reciprocating lateral motor started, and the lateral movement trimmed the edges. After the entire cutting process was completed, the fixed platform returned to its original position. Forklifts then picked up the cut plates and stacked them. Simultaneously, another set of uncut plates was replaced to repeat the above operation. This cutting method is inefficient and significantly limits production efficiency. Moreover, due to the vibration of the lateral cutting machine during operation and the accumulation of cutting dust, the cutting accuracy decreases, and the diagonal size deviation becomes increasingly larger.

With the development of technology, manufacturers of fireproof board equipment have increasingly higher requirements for production efficiency. The daily output has gradually increased from 800 sheets to 2000–3000 sheets. The traditional reciprocating four-sided cutting machine has far failed to meet the demand. To keep up with the development of the industry, it has been gradually upgraded and improved to an L-shaped cutting machine (Fig. 3.15).

The L-shaped cutting machine adopts a conveyor line structure. Both ends are equipped with lifting platforms and hydraulic cylinder pushers, which are used for storing and arranging the board materials. The loading platform pushes the neatly stacked billets from the fixed-height hydraulic cylinder pushers onto the transmission belt. The lifting platform's each lifting height is controlled by an infrared sensor to ensure that the total thickness of different thicknesses of the plates pushed each time remains unchanged, without affecting the subsequent cutting process.

The billets entering the transmission belt are conveyed by the belt, and then cut and trimmed by the fixed transverse cutting machine. When they touch the longitudinal limiters, they are pushed longitudinally by the hydraulic cylinder pusher and enter the longitudinal transmission belt. There, they are cut and trimmed by the cutting saw fixed in the longitudinal direction. After the process is completed, they are transported to the plate stacking lifting platform, where they are neatly stacked by the hydraulic cylinder pusher. After stacking to a certain height, they are transported away by a forklift and packed.

The L-shaped cutting machine ensures the continuity of the entire cutting process, essentially achieving uninterrupted production and greatly enhancing production efficiency. Additionally, the fixed cutting unit, combined with the configuration of multiple hydraulic cylinder pushers, guarantees the transportation stability and cutting accuracy of the billets during the cutting process.

3.6.2 Leveling process

During the curing process of magnesium-based cementitious materials, volume expansion often occurs, resulting in a thickness that is 1–2 mm greater than the preset thickness. To enhance dimensional accuracy and meet diverse usage requirements, leveling treatment must be performed on the initially formed sheets. Common leveling methods include sanding and flattening. Typically, in products such as glass-magnesium sheets and hollow magnesium-sulfate oxide sheets, sanding machines are used for sanding; for products like fireproof door core sheets, press machines are primarily employed for flattening, while in special cases, engraving machines are used for carving and leveling.

3.6.2.1 Sandblasting

Traditional sandblasting of sheet materials often draws inspiration from woodworking processing tools and utilizes wide belt sanding machines for sanding and leveling. The wide belt sanding machine is a type of machine that uses a group of parallel shaft rollers to cut the circular sand belt under tension [3] (Fig. 3.16).

Figure 3.16 Sanding machine.

Its main functions are as follows.

1. Thickness control sanding is a process designed to enhance the thickness accuracy of sheet materials. This is employed, for instance, when hollow magnesium oxide sheets require thickness control sanding prior to double-sided bonding.
2. Surface sanding is a process performed on the surface of a sheet to uniformly remove a layer of material, enhancing surface quality and eliminating burrs left from previous processes, resulting in a smooth and aesthetically pleasing sheet surface. This process is also employed for veneering, dyeing, printing, and painting.
3. Sanding roughness refers to a sanding process for increasing the roughness of the backside of the decorative board to ensure the bonding strength between the veneer and the base material.

The broadband sanding machine is mainly composed of a conveying platform, a sand belt, a sand belt tensioning device (Fig. 3.17), and a sanding chamber. The conveying platform achieves reciprocating motion through the coordination of an electric motor and a transmission system, enabling the sand belt to perform grinding on the workpiece surface; the sand belt tensioning device can effectively adjust the tightness of the sand belt, ensuring uniformity and stability in the sanding process; the sand belt is a sanding tool that coats granular abrasive onto a flexible strip-shaped substrate. During operation, the sand belt rotates at high-speed, and each abrasive particle on its surface can be regarded as a tiny tool, participating in the sanding process of the workpiece.

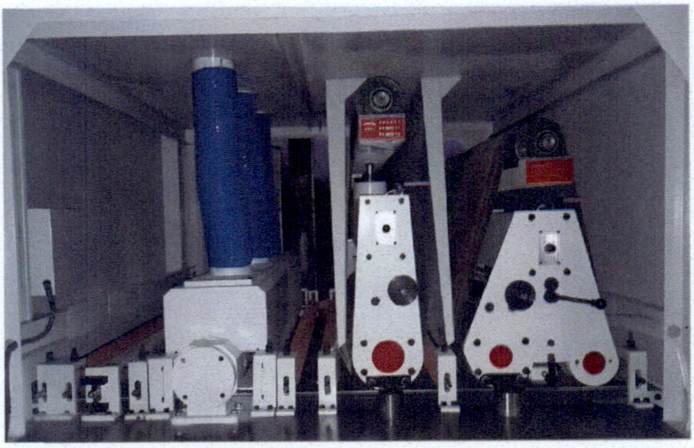

Figure 3.17 Abrasive belt tensioning device.

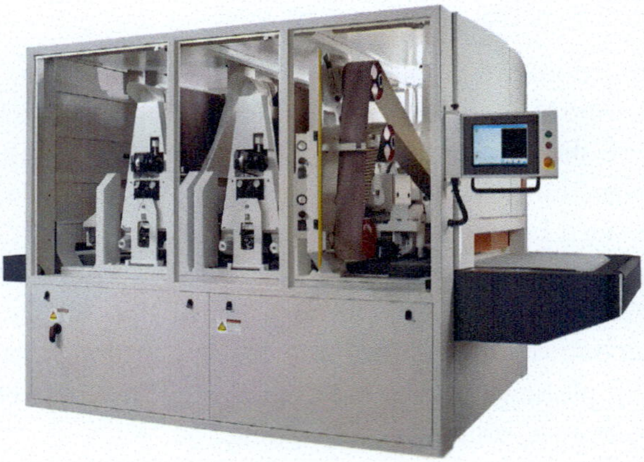

Figure 3.18 Powerful vacuum cleaning device.

In addition to installing the fixed sand belt tensioning device and the sand belt, the sanding room is also equipped with a powerful dust collection device (Fig. 3.18), which will promptly remove a large amount of debris generated during the sanding process. Moreover, the sanding machine is also equipped with a cleaning discharge roller, which can remove the remaining dust after the grinding is completed, ensuring the safety of the operators and the equipment.

Sandblasting is not only for making the surface smooth and uniform in thickness, but also for enhancing the strength of the board surface, enabling the board to adapt to various veneering processes. However, during the sandblasting process, problems such as sand leakage or sand penetration often occur, especially with wide band sandblasting machines. Wear of the sand belt and loosening of the tensioning device can result in some areas of the board not being sandblasted or being oversanded, seriously affecting product quality. The demand for sandblasting in high-end customization is also very extensive, including irregularly shaped board pieces, board pieces with height differences, and sandblasted with nonwoven fabric, all of which have high requirements for sandblasting processes and equipment.

To address these issues, an upgraded version of the broadband sanding machine—the piano-key sanding machine—was introduced. The sanding belt frame of the piano-key sanding machine is composed of several sections, resembling the keys of a piano. This design enables it to adapt to different thicknesses of boards and various irregular parts, whether large plates or small components, and perform precise processing. In the past, the keys of the piano-key sanding machine only had 40–60 sections, while the new model on the market now has keys increased to 80 sections, and the fine sanding during processing is more precise.

The new piano-key sanding machine is also equipped with an automatic edge detection system. During the processing, there is no need for manual intervention, and the sanding belt will not deviate. The existing piano-key sanding machines include a conveyor belt on the frame, cross-sanding components, longitudinal sanding components, fine sanding components, and polishing rollers.

The difference between the new piano-key sanding machine and traditional sanding machines lies in the adoption of longitudinal and transverse sanding units and fixed-thickness sanding units.

The grinding direction of the transverse sanding unit is at a right angle to the forward direction of the workpiece, which can avoid uneven paint and oversanding of the wood.

The grinding direction of the longitudinal sanding unit is consistent with the forward direction of the workpiece and is often used in conjunction with the transverse sanding unit to avoid processing difficulties caused by different board materials.

The fixed-thickness sanding unit uses steel fixed-thickness sanding rollers, which are adjusted through a digital control device. Compared with rubber rollers, its advantage is that it can avoid oversanding of softwood when sanding and has an overall sanding belt compensation device to ensure the thickness of the parts.

Figure 3.19 Hydraulic press.

3.6.2.2 Pressing

In addition to the traditional sanding process, the pressing process is also commonly used in foamed products such as fireproof door core boards. The pressing of the door core board is generally divided into two types: planar pressing and linear pressing. Planar pressing usually employs a hydraulic machine of 100 tons or more (Fig. 3.19). Positioning blocks are placed on the operation platform to control the pressing size; the door core board that has been cured but has not yet reached the final strength is placed on the hydraulic machine operation platform, and the hydraulic machine is turned onto press, thereby obtaining the door core board with the size predetermined by the positioning blocks. According to special requirements, certain patterns can be placed on both sides of the template, and simple patterns can be pressed onto the door core board simultaneously during the pressing process.

Linear presses usually consist of two to four sets of rollers, and the pressing size gradually increases until the last roller reaches the final required size. When using a linear press for pressing, it is necessary to control the timing of the material's reaction. If it is too early, the strength will not be sufficient; although it is pressed flat, it may not necessarily be able to be demolded. If it is too late, the hardness will be high, and the linear press can only achieve the leveling effect with just a few rollers, which cannot reach the tonnage of a hydraulic press. Moreover, when using a linear press for pressing, it is recommended to line each side with an

ABS material template to prevent the hard squeezing of the rollers from damaging the internal structure of the door core plate and causing incomplete surface curing of the material, resulting in the paste adhering to the rollers.

3.6.3 Painting: simulation decoration

For the dried and cured handicrafts, apply calcium carbonate as the base coat with resin paint, then sand and polish them to a smooth finish. Next, spray on waterproof light-colored resin paint. After drying, perform various material appearance simulation decoration treatments. Then, spray on two to three layers of transparent resin varnish. For large products, apply waterproof paint to the inner walls for protection. The specific techniques are as follows.

3.6.3.1 Wood-like decoration

Apply wood-grain paint, imitation redwood paint, or other wood-finish coatings to the product, or apply wood-grain paper. Then coat the surface with a resin varnish to ensure a durable, smooth, and aesthetically pleasing finish.

3.6.3.2 Metal-like decoration

Spray metal colors on the polished product, such as copper gold and aluminum silver. For high-end products, modern high-brightness gold and silver plating can also be glued using tree resin, and plastic paper can be used to decorate the product surface (this method is only applicable to large, regular products). Then apply resin varnish for coating to make the product appearance color luxurious, beautiful, and lasting. For ancient products and Buddha statues, the gold and platinum decoration method using earth lacquer can be used; that is, apply the earth lacquer for coating, and before it dries, stick the gold foil and silver foil on the carved artworks.

3.6.3.3 Stone pattern simulation technique

3.6.3.3.1 Single-color stone decoration method

Mix white marble powder, black jade powder, green stone powder, and red stone powder with transparent resin to ensure even blending. Then apply this mixture onto the product. While it is still wet, press stone powder of the same color onto the surface of the product using the dry bonding method. After drying, remove any loose stone powder. This process results in artworks that appear identical to single-color stone carvings, such as those made from white marble, black jade, green jade, and red stone.

3.6.3.3.2 Multicolor stone pattern decoration method

This technique is suitable for creating relief sculptures and large flat products. Mix magnesium oxide with colored stone powder and add color to the same color. Use magnesium chloride solution to prepare color pastes of different shades. Then apply several color pastes onto the mold walls of the product to form regular and irregular stone art textures. Next, press a fiber plastic fabric onto it. By using the squeezing effect during pressing, the color pastes will move slightly, forming natural and deep stone cutting surface patterns. When the color matching is appropriate, the stone patterns can be elegant, realistic, and natural and beautiful.

3.6.3.3.3 Granite-like technique

Apply white resin paint to the surface of the product. After it dries, apply black and gray in the form of raindrops onto the product. Then, apply fine silver powder dots for further decoration. This creates silver reflective points that resemble the internal mica crystalline sections of granite, giving a more realistic effect. The size and density of the dots depend on the thickness of the imitation stone specimen (if there are any mistakes in the dot application, some dots may be too large. In such cases, add white dots for division). After drying, apply a transparent resin topcoat to ensure the color lasts and the surface is firmly attached.

3.6.3.4 Ceramic-like decoration technique

Polish the product to achieve a smooth finish, remove dust, and apply a waterproof resin paint layer first. Once dry, apply the prepared colored glaze paint onto the product to create a flowing color effect, forming a ceramic-like artistic sculpture. The color glaze formula is simple. Once you master the formulation principles and process methods, you can handle it confidently and operate freely. Below are detailed descriptions of the formulation principles and process methods.

3.6.3.4.1 Mixed glaze technique

Select several colors that are not easily mixed or dissolved with each other, such as "mixed colors of varying intensity," "the mixture of oil painting colors and varnish colors,", "the mixture of different types of colors."

These will all result in uneven point-like patterns, and by utilizing the flow of the lacquer glaze, a natural pattern effect can be achieved. If the process is operated properly, it can produce sculpture art products with the same effect as the fired ceramic glaze.

3.6.3.4.2 Glaze scatter point technique

Apply the glaze to the product according to the design, and then quickly sprinkle different colored glazes onto it, making them flow simultaneously with the previously applied glaze, creating a natural, gradated point-like glaze effect.

3.6.3.4.3 Transparent glaze technique

This method is most suitable for relief-style incised and engraved products. First, coat the product with a white or other light base color, then mix transparent resin paint with transparent colors to make a glass-like, thin, and runny transparent glaze paint (note the concentration and control the flow). Quickly apply the glaze paint onto the product, due to the flow of the paint surface, the paint in the recessed lines of the relief is thick and dark, while the paint on the convex lines and the flat surface is thin and light, thus forming a contrast of depth, clear patterns, and natural elegance and simplicity of the glaze art product.

3.6.3.4.4 Under painting with lacquer technique

Polish the product to the smoothness of a ceramic product, then spray white resin paint. After drying, perform detailed coloring such as printing ink drafts, outlining, and filling colors according to the design. Characters, animals, flowers, birds, and landscapes can be drawn on the product, and then spray transparent paint on the picture to form an underpainted picture, which can maintain the color permanently, similar to the artistic effect of glaze under painting on ceramics.

3.6.3.4.5 Color printing method

The noncolored parts of the product are covered by a specially designed "shell mold cover," and then the colors are sprayed onto the product using a spray gun. Depending on the direction of spraying, the colors can produce a uniform gradation of intensity. After drying, the points and shading can be drawn with a pen, and a colorful printed product can be obtained. This method is best for decorating floral paintings.

Figure 3.20 Spray painting of magnesium clay flower pot.

The spray painting of magnesium mud flower pots mainly uses the traditional high-pressure air pump to connect the spray gun. The spray gun is filled with paint, and the air pressure sprays the atomized liquid onto the surface of the flower pot (Fig. 3.20). According to the color depth of the flower pot and the color overlay, the number of spray painting layers is selected. For pure color flower pots, generally one to two colorings are sufficient. The choice of spray painting is relatively broad, mainly using exterior wall paint, water-based paint, and real stone paint. The selection of various paints depends on the final positioning of the flower pot. The recommended paints can all be used for coloring. Generally, exterior wall paint is slightly matte, with low chroma and unsaturation, which is more suitable for ordinary flower pots. Water-based paint has a wide variety of colors, high saturation, and bright colors and is more suitable for midrange flower pots. True stone paint is the most complex, often requiring three to four layers of spraying, with a realistic imitation stone effect and excellent waterproof performance, and is more suitable for high-end flower pots.

3.6.4 Coating

Nowadays, many environmentally friendly and green building materials have emerged both domestically and internationally. Among them, magnesium-based cementitious materials have stood out in the field of green building materials due to their excellent physical properties. This material is mainly used in partition walls, fireproof boards, decorative boards, and so forth. If used as decorative materials, additional processing is required, such as applying latex paint, pasting wallpaper, and laying ceramic wall tiles, to complete the final decoration.

The common postprocessing methods mainly include the following three types.

1. Facing method, materials using this method are numerous, such as decorative veneers (thin wood), high-pressure melamine resin laminated board (fireproof board), low-pressure melamine impregnated film paper, prepainted paper, thin sheet paper, PVC film, softwood, metal foil, and textiles.
2. Coating method, including coating, direct printing, and transfer printing.
3. Mechanical processing method, such as molding, milling, laser engraving, hand engraving, drilling, slotting, and brushing.

The three methods can be applied individually or in combination. Typically, magnesium decorative panels undergo secondary processing before being transformed into various downstream products such as furniture, flooring, and doors. In some cases, panels are directly processed into products and then receive secondary treatments such as painting.

The magnesium decorative panels that undergo secondary processing have improved surface decorative properties, along with enhanced waterproofing, resistance to deformation, UV aging, wear, heat, stains, scratches, and contamination. Thus these magnesium decorative panels can meet various usage requirements, enhance their value, expand their application scope, and bring better economic benefits to the manufacturing enterprises.

With the technological development in recent years, the high-energy consumption and high-pollution natural thin wood veneer and high-pressure long-term impregnated paper veneer have gradually transformed towards more environmentally friendly artificial thin wood and low-pressure short-term impregnated paper veneer. The short-term sheet veneer has been widely applied in the decoration industry due to its precise, fast production process and economic practicality.

Low-pressure short-term impregnated paper generally uses melamine impregnated paper. Melamine impregnated paper is also known as "methylated" paper. It is a plain paper or printed decorative paper that is impregnated with amino resins (melamine formaldehyde resin and urea formaldehyde resin) and dried to a certain extent, with a certain resin content and volatile matter content. After heat pressing, it can be bonded to each other or with magnesium building materials.

In addition to applying the melamine impregnation paper process to decorate the base board with a coating, the light-curing coating method can also be adopted. First, use a sanding machine to sand both sides of the magnesium board to make it smooth; after the sanding is completed, use a vortex dust removal machine to remove the surface dust and static electricity from the board. Apply a layer of adhesion primer to the surface of the magnesium board, and after UV light curing, apply a layer of sealing primer, continue UV light curing, and then sand again. After the sanding is completed, use the vortex dust removal machine to remove the surface dust and then apply a layer of cover color primer by roller coating.

After the primer is applied to the sheet, it undergoes the first infrared heating irradiation by a 65°C infrared leveling machine. After the interior panel is irradiated by infrared light, it undergoes roller coating of the film and adhesive. After the film and adhesive are roller coated, it is controlled by a 65°C infrared leveling machine for temperature control irradiation, and then UV light curing is carried out. After completion, another layer of paint is roller coated.

The secondary processing of decorative panels has evolved from thin wood veneer to the predominant use of adhesive film paper, followed by thin wood, artificial film adhesive paper, and finally entered a new era characterized by direct painting and application. Thin wood veneer emphasizes "naturalness," showcasing the beauty, authenticity, and distinct cross-section patterns of trees; artificial film veneer emphasizes "protection," enhancing the fire resistance, wear resistance, and corrosion resistance of panel products; direct painting emphasizes "decoration," being vibrant, colorful, and aesthetically pleasing.

References

[1] Yujie Z. Research and production practice on the maintenance mechanism and technology of magnesium magnesium products. Brick and Tile 2006(05):54–7.

[2] Shandong Magnesium Jiatu New Material Technology Co., Ltd A microwave additive for rapid solidification molding of foam boards thermal system CN116117980A [P/OL]. 2022-05-16 [2400-03-11].

[3] Bo Z. Working principle of broadband sanding machine for artificial board (continued). China Artificial Board 2013;20(04):25–8. +31.

CHAPTER 4

Magnesium-based cementitious material products

Since the introduction of magnesium-based materials in China, magnesium products have undergone a series of developmental stages, progressing from agricultural and forestry products, to building insulation materials, and finally to specialized and medical-grade materials. In the initial phase of magnesium-based material development, the related theories were not yet mature, and practitioners primarily focused on producing low-value-added agricultural and forestry products such as greenhouse poles, biogas tanks, and manhole covers. The lack of mature theories and outdated production techniques once hindered the advancement of magnesium-based materials.

As magnesium-based materials gained prominence in the building materials and insulation industries, the production technology and processes of magnesium-based products flourished, leading to the development of many products such as door core boards, fireproof boards, homogeneous boards, and others. Magnesium prefabricated building modules have become the pinnacle of the magnesium-based building materials field. The growing aesthetic demands of people have driven the advancement of magnesium imitation wood strips, magnesium imitation ancient building materials, and magnesium handicrafts.

In recent years, magnesium-based materials have continued to evolve toward high-value-added special materials, medical materials, and new energy fields, resulting in products such as salt-resistant magnesium photovoltaic pipe piles, antibacterial and antiseptic medical purification boards, and convenient and rapid bone repair products. These new products not only promoted the growth of the magnesium-based materials industry but also represented significant breakthroughs in the domestic materials field.

4.1 Artifacts

In modern life, handicrafts are increasingly integrated into people's home environments and gradually extend to more public spaces such as squares,

Magnesium-Based Cementitious Materials
DOI: https://doi.org/10.1016/B978-0-443-54075-2.00007-2

roads, and buildings. They not only have strong practicality but also possess ornamental and artistic qualities. Previously, handicrafts were predominantly made through ceramic firing, which not only polluted the environment and wasted resources but also incurred high costs. Finding more environmentally friendly and energy-saving materials to replace the original ones or improving production techniques has become a consensus among people. The emergence of magnesium materials precisely meets these current demands. Magnesium cement, as a new type of material, has advantages such as low cost and environmental friendliness, and has become the preferred raw material for many manufacturers.

In coastal regions such as Guangdong and Fujian, there are factories producing handicrafts using magnesium cement. The typical features of these factories include a multitude of small enterprises (including foreign-funded ones), a diverse range of products, and significant export volumes. Most operate on an order-based production system. The high export volume can be attributed to the fact that since the 1990s, from the Pearl River Delta to the Yangtze River Delta region, thousands of factories using resin to produce handicrafts have emerged, almost all of which are engaged in export business. Later, due to the continuous increase in resin prices and environmental protection requirements, many foreign investors have long wanted to find an alternative raw material to produce handicrafts. Using magnesium cement to produce handicrafts has the characteristics of being green, environmentally friendly, low-carbon, lightweight, high strength, and having realistic wood and stone imitation effects. Moreover, the raw material price is only 1/10 of that of resins. Some factories that used resins as raw materials to produce handicrafts and some newly established enterprises have chosen to use magnesium cement to produce handicrafts.

From the perspective of handicrafts, more magnesia cement products can be produced by leveraging its characteristics, such as being environmentally friendly, nontoxic, nonflammable, lightweight, and high strength, with wood-like and stone-like effects. These can be combined with modern surface treatment technologies and new decorative techniques. Common examples include magnesium clay flower pots, Mount Taishan stone dare, decorative reliefs, and ornaments.

4.1.1 Magnesium clay flower pot

The magnesium mud flower pot is fabricated by blending light-burned magnesium oxide with brine at a specific ratio, incorporating a defined

Figure 4.1 Magnesium clay flower pot.

Table 4.1 Production ratio of magnesium clay flower pots (dynamic ratio) (unit: kg).

Light calcined magnesia	Brine	Modifier				Saw powder	Fly ash	Quartz sand
		4#	5#	12#	16#			
100	90	0.6	0.6	0.6	1	15	30	10

quantity of modifiers, reinforcing materials (such as short fibers), and suitable auxiliary materials (like fly ash, sawdust, etc.). The well-mixed slurry is then poured (or coated) into a mold cavity lined with glass fiber cloth, after which it solidifies and is demolded following curing (as shown in Fig. 4.1).

4.1.1.1 Production process

1. Raw materials and mold preparation: Light-burned magnesium oxide, brine, modifiers, glass fiber cloth, auxiliary materials, etc., detachable mold;
2. Slurry preparation: According to the ratio (Table 4.1), pour the brine and light-burned magnesium oxide into the mixer and stir, add modifiers separately, then add various auxiliary materials such as sawdust, fly ash, and quartz sand, stir evenly, the slurry body has certain fluidity and can adhere to the surface of the mold, facilitating operation;
3. Manual molding: Apply a layer of slurry on the inner wall of the mold that has been coated with a release agent. Place glass fiber cloth, then apply another layer of slurry. According to the size and thickness requirements of the product, add thickness and fix at the corner positions;
4. Mold closing and curing: After the slurry is applied, close the mold. Under suitable curing conditions, wait for solidification and formation;

Figure 4.2 The finished product after spraying.

5. Demolding and surface treatment: After the flower pot hardens, demold it. Repair the corners or fill in the uneven areas to obtain the original body;
6. Spray color on the raw body and store the finished product (Fig. 4.2) for maintenance.

4.1.1.2 Advantages of magnesium mud flower pots

1. Environmentally friendly. The magnesium mud flower pots are made of magnesium-based cementitious materials, which are free from harmful substances and exhibit high safety standards. As demand for green and environmentally friendly products continues to grow and living standards improve, people's pursuit of a higher quality of life is also increasing. Magnesium mud flower pots are pollution-free, as they do not require firing during formation, meeting the current and future needs of consumers.
2. Lightweight and high strength. The magnesium mud flower pots have the characteristics of light weight and high strength. The pure magnesium cement slurry strength can reach over 80 MPa, and the strength can still reach over 20 MPa after adding auxiliary materials. Compared with cement flower pots, when the size is the same, the weight of magnesium mud flower pots is only half of that of cement flower pots, and it has an absolute advantage in transportation.
3. Cost savings.
   a. Abundant magnesite resources result in low raw material procurement costs.

      The raw materials include magnesium oxide, magnesium chloride, GX series modifiers, fly ash, river sand, wood chips, mesh fabric,

Table 4.2 Cost calculation of magnesium mud flower pots.

Magnesium oxide	Brine	Modifier	Fly ash	River sand	Saw powder
2 kg 1.6 RMB	2 kg 0.6 RMB	4.8 g 0.1 RMB	0.6 kg 0.12 RMB	0.2 kg 0.05 RMB	0.2 kg 0.2 RMB

molds, etc. Based on the average market purchase prices, the estimated cost of each magnesium mud flower pot can be controlled within 0.8–1.5 RMB per kilogram, as shown in Table 4.2.

b. The process that does not require high-temperature firing results in lower production costs and reduced carbon emissions.
Most magnesium cement is produced by hand, relying more on human labor than mechanical equipment. Under normal production output, each person can produce 200 flower pots with a size of 25 × 25 cm per day. From small family workshops to large production enterprises, the number of workers varies from 1 to multiple people.

c. The fast molding and demolding speed of the product results in lower time costs.

4.1.1.3 Current situation and development trends of magnesium clay flower pots

As more enterprises invest in the production of magnesium clay flower pots, some begin manufacturing without fully understanding the reaction mechanism of magnesium cement. Initially, the products may appear flawless, but their internal quality varies significantly. Some manufacturers produce pots that are moisture-resistant, stable against reoxidation, and resistant to deformation and cracking; however, many products start to reoxidize, crack, deform, or even corrode and damage containers during transportation. The losses incurred by these enterprises range from tens of thousands to millions or even tens of millions of yuan. Simultaneously, this situation brings serious negative impacts to the magnesite products industry. The lesson is deeply painful. While the appearance of the products may seem fine, their internal quality is highly inconsistent. It is precisely due to this deceptive nature of magnesium cement that some enterprises disregard ingredient guidelines, use unqualified raw materials, and incorporate additives of unknown composition, which further exacerbates the serious negative impacts on the magnesite products industry.

Figure 4.3 The magnesium clay.

4.1.1.4 Common problems and analysis

Regarding the common problems that are widespread in the market for magnesium clay flower pots, Shandong Meijiatu New Materials Technology Co., Ltd.(MJT) has provided corresponding solutions for the industry personnel for reference.

4.1.1.4.1 Cracking

The cracking of magnesium mud flower pots (Fig. 4.3) is the most obvious manifestation that everyone can observe. One common problem with magnesium cement handicrafts is that the location, time, and pattern of cracking vary greatly among different products. Therefore it needs to be analyzed from multiple perspectives and corresponding solutions should be provided based on specific techniques and production methods.

Qualified raw materials Active and reasonable magnesium oxide is the foundation for ensuring the quality of magnesium cement. The magnesium oxide used in magnesium mud flower pots is generally made of 80 powder and 85 powder, with an activity of 55%–65%, which provides the best effect. Chloride magnesium can be in the forms of halite flakes, halite particles, halite powder, and industrial brine. Usually, magnesium mud flower pots use halite flakes more often, which can achieve sufficient reaction and better strength. The quality of raw materials is a prerequisite condition. It means that if there is a problem with the raw materials, it is equivalent to the concept of 100–1 = 0. For example, if the flower pot cracks, the first consideration factor is the quality of the raw materials. Magnesium oxide or halite flakes can be tested chemically to determine their activity or content, which is an intuitive and effective judgment method. If the raw materials are fine, then other factors should be considered.

Reasonable proportioning Even with high-quality raw materials, without proper proportioning, it is difficult to produce qualified products. The strength will naturally fail to withstand the test, and thus crack. For example, using 10 parts of 85 light-burned magnesium oxide with an activity of 65%, a scientific proportioning is required. Ten parts of brine (25°Bé) need to be added to achieve the scientific proportion, so that the molecules can effectively combine and the crystal structure can be stable. If the concentration of the brine is too low, the chloride ions in the brine are exhausted, the active magnesium oxide does not fully react, resulting in a large amount of excess magnesium oxide, and the crystal structure lacks water participation, leading to insufficient construction and instability. The large amount of excess magnesium oxide will undergo a secondary reaction when exposed to water later, causing the magnesium mud flower pot to contract and expand, and there is a risk of cracking.

At the same time, it is necessary to adapt to the dynamic proportioning according to the seasons. Due to the magnesium mud flower pot being very sensitive to environmental temperature, usually in summer, the amount of brine is increased, the concentration is reduced, while the ratio of auxiliary materials remains unchanged. For example, using 10 kg of magnesium oxide (85 powder), at the highest temperature, 10–12 kg of brine (21°Bé) can be used; as the temperature gradually decreases, the amount of brine gradually decreases, the concentration gradually increases, and the minimum concentration should not be lower than 8.5 kg, and the specific gravity should not exceed 26°Bé.

Appropriate curing conditions During the solidification process of the magnesium mud flower pot, it needs to be maintained under certain temperature and humidity conditions. Generally, after 4–6 hours, the mold has completely solidified and can be demolded. After the magnesium mud flower pot is demolded, it also needs to be maintained because there is still a small amount of moisture inside the mold, and the crystal structure needs to be stabilized, thus requiring maintenance. The maintenance time is usually more than 7 days in summer and more than 15 days in winter. The maintenance environment should be an indoor environment without wind or direct sunlight.

If the temperature is too high, it will solidify over a period of time, resulting in a significant loss of moisture. You can observe a phenomenon. For example, the dry weight of the flower pot after demolding is 5 kg. If only 5 kg of magnesium slurry is used, after demolding, a dry weight of 4.8–5 kg of the flower pot can be obtained. The reason is that the moisture in the slurry has not evaporated. The moisture combines with magnesium oxide and undergoes a chemical reaction to form a crystalline structure. Among them, the $5Mg(OH)_2{\cdot}MgCl_2{\cdot}8H_2O$ (5·1·8) phase is the main representative. However, if the curing speed is too fast, the flower pot will have unstable crystal structure due to the evaporation of moisture caused by the high-temperature, and the amount will be small. Without the participation of moisture, it will lead to the completion of the structure being weak and unstable, prone to carbonization, prone to knocking off corners, short lifespan, and even cracking when demolding. Therefore the optimal curing time for magnesium slurry flower pots is usually 4–6 hours, and the slowest curing time should not be lower than 24 hours. If the temperature is too low, the curing process is too slow, and due to the long time, the crystalline structure cannot be formed. The loss of moisture mainly relies on evaporation, and the magnesium oxide loses its activity, thereby affecting the strength, causing cracking, products being prone to deformation, reverse leaching, reverse alkalization, low strength, and short lifespan.

Scientific modification The solidification of magnesium mud flower pots is also a chemical reaction process, and chemical reactions cannot be carried out without modifiers. Modifiers can assist magnesium mud in optimizing crystals and enhancing the strength and stability of the magnesium mud structure. MJT has launched a series of chemical modifiers, which can be combined and matched according to the performance requirements of the produced products. Generally speaking, the modifiers used for magnesium mud flower pots include halogen-resistant enhancers (GX-4#), defoaming and color-retaining agents (GX-5#), toughening and solvent-reducing agents (GX-12#), waterproof agents (GX-13#), and curing agents (GX-16#). These are all common types that are applicable throughout the year. Seasonal types include promoting coagulation to increase strength modifiers (GX-0#), antisetting and antihalogen agents (GX-1#), and promoting setting and antihalogen agents (GX-8#).

At the same time, auxiliary material modification is also an indispensable part of the production process. The addition of solid waste such as fly ash, silica fume, talc powder, ordinary stone powder, and calcium powder can not only reduce the amount of raw materials but also utilize their potential activity and filling properties to increase the compactness and strength of the magnesium mud structure, thereby improving the water resistance of the product. For example, adding about 30% more fly ash or talc powder, calcium carbonate powder, etc. can increase toughness without affecting strength, and also save costs. Adding low density fillers such as wood chips and rice bran can significantly reduce the product quality. Based on production experience, these auxiliary materials can also increase the toughness of the product, thereby reducing the risk of cracking of the flower pot due to excessive brittleness. For example, adding 10%–15% by dry weight of rice bran or wood chips to the magnesium mud slurry can effectively increase the toughness of the magnesium mud flower pot, reduce the product density, and make the flower pot light and strong. Similarly, if the rigidity of the product is required to be high, adding 5%–10% of 40–70 mesh fine sand can form a framework inside the magnesium mud flower pot, improving rigidity and antideformation performance.

Generally, the total amount of recommended additives should not exceed 60% of the mass of magnesium powder. The slurry of magnesium mud exhibits strong gelation properties and excellent compatibility with additives. It is possible to incorporate additives up to 200% of the magnesium mass. Typically, excessive addition of additives increases the risk of cracking. However, at the same time, the amount of additives should not be less than 35% of the magnesium mass. Because magnesium mud products are brittle materials, too low addition of additives can also cause cracking. Adding an appropriate proportion of additives can enhance the toughness of the product, avoid cracking issues, and has no impact on the product strength.

4.1.1.4.2 Surface pores

The formation reason of surface pores (Fig. 4.4) is that air is mixed into the slurry during the stirring process. The more viscous the slurry is, the less likely the gas is to be expelled. Solutions: Firstly, when making the magnesium mud flower pot, the surface should be kept thin and uniform, and the surface should be repeatedly brushed until the internal gas is expelled, and then the mold can be removed to reduce the formation of pores; Secondly, it is advisable to use water-based release agents, as oil-based release agents are not easily combined with the magnesium mud slurry and will present unevenly sized aggregated pores when removing

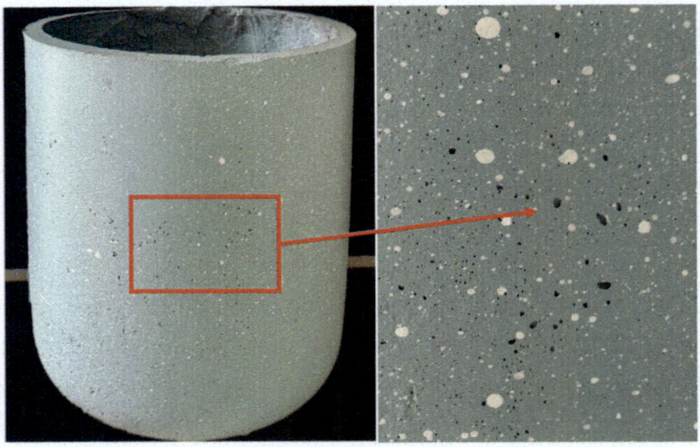

Figure 4.4 Air pores on the surface of the magnesium clay pot.

the mold; Thirdly, when using filling additives, it is advisable to choose additives with high fineness and no soil, as the finer the slurry is, the easier the pores are to be expelled, and the surface of the product will be smoother.

Therefore qualified raw materials, reasonable ratios, scientific modification, meticulous processes, and appropriate curing are necessary conditions for making high-quality magnesium mud flower pot handicrafts.

4.1.2 Taishan stone guardian

The culture of "auspiciousness and peace" embodied by the Taishan Stone Guardian (Fig. 4.5) reflects the common aspiration for peace and harmony among people, and exemplifies the humanistic spirit and cultural creativity of the Chinese nation. It holds not only aesthetic and artistic value but also stands as a significant historical relic, uniquely bearing witness to the vitality of the Chinese nation's cultural traditions. In today's society, the demand for the Taishan Stone Guardian is growing. However, due to restrictions on stone mining in its original production area, the traditionally carved Taishan Stone Guardian cannot meet current needs. Consequently, people have begun using magnesium-based cement materials to produce the Taishan Stone Guardian, allowing this cultural heritage to be better promoted and preserved. The production process and other aspects of the magnesium-based Taishan Stone Guardian are similar to those of magnesium clay flower pots, which will not be further elaborated here.

Figure 4.5 Stone Guardian of Mount Tai.

4.1.3 Magnesium-based cementitious material 3D-printed products

Magnesium-based cementitious materials 3D printing technology refers to the process of using digital model files as the basis, replacing the original adhesive materials such as powdered metals or plastics with magnesium-based cementitious materials, and constructing products through layer by layer printing. This application has expanded from early fields such as mold manufacturing and industrial design to direct manufacturing of some products, including 3D-printed components, handicrafts, and housing applications (Fig. 4.6).

4.1.3.1 Working principle

This process operates similarly to the principle of ordinary printing. The printer is equipped with a flowable magnesium gel material as the "printing medium." After connecting to a computer, it is controlled by the computer to deposit the magnesium gel material layer by layer. Once solidified, the design on the computer is ultimately transformed into a physical object.

4.1.3.2 Process flow

1. Use computer drawing software to draw the graphics and connect the computer to the 3D printer.
2. Prepare the appropriate slurry according to the required fluidity of the 3D printer and the mechanical properties required for the product.

Figure 4.6 3D-printed artifacts using magnesium gel materials.

3. Start the 3D printer to print. Depending on the specific situation, the height of the printed product should be considered. When the height exceeds a certain size, intermittent printing should be carried out. First, let the slurry solidify to a certain strength, and then continue printing.
4. After printing is completed, it should be cured under a humid temperature condition, and then the product is formed.

In summary, the integration of 3D printing technology and magnesium gel materials will bring greater application value to society and have a broad market prospect.

4.1.4 Magnesium clay oven

The manufacturing process and procedure of the magnesium mud oven are similar to those of the magnesium mud flower pot. Both are made by mixing magnesium oxide with brine in a certain ratio, adding chemical modifiers, and incorporating a certain proportion of auxiliary materials. The reaction mechanisms are similar, but the usage environments are different. The magnesium mud flower pot typically focuses on water resistance, with the main research direction being how to control back-salt and a series of issues encountered after exposure to water. In contrast, the oven is quite different. The oven needs to withstand temperatures above 400°C for heat conduction, while also avoiding a series of problems caused by excessive temperature differences.

The common issue with the oven is that, during operation, temperature differences develop after exposure to high heat, leading to cracking after multiple uses. During transportation, particularly in long-distance international shipping, cracking may occur. After use, while in storage, surface peeling and carbonization from water exposure and moisture absorption can affect its aesthetic appearance.

To address these issues, when producing the oven products, it is necessary to reasonably control the formula ratio, and add suitable auxiliary materials and chemical modifiers.

Typically, auxiliary materials for ovens are primarily high-kaolin, which has extremely high refractoriness and can withstand temperatures above 1250°C. It is a key material used in insulators and ceramics. Among these materials, calcined high-kaolin is considered the best and the most expensive. Adding 20%–25% of high-kaolin can inhibit the alkali-aggregate reaction of magnesium cement, provide stronger antishrinkage capability, and serve as an excellent auxiliary material for ovens.

Due to the characteristics of the oven, the use of the MJT series modifiers is more stringent. Usually, adding 5% of the modifier to the slurry is required to extend the lifespan of the oven and reduce its brittleness. By combining the antihalogen reinforcing agent (GX-4#), the toughening and solvent-reducing agent (GX-12#), and the organic modifier (GX-16#) developed by MJT, various problems encountered during the oven's use can be comprehensively and efficiently solved, and the defects of the material itself can be improved.

The magnesium mud oven reinterprets magnesium-based cementitious materials, offering infinite possibilities in texture, shape, and color, thus giving the products excellent practicality and decorative properties (Fig. 4.7).

4.2 Imitation of ancient architecture

Imitation ancient buildings, as treasured gems of traditional Chinese architectural culture, possess unique aesthetic appeal and profound historical depth, making them increasingly favored and esteemed in modern society. They serve not only as witnesses to history but also as bearers and perpetuators of culture.

In modern urban planning, imitation ancient buildings have become a distinctive landscape. Through meticulous design and construction, these structures can seamlessly integrate into the urban environment, creating a striking contrast with surrounding modern buildings and adding a unique

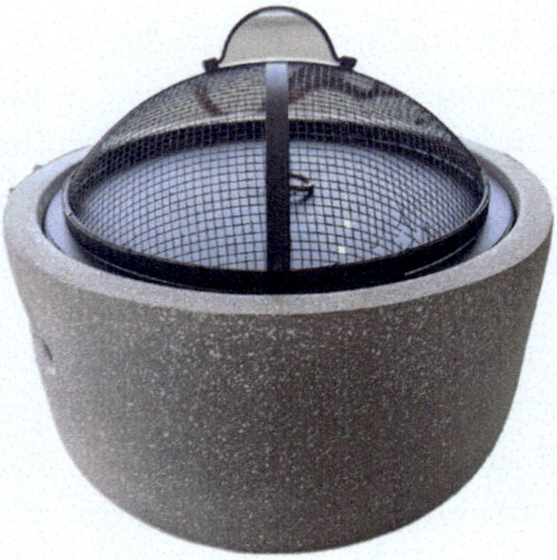

Figure 4.7 Magnesium clay oven.

flair to the city. They not only enhance the city's overall aesthetic appeal but also provide a space for people to experience history, offering a place of tranquility and belonging amid fast-paced modern life.

In the development of tourist attractions, imitation ancient buildings play a crucial role. By recreating the architectural style and cultural atmosphere of ancient times, they offer tourists an immersive time-traveling experience. Whether it is ancient temples, courtyards, or streets, imitation ancient buildings enable tourists to feel as if they are in an ancient world, experiencing the characteristics and cultural charm of that era. This immersive experience not only deepens people's understanding of traditional culture but also adds unique charm to tourist attractions, attracting more visitors to explore (Fig. 4.8).

However, the construction and preservation of traditional-style buildings also face challenges. While pursuing economic benefits, we must not overlook the respect and protection of historical culture. When building traditional-style structures, we should focus on maintaining their original historical appearance and cultural essence, avoiding excessive commercialization and overdevelopment. At the same time, we should also enhance the maintenance and upkeep of these buildings to ensure their long-term preservation, leaving valuable historical and cultural heritage for future generations.

In conclusion, traditional-style buildings, as a vital part of traditional Chinese architectural culture, maintain irreplaceable value and significance in modern society. We should cherish these precious cultural heritage sites and, through thoughtful construction and protection, allow them to continue shining in contemporary society, leaving more historical memories and cultural treasures for future generations.

Figure 4.8 Traditional-style arch tecture—temple.

4.2.1 Composition of magnesium-based antique-style building materials

4.2.1.1 Antique grey bricks

Antique grey bricks (Fig. 4.9) are one of the commonly used materials in antique buildings. With their unique texture and historical feel, they add an air of simplicity and elegance to the buildings. Antique grey tiles are hard in texture, slip-resistant and wear-resistant, and are often used to pave the floors and walls of courtyards, streets, temples, and other places, creating an atmosphere of antiquity and solemnity.

In traditional production techniques, antique grey bricks mostly use clay as the main raw material and are formed through high-temperature firing. Although the traditional clay firing process gives antique grey bricks their unique charm, it is inevitable that some consumption and emissions will occur during the production process. For example, a large amount of fuel is needed during the firing process, which not only increases production costs but may also cause certain pollution to the environment. In addition, the waste gas, wastewater, and solid waste generated during the firing process also need to be properly treated to prevent adverse effects on the environment. Moreover, the fired antique blue tiles are brittle, and due to knocks and bumps during transportation, a large number of corners and edges may break off, which increases the usage cost.

Figure 4.9 Magnesium-colored antique-style grey bricks.

The new antique-style grey bricks are made using magnesium-based cementitious materials as the binder, with a large amount of fly ash, mineral powder, and slag as fillers. Through proper proportioning and scientific modification, they are pressed into shape by a press machine. During the production process, this product can handle a large amount of solid waste, without additional energy consumption or waste generation. It may be regarded as an environmentally friendly product.

4.2.1.2 Antique-style floor tiles

Antique-style floor tiles (Fig. 4.10) are an indispensable part of antique-style buildings. With their unique texture and color, they bring a warm and rustic atmosphere to the interior space. Similar to the antique-style blue bricks, antique-style floor tiles also strive to accurately reproduce natural materials, emphasizing texture and historical sense.

Traditional antique floor tiles are prone to moisture absorption, deformation, and have high maintenance costs. They also have certain environmental issues such as formaldehyde emission. To overcome these problems, modern antique floor tiles adopt more environmentally friendly and durable magnesium-based cement materials. By adding coarse aggregates like river sand and stones, the product's rigidity is enhanced while warping deformation can be controlled and the water permeability is increased.

Figure 4.10 Example of fully laid magnesium imitation antique floor tiles.

Moreover, in the manufacturing process of magnesium-based antique floor tiles, the mold pouring molding technology plays a crucial role, enabling each tile to present a unique shape and texture (Fig. 4.11).

Designers will design various mold patterns based on market demands and consumers' aesthetic preferences, ranging from classical floral and geometric patterns to modern abstract art, covering everything. These molds not only give the floor tiles unique shapes but also endow them with rich cultural connotations and artistic value.

In addition to mold design, the pouring and shaping technology is also a key step in the production of magnesium imitation antique floor tiles. During the pouring process, workers need to precisely control the proportion of raw materials, the uniformity of mixing, as well as the temperature and humidity during curing, to ensure the quality and performance of the floor tiles.

In summary, magnesium imitation antique floor tiles, with their unique mold pouring and shaping technology, not only possess high artistic and

Figure 4.11 The unique shape and texture of the magnesium imitation antique floor tiles.

aesthetic value, but also have excellent physical properties and durability, and are an indispensable part of antique buildings.

4.2.1.3 Antique grey tiles

Antique grey tiles (Fig. 4.12) are an important material used for roofing in antique buildings, and their unique shape and texture contribute to the distinctive style of these structures. Traditional antique grey tiles are mostly made by firing natural materials such as clay or shale, and have good waterproof performance and durability. However, the traditional firing process has problems such as high energy consumption and heavy pollution, and also causes the finished grey tiles to be brittle and prone to damage.

To address these issues, lightweight and high strength magnesium-based cementitious materials are used as the base material. By adding solid waste materials such as fly ash, molds are used to pour the material into molds to form magnesium imitation ancient tiles. These imitation ancient tiles not only have the waterproof performance and durability of traditional tiles, but are also lighter, easier to install, and more environmentally friendly and energy efficient (Fig. 4.13).

4.2.1.4 Imitation stone

Imitation stone is one of the commonly used materials in ancient-style architecture. With its hard texture and unique patterns, it adds a rustic and elegant atmosphere to the building. Imitation stone can be used for

Figure 4.12 Magnesium imitation ancient blue tiles.

Figure 4.13 Magnesium imitation ancient blue tiles with unique design and convenient installation.

decorating the building's facade, flooring, steps, etc., creating an ancient and solemn atmosphere. At the same time, imitation stone offers excellent durability and weathering resistance, allowing it to maintain its beauty and stability over extended periods.

Figure 4.14 Magnesium-based antique stone material.

Magnesium imitation stone (Fig. 4.14) is generally formed by pouring into an open-close mold. For large-scale stone shapes, foam boards, extruded boards, and other lightweight aggregate materials are filled in the core of the mold to form an empty cavity structure, reducing the problem of excessive reaction temperature and expansion cracking caused by the accumulation of large-volume slurry. To achieve the effect of natural marble or white marble, a certain proportion of quartz stone is added to the slurry. After the imitation stone is fully cured, it undergoes multiple rounds of fine polishing, and can completely pass as the real thing.

4.2.1.5 Other ancient-style materials

Magnesium-based cementitious materials serve as the base material. Depending on different performance requirements, by adjusting the material ratio and conducting reasonable modification, and through various production processes such as manual pasting, mold pouring, and equipment extrusion, a wide variety of ancient-style building materials can be obtained. For example, common ancient-style reliefs (Fig. 4.15), ancient-style wall screens (Fig. 4.16), ancient-style ornaments (Fig. 4.17), etc.

Figure 4.15 Magnesium-based antique relief sculpture.

Figure 4.16 Magnesium-based antique-style wall paneling.

4.2.2 Construction process of traditional architecture

The construction of ancient-style buildings (taking ancient-style temples in the north as an example) usually involves the following main steps.

1. Design and planning.

   Before starting any construction work, a detailed floor plan and construction schedule must be developed.

Figure 4.17 Magnesium imitation antique decorative piece.

2. Site selection and surveying.
   When selecting the site, factors such as terrain, climate conditions, and transportation convenience should be considered to ensure the stability and practicality of the building. At the same time, soil surveys and environmental assessments should be conducted to ensure the safety and feasibility of the site selection.
3. Material preparation.
   Choose materials that conform to the characteristics of ancient architecture. According to the requirements of the drawings, various ancient-style components should be prefabricated in factories in advance, and the materials should undergo quality inspection and processing to ensure their quality meets the standards.
4. Foundation construction.
   First, level the ground and excavate the foundation pit. Then, carry out foundation pouring and reinforcement to ensure the stability and load-bearing capacity of the building. During the foundation construction process, underground pipelines should be laid and inspected to ensure the normal operation of the water supply, drainage, and ventilation systems of the building.

Figure 4.18 Structure construction.

Figure 4.19 Roofing and waterproofing treatment.

5. Main building construction.
   Build the structure according to the design drawings, including walls, roofs, stairs, etc. (Fig. 4.18).
   Then, complete the building's topping-out and waterproofing treatment to ensure the airtightness and waterproof performance of the building (Fig. 4.19).

Figure 4.20 Double-layered structure wall.

Figure 4.21 Installing variously shaped ridge tiles.

For the ancient-style temple buildings in the north, most of them adopt a double-layer wall design. That is, a layer of red brick wall is built to provide load-bearing support, and on the outside of the red brick wall, a layer of ancient-style bricks is built to achieve insulation and decorative purposes. Finally, special lime mortar is used for jointing (Fig. 4.20).

6. Decoration techniques.

This is one of the highlights and features of ancient-style buildings. Through meticulous carving, painting, and the use of variously shaped floating sculptures and ridge tiles with decorations, the buildings present the style and characteristics of ancient architecture (Fig. 4.21).

In conclusion, magnesium-based antique building materials, with their unique texture and historical charm, bring a more realistic and rustic visual effect to antique buildings. At the same time, by adopting environmentally friendly and energy-saving production processes and materials, magnesium-based antique building materials not only have better performance and aesthetics, but also better meet the requirements of sustainable development, injecting new vitality into the inheritance and development of traditional architectural culture.

4.3 Magnesium imitation wood strip board

Wildfires are closely linked to human survival, the economy, and ecosystems, and have long been a major concern. With the acceleration of global warming, the urgency of preventing wildfires in the future has become increasingly critical.

According to satellite monitoring data from NASA, the severe wildfires in California in 2018, caused by severe drought, burned an area of approximately 800,000 hectares. In 2019, the Amazon rainforest suffered fires that burned an area of approximately 1.8 million hectares (Fig. 4.22).

According to CNN, in 2019, more than 5.9 million hectares of land in Australia were burned. The area was nearly half of that of England. Among the states in Australia, New South Wales was the most severely affected. As of January 3, 2020, more than 130 wildfires had occurred across New South Wales. Approximately 3.6 million hectares of land were burned, over 1300 houses were damaged, and thousands of residents were

Figure 4.22 Map of Amazon rainforest fire burns.

Figure 4.23 The Australian fire crisis.

displaced. This widespread fire caused severe damage to the local ecosystem. According to Minister Sussan Ley of the Australian Federal Environment Department, this fire is estimated to have caused the death of one-third of koalas in New South Wales and the complete destruction of one-third of the habitats of various species (Fig. 4.23).

According to incomplete statistics in China, in the summer of 2022, multiple districts and counties in Chongqing Province experienced more than 10 forest fires. Under the conditions of continuous 40°C+ high temperatures and drought, severe forest fires were triggered, with black smoke spreading and the fires raging. Since the Spring Festival of 2024, multiple forest fires have occurred successively in Guizhou Province. Especially from February 10th to 21st, the fires broke out in an explosive manner and caused casualties, seriously threatening ecological security and the safety of people's lives and property. In forest fires, the firebreaks have played an important role.

The raging fires caused huge casualties and property losses in the area. How to deal with forest fires has been a matter of great concern for the local government. Among them, promoting the construction of forest fire prevention isolation systems has become one of the preferred solutions. The use of fire-resistant, environmentally friendly, and high strength inorganic magnesium imitation wood board as the isolation belt emerged as a result, aiming to minimize the harm of forest fires to the greatest extent.

Products crafted using wood imitation techniques achieve a visual appearance indistinguishable from genuine wood products, making it

Figure 4.24 Magnesium imitation wood strip panel.

difficult to discern any difference. In terms of lifespan, however, inorganic wood imitation far surpasses real wood. It will not corrode even after prolonged exposure to wind and sunlight, nor will it become infested by insects. This approach is both environmentally friendly and addresses the issue of wood shortage.

Magnesium wood imitation board, as the name implies, replicates the appearance and texture of wood visually, yet fundamentally, it is constructed from lightweight, high strength magnesium cement materials that are fire-resistant and environmentally friendly. It is processed through specialized techniques to form strip boards featuring varied colors, textures, and aesthetic qualities (Fig. 4.24). This building material holds significant potential, combining practicality with the achievement of a natural visual effect. Through appropriate construction methods, it can contribute to creating architectural spaces that harmonize nature with functionality.

4.3.1 Characteristics of magnesium-infused wood-look panel

1. Fire resistance.

   The material used for this board is inorganic magnesium-based material, which inherently has fire resistance.

2. Lightweight and high strength.
   The material used for this board has better strength than other inorganic materials (such as cement, gypsum, etc.). Under identical strength requirements, it can be made lighter.
3. Easy to shape.
   The magnesium-based cementitious material used for this board is well-mixed and exhibits good fluidity. Different molds can be used to produce boards of different shapes.
4. Compatible with different colors.
   The material slurry of the magnesium-based cementitious material turns white after curing. Different pigments can be mixed to produce boards of various colors.
5. Environmental friendliness and economic efficiency.
   (a) The raw materials for magnesium-based imitation wood strips primarily consist of calcined magnesium oxide, brine, and other solid waste materials. The firing temperature of calcined magnesium oxide is lower than that of cement, and it requires less amount at the same strength, making it more energy-efficient and environmentally friendly. Brine is a by-product of the salt industry, while fly ash and other solid wastes are waste from the thermal power industry. During the production of imitation wood strips, these by-products and solid wastes are extensively used, transforming waste into valuable resources while also yielding certain economic and environmental benefits. (b) Magnesium-based imitation wood strips exhibit stable performance, are noncombustible, corrosion-resistant, and durable, making them excellent decorative materials suitable for a wide range of indoor and outdoor applications.
6. Long service life.
   The strips made from magnesium-based cementitious materials, after reasonable formulation and modification, have a longer lifespan than wood strips, significantly saving wood.

4.3.2 Production process

4.3.2.1 Preparation work

1. First, issue the molds as per the requirements. The molds include silicone molds, resin molds, polymer plastic molds, etc.
2. First, select a release agent suitable for magnesium materials. When applying the release agent on the inner side of the mold, do it evenly

and avoid applying too much. Place the bamboo strips, reinforcing materials, etc. that have been prepared inside the mold cavity.

3. Pour the prepared magnesium oxide, brine, modifier, pigment, auxiliary materials (such as sawdust, etc.) into the mixing container in the prescribed sequence and proportion. Mix them evenly. Refer to the ratio in Table 4.3.

4.3.2.2 Forming

1. Pour the evenly mixed slurry into the prepared mold. First, scrape off a layer of slurry and insert bamboo strips and mesh fabric; then, place another layer of slurry, level it off, and use the vibration platform to vibrate to remove air and ensure there are no air holes on the surface.
2. Curing: Place it in the curing room for curing.
3. Demolding: After the solidification and formation is complete, remove it from the mold at the right time.
4. Storage: Repair the edges and corners, and pack and store.

4.3.3 Common problems and analysis

1. Pay attention to the quality of the raw materials.

   The quality of magnesium oxide is crucial. If only the price is considered while the quality is ignored, using low-quality magnesium oxide will lead to problems such as low strength and slow curing; at the same time, the local climate temperature should also be taken into account. For example, if the temperature is high in summer, the reaction between magnesium oxide and brine will be fast, and a retarder like MJT needs to be added for intervention; otherwise, the "burning board" phenomenon will occur.
2. The color is not uniform.

   Due to different requirements, when making products of different colors, problems such as appearance with alkaline residue and whitening, and inconsistent color of the color paste (Fig. 4.25) may occur, which requires process improvement.
3. The mold sticks and peels off during demolding.

   The demolding time needs to be controlled properly. If demolding is too early, the strength is not sufficient, and the product will stick to the mold during demolding, resulting in defects; if demolding is too late, the product will stick to the mold and is difficult to demold, consuming manpower.

Table 4.3 Reference ratio (unit: kg).

Light calcined magnesia	Brine	Modifier			Saw powder	Fine sand	Fly ash	Pigment
		4#	12#	13#				
100	120	0.8	0.6	1	20	15	5	Proper amount

Figure 4.25 Different-colored panels.

4. Warping and deformation.
   Since magnesium imitation wood strips belong to magnesium products with large range and small cross-sectional dimensions, warping and deformation are prone to occur. The main reasons are: during the curing period of the product, the surface moisture evaporates significantly and loses water too quickly, while the bottom moisture is not easy to evaporate, and the contraction of the two is inconsistent, resulting in warping and deformation; under high-temperature conditions, the moisture reaction of the magnesium cementitious material is intense. If heat dissipation is not timely, thermal stress will be unevenly distributed in each part of the product, and the upper and lower surfaces of the product will not have balanced heat dissipation, causing the strips to warp and deform; improper placement during curing can also cause warping and deformation, etc.

4.3.4 Outlook and prospects

There are many houses in both domestic and foreign regions that are constructed with pure wooden structures. This method has limitations in terms of historical technology and regional resources. Although it is lightweight and convenient, its biggest drawback is that it is not fireproof. In case of a fire, it will cause significant losses of both human lives and property. For example, the ancient city of Weishan in Baiyin,

Yunnan Province, has a history of over 600 years. Due to the fire disaster, this historical and cultural ancient city in Yunnan suffered heavy losses. In foreign countries such as New Zealand, the foundation filling part of the prefabricated houses in some countries has foam-like materials with shock absorption and buffering functions, while the ground and roof wall panels mostly use wooden structure boards filled with wool in the middle. They also face the risk of fire. If combined with inorganic magnesium-based cementitious materials, which are environmentally friendly, pest-resistant, antimold, lightweight, and high strength, and used to form an enhanced fire-resistant layer in wooden structures, it could become a new application.

4.4 Nonmetallic magnesium photovoltaic pipe piles

4.4.1 Background of the photovoltaic industry

Photovoltaics, as a key component of global new energy, is gradually changing the global energy landscape. Since China surpassed Germany in 2013 and became the world's largest market for photovoltaic applications, China's photovoltaic industry has been steadily growing. According to data from the International Energy Agency, by 2023, global solar production expenditures (approximately 380 billion US dollars) will exceed oil production expenditures (about 370 billion US dollars). China holds a significant share in this regard, with 80% of the world's solar panels, 85% of solar cells, and 97% of solar silicon wafers being produced in China.

Today, China's photovoltaic industry has entered the trillion-dollar market. In 2022, the total annual output value of the photovoltaic industry exceeded 1.4 trillion US dollars; if including auxiliary materials and equipment, the total output value exceeded 2.2 trillion US dollars. As of the end of 2022, the total market value of 138 photovoltaic listed companies reached 3.8 trillion US dollars, approximately accounting for 5% of the market value of China's A-share market.

4.4.2 Photovoltaic application branch

4.4.2.1 Solar-PV agriculture integration

Solar-PV agriculture, also known as photovoltaic agriculture, utilizes the environmental-friendly and zero-emission advantages of solar power generation and combines it with advanced technology greenhouses

Figure 4.26 Photovoltaic and agricultural greenhouses.

(including planting greenhouses and breeding greenhouses). On the sunny side of the greenhouse or by covering the entire area with photovoltaic solar power generation equipment, it can generate electricity while providing suitable living environments for crops, edible fungi, and livestock breeding, thereby creating better economic and social benefits. Common forms include photovoltaic agriculture planting greenhouses, photovoltaic breeding greenhouses, etc. (Fig. 4.26).

4.4.2.2 Water-solar integration

Water-solar integration can be further divided into three types: fishery-solar integration, offshore solar power generation, and coastal tidal flat solar power generation. Fishery-solar integration involves building solar power stations on fish ponds, allowing for both fish farming and electricity generation simultaneously. Offshore solar power generation takes place in areas along the coast that do not damage the ecological environment, where fixed or floating solar power stations are constructed to enhance the efficiency of marine use. Coastal tidal flat solar power generation occurs in coastal tidal flat areas, where the ecological environment is restored and protected while using vacant land to build solar power stations (Fig. 4.27).

4.4.2.3 Lin Guang complementarity

Lin Guang Complementarity refers to generating electricity above the solar photovoltaic panels and planting shade-tolerant crops below, thereby

Figure 4.27 Photovoltaic power generation and tidal flat fish ponds.

Figure 4.28 Photovoltaic station.

making full use of the land. Currently, this method is primarily applied to forested areas that were previously unused and suitable for afforestation, utilizing land resources that originally had a coverage rate of less than 30% (Fig. 4.28). It does not matter whether the trees are tall or short (including dwarf shrubs).

4.4.3 Composition of the photovoltaic industry chain

The photovoltaic industry chain refers to the entire process from silicon materials, photovoltaic components to photovoltaic power generation and applications. In 2019, the installed capacity of photovoltaic power generation in China reached 204.5 GW, and the scale of the entire photovoltaic industry chain has exceeded 1 trillion yuan. The rapid development of this industry has driven the development of a series of surrounding industries, such as photovoltaic cell sheets, inverters, pipes, and supports. The photovoltaic industry chain has continuously extended and has formed a complete industry. This section only introduces the pipes related to magnesium-based materials.

Photovoltaic pipe piles are structural components installed on the ground to support and fix photovoltaic cell panels, with the aim of improving power generation efficiency. According to different classification methods, there are many types of photovoltaic pipe piles (Fig. 4.29).

Based on the material, photovoltaic pile foundations mainly consist of steel and concrete. The steel pile foundation is composed of angle steel, steel plates, and screws, and it has the advantages of being lightweight and easy to install, making it suitable for short-term use or locations where long-term maintenance is not required. The concrete pile foundation is made up of reinforcing bars, concrete, and formwork, and it has high stability and a long service life, making it suitable for places that require long-term use or have high requirements for structural strength.

Figure 4.29 Photovoltaic cement pile.

According to the installation method, photovoltaic pipe piles can be divided into embedded type and nonembedded type. Embedded pipe piles are buried underground and secured to the concrete foundation through expansion bolts, etc., making them suitable for use in areas where the ground consists of hard soil or rock. Nonembedded pipe piles are installed on the ground, with photovoltaic panels fixed to the pipe piles via supports and fasteners, and are suitable for areas where the ground is soft soil or where ground settlement needs to be considered.

The types of prestressed pipe piles are classified according to the effective prestress value or mechanical properties of the concrete in the pile body. There are mainly four types: A type, AB type, B type, and C type. Their effective prestress values are 4 , 6 , 8 , and 10 MPa respectively. Different types of prestressed pipe piles have different amounts of reinforcing bars used, mechanical properties, and prices. Because prestressed pipe piles are mainly used to bear longitudinal pressure, their bending resistance performance must meet the requirements for lifting and stacking of the pipe piles.

4.4.4 Advantages of the new nonmetallic magnesium photovoltaic pipe piles

Due to their flat terrain and high light intensity, coastal areas and near-sea regions have become popular locations for new energy development. They offer significant development potential, high comprehensive benefits, and are environmentally friendly. However, because these areas have severe soil salinization and high concentrations of Cl^- and SO_4^{2-}, they have strong corrosive effects on metals and traditional prestressed concrete photovoltaic pipe piles and supports. Ordinary photovoltaic pipe piles and supports cannot meet the requirements for service life and safety in such environments.

Looking to the future, with the development of national policies and the photovoltaic industry, offshore photovoltaics will become an important direction in photovoltaic design. Moreover, as the photovoltaic industry develops, the load of multimodule assembly is too large, making installation rather troublesome. Therefore the durability and lightweight of photovoltaic supports will become a development trend.

In order to develop a photovoltaic pipe pile and support with stable structure, high durability, and light weight, researchers based on actual construction projects and prepared nonmetallic magnesium photovoltaic pipe piles. Starting from the wind loads, snow loads, self-weight loads, and earthquake loads borne by the photovoltaic support, through calculations, the strength verification of key components and nodes in the structural design of the pipe pile and support was carried out. At the same time, through wind tunnel tests and other methods, the feasibility of the nonmetallic magnesium photovoltaic pipe pile in practical applications was verified.

4.4.5 Selection and proportion of raw materials for nonmetallic magnesium photovoltaic pipe piles

4.4.5.1 Raw materials

4.4.5.1.1 Magnesium oxide

The performance indicators of calcined magnesium oxide need to comply with the relevant provisions of "Calcined Magnesium Oxide for Magnesite Products" WB/T 1019–2002, as shown in Table 4.4.

4.4.5.1.2 Magnesium chloride

The performance indicators of magnesium chloride must comply with the relevant provisions of "Industrial Magnesium Chloride for Magnesium Oxide Products" WB/T 1018–2002, as shown in Table 4.5.

4.4.5.1.3 Modifier

The modifiers are listed in Table 4.6.

4.4.5.1.4 Stone properties

The stone properties indicators are shown in Table 4.7.

4.4.5.1.5 Properties of sand

The properties of sand are listed in Table 4.8.

4.4.5.2 Production ratio

The raw materials selected for nonmetallic magnesium-based photovoltaic pipe piles are different from those of traditional prestressed concrete pipe piles. Therefore their material ratio and mixing sequence also differ. Taking the pipe pile with a production specification of PHC-400-AB-11 meters as an example, the reference ratio for calculation is shown in Table 4.9.

Table 4.4 Performance indicators of lightweight magnesium oxide.

Level	Magnesium content/%	Active magnesium oxide content/%	Fine degree/mesh	Active calcium oxide content/%	Dry loss on ignition/%
Premium product	≥85	≥65	≥200	≤ 1.5	1~9

Table 4.5 Performance indicators of magnesium chloride.

$MgCl_2$	NaCl	KCl	$CaCl_2$	SO_4^{2-}	Appearance
≥45.0	≤1.5	≤0.7	≤1.0	≤3.0	White, off-white or yellowish-brown, in fragment or granular form. It must not deliquesce and must not present as large crystalline masses.

Table 4.6 Modifier.

Type	Name	Addition amount/‰	Appearance
GX-4#	Halogen-resistant enhancer	5–10	Deep reddish-brown viscous liquid
GX-5#	Defoaming and color-retaining agent	10	Milky white liquid
GX-12#	Toughening and solvent-reducing agent	10–20	Orange-red liquid
GX-16#	Curing agent	5–10	Pink liquid

4.4.6 Nonmetallic photovoltaic tube pile forming and curing process

There are two traditional forming processes for prestressed concrete pipe piles: the pretensioning method and the posttensioning method. The pretensioning method produces prestressed pipe piles by using the pretensioning process and the centrifugal forming process. These are slender concrete prefabricated components, mainly hollow cylinders, consisting of a cylindrical pile body, end plates, and steel sleeves. The posttensioning method forms prestressed pipe piles by applying prestress through the reserved holes after the concrete in the pipe pile begins to set. The nonmetallic magnesium photovoltaic pipe piles are new types of pipe piles that are formed by adapting the forming process of pretensioning prestressed pipe piles, adjusting the inorganic cementitious material ratio, and through natural curing or low-temperature steam curing (Fig. 4.30).

Table 4.7 Stone properties indicators.

Variety	Grain composition	Bulk density/ (kg/m^3)	Apparent density/ (kg/m^3)	Compact density/ (kg/m^3)	Content of needle-like and flake-like particles/%	Crushing value/%	Clay content/%
Basalt	5–16	1500	2910	1680	2	3.2	0.7

Table 4.8 Properties of sand.

Variety	Type	Fineness modulus	Clay content/%	Compact density/ (kg/m^3)	Bulk density/ (kg/m^3)	Apparent density/ (kg/m^3)
Natural sand	Two zone medium gravel	3.0	1.8	1720	1560	2690

Table 4.9 Reference ratio for pipe piles (unit: kg).

Light calcined magnesia	Brine	Modifier				Pebble	Sand
		GX-4#	GX-5#	GX-12#	GX-16#		
500	380	5	2.5	5	2.5	1300	700

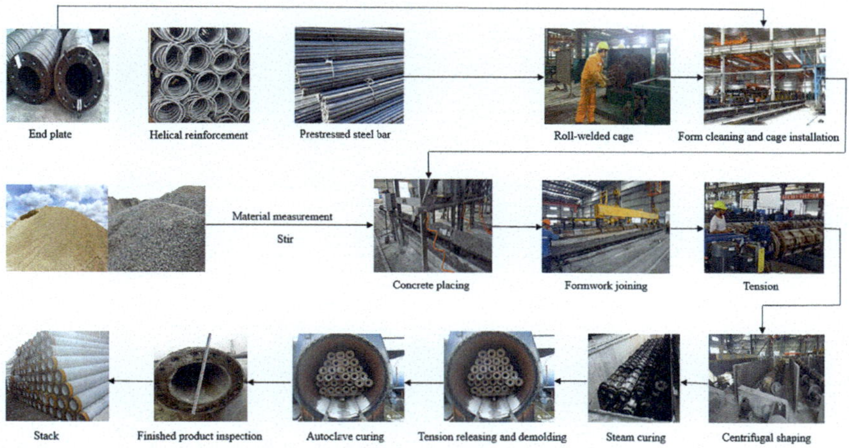

Figure 4.30 Production process of nonmetallic magnesium-based photovoltaic pipe piles.

4.4.6.1 Binding of the steel cage

Binding the steel cage for pipe piles is performed using mechanized operations, which are rapid and efficient. Typically, a worker requires 2–3 hours to bind the steel cage needed for one day. The worker places the processed main reinforcement bars into it, and the disc on the small vehicle can stretch the main bars. The vehicle runs on lateral guide rails, and it can adapt regardless of the length of the pipe pile. The stirrups are placed on the disc in the binding machine, and the vehicle runs at a certain speed, and the steel bars on the disc wrap around the main bars in a circle, forming the steel cage (Fig. 4.31).

4.4.6.2 Insertion into mold

After the reinforcement cage is tied-up, it is placed into the prefabricated mold using a truss crane (Fig. 4.32). Both ends of the reinforcement cage

Figure 4.31 Rolling welding cage.

Figure 4.32 Clear mold in cage.

have nuts, and after being placed in, it is fixed with a ring-shaped steel plate, which has corresponding bolts for the nuts. The tensioning end and the other end are different in appearance, and the gap in the mold is the tensioning gap. Finally, the tied-up reinforcement cage is lifted to the concrete pouring area using the truss crane.

Figure 4.33 Pouring the slurry of magnesium-based cementitious material.

4.4.6.3 Pouring of magnesium-based cementitious material slurry

In the pouring area, there are corresponding pouring vehicles running along the side guide rails of the operation platform. The amount of magnesium-based cementitious material slurry for one pile is controlled to be within one vehicle. The small vehicle runs on the hollow semiopen mold, and the slurry that leaks from the bottom gradually fills the mold (Fig. 4.33).

After the pouring is completed, the other half of the steel mold is lifted in for closing and sealing (Fig. 4.34).

4.4.6.4 Tensioning

Move one end of the pipe pile into a specific machine for tensioning. The specific tensioning amount depends on the design requirements of the pipe pile (Fig. 4.35).

4.4.6.5 Centrifugal molding

Place the tensioned pipe pile into a centrifugal slot for centrifugal operation. This process is divided into four stages: slow speed, low speed, medium speed, and high speed. This allows the magnesium-based cement

Figure 4.34 Closing the mold.

Figure 4.35 Tensioning reinforcing bars.

slurry to be more compact. The centrifugal time is approximately 15 minutes, depending on the centrifugal speed. The cavity in the middle of the ring-shaped steel plate is blocked with a foam board to facilitate centrifugal molding (Fig. 4.36). After forming, the foam board can be pulled out, facilitating the removal of excess slurry and maintaining the interior of the pile as flat.

Figure 4.36 Centrifugal molding.

4.4.6.6 Maintenance

Place the nonmetallic magnesium photovoltaic pipe piles that have undergone centrifugal forming into the maintenance pool. Use normal temperature maintenance or low-temperature steam maintenance. The steam maintenance pool is a large concrete pool. Place the centrifugally formed pipe piles in the pool, cover with a concrete cover plate, and maintain under natural conditions, or appropriately heat to 40°C–50°C during low-temperature maintenance (Fig. 4.37).

4.4.6.7 Demolding

After normal natural maintenance or heating maintenance for 8–12 hours, lift the steel mold. Remove the fixing bolts on both sides, and dismantle the model using the truss crane (Fig. 4.38).

4.4.6.8 Inspection and storage

The fabricated prestressed nonmetallic magnesium photovoltaic pipe piles, after passing the inspection, are neatly placed in the storage area according to their type, length, etc. (Fig. 4.39).

Figure 4.37 Steam curing at normal temperature or low-temperature.

Figure 4.38 Demolding.

Figure 4.39 Storage of pipe piles.

4.4.7 Construction method of nonmetallic magnesium photovoltaic pipe piles

Photovoltaic power generation involves installing photovoltaic panels in open areas to convert sunlight into electricity. These panels are mounted on pipe piles and distributed across various terrains, including land, mudflats, mountains, and grasslands. The construction methods for the pipe piles vary depending on the site conditions and geological characteristics.

4.4.7.1 Ponds

For example, fish ponds and shrimp ponds, the water depth is approximately 2–4 m. Here, installing photovoltaic pipe piles can be done using a pile-driving vessel along with a tugboat (responsible for transporting the piles). The pile driver usually uses a hydraulic pile hammer. If the water depth is deeper, the method of sinking the piles with a vessel and an excavator's vibration hammer can also be used.

4.4.7.2 Mudflats

The water depth is approximately 0–2 m. Here, installing photovoltaic pipe piles can be accomplished using an excavator equipped with a vibration hammer.

4.4.7.3 Hilly and hard soil areas

For installing photovoltaic pipe piles here, the process typically involves using a spiral photovoltaic drilling machine to create the holes, followed by using an excavator equipped with a vibration hammer. Spiral drilling can be performed with a dedicated spiral drilling machine or by configuring a hydraulic spiral drill on the excavator.

4.4.7.4 Rock slope

In some mining rehabilitation projects, bare rocky slopes are also the distribution area for photovoltaic equipment. The terrain here is hard and solid, so the method of using a drilling hammer combined with an excavator equipped with a vibration hammer can be adopted to construct the photovoltaic foundation piles. The drilling holes are made by the drilling hammer, and then the prefabricated piles are driven in with the vibration hammer.

4.4.8 Advantages and application prospects of nonmetallic magnesium photovoltaic pipe piles

The application of nonmetallic magnesium-based photovoltaic pipe piles in the field of photovoltaic power generation has the following advantages.

1. High strength.
   Magnesium-based cementitious materials have high strength, which enables the photovoltaic pipe piles to have good performance in terms of bearing the weight of photovoltaic equipment and wind loads. Through calculation, the physical and chemical performance indicators of magnesium-based photovoltaic pipe piles of the same age can be comparable to those of C80 concrete prestressed pipe piles.
2. Excellent corrosion resistance.
   Magnesium-based cementitious materials have good corrosion resistance. Especially in the required mixing agent halogen water for self-curing reaction, there are Mg^{2+} and Cl^{-} ions. When used in the harsh environment of tidal flats and saline-alkali lands, it is beneficial to the development of their strength and prolongs the service life of photovoltaic pipe piles.
3. Environmental performance.
   The production and construction process of nonmetallic magnesium-based photovoltaic pipe piles cause less environmental pollution and is in line with the concept of green development.

4. Convenient construction.
   By adopting advanced molding techniques and equipment, the production efficiency of photovoltaic pipe piles is improved, and the construction difficulty and cost are reduced.
5. Strong adaptability.
   Nonmetallic magnesium-based photovoltaic pipe piles can adopt different construction methods according to different geological conditions and site environments, and have strong adaptability.
6. Beautiful generous.
   Magnesium-based cementitious materials have good surface smoothness, making photovoltaic pipe piles have better aesthetic performance.

In summary, nonmetallic magnesium-based photovoltaic pipe piles, with their unique advantages, have broad application prospects in the photovoltaic power generation industry. With the continuous advancement of technology and the continuous expansion of the market, the application fields of nonmetallic magnesium-based photovoltaic pipe piles will be further expanded, injecting new vitality into the development of the photovoltaic power generation industry.

4.5 Nonmetallic magnesium-based photovoltaic support structures

At present, the commonly used photovoltaic supports in China include concrete supports, steel supports, and aluminum alloy supports. Concrete supports and steel supports (Fig. 4.40) are used in large-scale photovoltaic power stations due to their high stability. Aluminum alloy supports, being

Figure 4.40 Photovoltaic support structure.

lightweight, aesthetically pleasing, and durable, are generally applied to the roofs of civilian buildings and sunrooms. However, due to their high cost and low bearing capacity, they cannot be used in large-scale solar power stations. The air humidity in tidal flats and offshore areas is high, and the contents of Cl^{-} and SO_4^{2-} are relatively high, which have strong corrosive effects on the structures of metal and concrete photovoltaic supports. The traditional coating solution is costly. Therefore a new type of photovoltaic support that is lightweight, high strength, corrosion-resistant and low cost is needed.

4.5.1 Advantages of the new nonmetallic magnesium-based photovoltaic support structure

The nonmetallic magnesium-based photovoltaic support structure is made of magnesia cement as the binding material, supplemented by bamboo fibers, glass fiber cloth, etc. as reinforcing materials, and combined with inorganic and organic filling materials. It has the advantages of light weight, high strength, corrosion resistance, strong bearing capacity, and low cost.

4.5.2 Selection and proportion of raw materials for nonmetallic magnesium-based photovoltaic supports

4.5.2.1 Raw materials

The requirements for magnesium oxide, magnesium chloride, and modifiers are the same as those for nonmetallic magnesium-based photovoltaic pipe piles.

4.5.2.2 Production ratio

The raw materials selected for nonmetallic magnesium-based photovoltaic supports are different from those of traditional concrete supports and steel supports. As a new type of material and product, their ingredient ratios and mixing and discharging sequences are also different. Taking the support with a production specification of PHC-400- AB-11 m 2 × 6 as an example, the reference mix ratio of the slurry is shown in Table 4.10.

4.5.3 Nonmetallic photovoltaic support forming and curing process

1. Preparation and molding
   Evenly spread a layer of slurry in the impregnation mold according to the designed dimensions. Place the central aggregate and carbon fiber

Table 4.10 Reference mix ratio of slurry (unit: kg).

Light calcined magnesia	Brine	Modifier				Carbon fiber	Silica fume
		GX-4#	GX-5#	GX-12#	GX-16#		
100	85	1	0.5	1	0.5	0.5	1

fabric for impregnation. Cut the glass fiber fabric and nonwoven fabric in advance. Prepare the molding mold according to the designed dimensions. First, lay the slurry, then place the central molding aggregate, cover with the impregnated carbon fiber fabric, then cover with the glass fiber fabric and nonwoven fabric. Treat the surface until the corresponding size is achieved, and then place the mold for curing and molding.

2. Maintenance
   Place the formed nonmetallic magnesium photovoltaic support with the mold into the curing chamber and carry out curing at normal temperature or low-temperature steaming. Under natural conditions or when at low-temperature, appropriately raise the temperature to 40°C–50°C for curing.
3. Demolding
   After normal natural curing or heating curing for 8–12 hours, remove the model. Continue the curing under natural conditions.
4. Drilling
   After the fabricated support structure has passed the inspection, it will undergo sanding and drilling according to the drawings.
5. Installation
   The photovoltaic support structure consists of rear column, front column, cross beam, diagonal brace, diagonal beam, rear pull rod, connecting parts, etc. It is connected by bolts. The photovoltaic module is connected to the support cross beam through the installation hole and the support beam (Fig. 4.41).

4.5.4 Advantages and application prospects of nonmetallic magnesium-based photovoltaic supports

The application of nonmetallic magnesium-based photovoltaic supports in the field of photovoltaic power generation has the following advantages.

1. High strength.
   Magnesium-based cementitious materials have the characteristics of light weight and high strength, which enables the photovoltaic supports to have excellent performance in bearing the weight of solar photovoltaic panels and wind and snow loads. Through calculation, the flexural strength of the nonmetallic magnesium-based photovoltaic supports is 142 MPa, and the compressive strength is 84.5 MPa.

Figure 4.41 Nonmetallic magnesium-based photovoltaic support structure.

2. Excellent corrosion resistance.
   The nonmetallic magnesium-based photovoltaic support has good resistance to salt brine. Salt Lake brine contains Cl^-, SO_4^{2-}, Mg^{2+}, and other ions. When the chloromagnesium cement is in the salt brine environment, these ions will invade the interior of the cement and react with it, forming cross-shaped needle-like structures. This enhances the cement strength, thereby influencing the composition and content of the hydration products of the system as well as the internal structure.
3. Environmental performance.
   The production and construction process of the nonmetallic magnesium-based photovoltaic support causes less environmental pollution and is in line with the concept of green development.
4. Convenient construction.
   The use of prefabricated production technology improves the production and installation efficiency of the photovoltaic support, which reduces the construction difficulty and cost.
5. Beautiful generous.
   Based on practicality, the magnesium-based cementitious material products have good aesthetics, making the photovoltaic support a beautiful landscape.

In summary, the nonmetallic magnesium-based photovoltaic support has the above advantages and has a wide application prospect in the photovoltaic power generation industry.

4.6 Modular structure homogeneous board

4.6.1 Development status of prefabricated construction abroad

In Europe, France has the longest history of prefabricated construction. It has a 130-year history of prefabricated building development and currently, the prefabrication rate in France has reached 80%. It mainly adopts a prestressed concrete prefabricated frame structure system, and the production and construction quality are very high. In Germany, prefabricated housing mainly adopts a laminated board, concrete, and shear wall structure system, and combines component prefabrication with concrete structures, which has good durability. Germany is the country with the fastest reduction in building energy consumption in the world, and in recent years, it has even proposed the development of zero-energy passive buildings. From significant energy savings to passive buildings, Germany has implemented prefabricated housing to implement it, and the prefabricated housing and energy-saving standards are fully integrated. The United States also has a nearly 100-year history of prefabricated building development and as early as 40 years ago, it legislated for industrialized buildings and issued relevant industry norms, paying attention not only to quality but also to aesthetics. Currently, affordable housing in the United States mainly adopts prefabricated buildings, and among every 16 people, one lives in a prefabricated building. In the 1960s, Australia proposed the concept of "rapid installation of prefabricated houses," and in 1987, high strength cold-formed thin-walled steel structure technology matured. In 1996, the Australian and New Zealand joint standard AS/NZS 4600 for cold-formed formed structural steel was released and implemented. After the release of the standard, Australia spent approximately 600 million US dollars each year to build 120,000 steel frame independent houses, accounting for 24% of Australia's total construction business output (Fig. 4.42).

Prestressed housing in countries like Europe and America has reached a relatively mature and complete stage. The major representatives in Asia are Japan and Singapore, both of which began to vigorously develop prefabricated buildings in the 1990s. They should have been 20 years ahead of China.

The development of prefabricated housing in countries such as Europe, America, and Japan roughly went through three stages: The first stage was the initial stage of industrialization, with the focus on establishing an industrial production (construction) system; the second stage was the development period of industrialization, gradually improving the quality

Figure 4.42 Development status of prefabricated construction abroad.

Figure 4.43 Resource-circular residential building.

and cost-effectiveness of products (housing); the third stage was the mature period of industrialization, further reducing the material consumption and environmental load of housing, and developing resource-circular housing (Fig. 4.43). The practice of these countries has proved that using industrial production methods is the fundamental way to achieve low energy consumption and low pollution in housing construction, and to achieve resource conservation, quality improvement, and efficiency enhancement.

4.6.2 Current situation of domestic prefabricated construction development

In 2016, the promotion and development of prefabricated buildings began to spread rapidly across various regions of China. The government vigorously promoted the prefabricated construction method. Currently, there are mainly three types of prefabricated buildings: prefabricated concrete buildings (PC), steel structure buildings, and wood structure prefabricated buildings. Among them, PC buildings have received the most attention. The PC building industry in China has taken shape to some extent. By the end of 2015, there were 104 PC building factories and 56 industrialization bases across the country, mainly distributed in coastal areas such as Shandong, Zhejiang, Jiangsu, and Shanghai, with fewer distribution in the inland regions. PC buildings are one of the representatives of low energy consumption and green buildings, and have outstanding performance in reducing energy consumption and conserving resources. Compared with the traditional pouring concrete production method, PC buildings have significant advantages in production efficiency and engineering quality. Secondly, the market capacity of PC buildings is huge: nearly 500 million square meters of new construction area, 3 trillion yuan in output value, and the construction of 1.5 million units of affordable housing. The development space of PC buildings in China is huge. According to the "Development Outline of Modern Construction Industry," during the "13th Five-Year Plan" period, prefabricated buildings should account for more than 20% of new buildings, and the construction of 40% of affordable housing should adopt the prefabricated construction method.

In recent years, people's living standards have been continuously improving, their demands have become increasingly diverse, comfort requirements have risen sharply, and the awareness of environmental protection and energy conservation has become deeply rooted in people's hearts. The previously available several types of prefabricated buildings in China could no longer meet people's demands. The PC buildings, which use cement as the base material, have too heavy quality, inconvenient transportation, and fail to meet the required insulation effect; steel structure buildings also fail to achieve the required insulation effect; and wood structure prefabricated buildings are not fireproof and have low safety. The prefabricated structure uniform board designed and introduced in this section utilizes environmentally friendly magnesium materials. Compared

Figure 4.44 Standard installation

with cement, it is lighter, stronger, and higher in quality. The high strength cold-formed thin-walled steel structure is used as the keel, and foam particles are added as insulation materials. The prefabricated building panels made by this method have the characteristics of high strength, good insulation effect, light overall quality, and standardized installation (Fig. 4.44), meeting people's various demands.

4.6.3 Modular structure homogeneous board

The prefabricated homogeneous board is a new type of board that combines functional materials such as magnesium-based cementitious materials and expandable polystyrene (EPS) particles, as well as modified materials, through certain special processes with prefabricated structural boards (made of wood or steel, etc.). It is then solidified and integrated with the structural boards (Fig. 4.45). Currently, this product is widely promoted in both domestic and international markets as prefabricated lower-level houses. This product is increasingly favored by customers.

4.6.4 Characteristics of homogeneous panels in prefabricated structures

4.6.4.1 Fireproof and thermal insulation

This board has the insulation property of organic materials and the fire resistance of inorganic materials.

Figure 4.45 Homogeneous plate of magnesium composite structure.

4.6.4.2 Lightweight and high strength

This board has better strength than other inorganic materials (such as cement, gypsum, etc.) and can be made lighter in weight.

4.6.4.3 Standardized assembly

This board can be prefabricated in the factory, which is easy to achieve standardization and can realize modular assembly, saving construction time on-site.

4.6.4.4 Excellent mechanical properties

The framework or interior uses steel structure (or other materials), and is fully combined with the modified inorganic cementitious materials. After solidification, it can leverage the advantages of both, allowing the board to bear greater loads.

4.6.4.5 Environmental safety

Magnesium-based cementitious materials, as an environmentally friendly and pollution-free material, are increasingly being applied in various industries.

4.6.5 Production process

4.6.5.1 Preparation work

1. Preassemble the structural framework and place it in the mold according to the position.
2. Select a demolding agent that is compatible with magnesium materials. Apply the demolding agent on the inner side of the mold.

3. Pour the prepared magnesium oxide, brine (or magnesium sulfate solution), modifier, and foam particles into the mixing container in the prescribed sequence and proportion. Mix them evenly. The reference ratio for the homogeneous plate of chlorinated magnesium composite material is shown in Table 4.11.
4. Use the preprepared foaming solution to foam with a foaming machine. Add the foam to the slurry and then carry out uniform stirring.

4.6.5.2 Molding

1. Pour the mixed slurry into the prepared mold and use the vibration pump to vibrate it to fill the mold and the framework (Fig. 4.46).
2. Curing: Place it in the curing room for curing.
3. Demolding: After the solidification and molding is completed, control the demolding time well.
4. Storage: Repair the edges and corners, and pack and store it.

4.6.6 Important notes and problem analysis

1. Quality control

   The quality of magnesium oxide plays a crucial role in the strength of the sheet material, so the raw materials must be of high-quality; the modifier also plays a very important role in it.
2. Adequate fluidity

   The slurry mixed in accordance with the proportion and formula should have sufficient fluidity. Otherwise, the operation will be inconvenient and defects such as holes in the plates will occur, which will affect the strength and appearance of the plates.
3. Control the demolding time

   In northern regions during winter, the temperature is low, and a curing room with appropriate curing conditions is needed. The temperature and time should be controlled properly. If the time is too short, the strength will not meet the requirements; if the time is too long, the phenomenon of burning the plates will occur.

4.6.7 Development prospects

Due to the advantages of homogeneous panels in prefabricated structures, their application scope will become increasingly extensive. Besides the common prefabricated houses, there will also be prefabricated hospitals, prefabricated hotels, and prefabricated apartments.

Table 4.11 Reference ratio for chlorinated magnesium homogeneous plate (unit: kg).

Light calcined magnesia	Brine	Modifier			Foam particles cube	Short-staple	Foaming agent	Quartz powder
		GX-1#	GX-4#	GX-13#				
100	120	0.6	0.8	1.5	0.8	0.5	0.2	20

Figure 4.46 Slurry pouring.

4.7 Magnesium-based prefabricated buildings

As the country has proposed to develop new construction methods and vigorously promote prefabricated buildings, it aims to increase the proportion of prefabricated buildings in new constructions to 20%–40% or more within about 10 years. Currently, low-rise prefabricated construction techniques have been widely adopted in some high-end villas and rural area development projects. With policy support and encouragement, low-rise prefabricated buildings have developed rapidly.

Traditional buildings mainly use materials such as cement, bricks, and steel bars. While promoting development, they also consume a large amount of land resources, water resources, etc., which aggravates environmental pollution and deteriorates the ecological environment. Traditional materials are limited by their material usage and structural aspects, and are extremely restricted in terms of energy conservation, earthquake resistance, etc. Prefabricated buildings, with their green, environmentally friendly, energy-saving, and emission-reducing characteristics, have gradually become a new trend in the construction industry. They move traditional building components to factories for standardized design and integrated production, which not only ensures good construction quality and short construction period, but also effectively reduces dust pollution and noise pollution.

The prefabricated wall panels are made with magnesium-based cementitious materials. By adding the modified agents developed by MJT, various

Figure 4.47 Factory-prepared magnesium structure crossbeam supports and components.

waste materials such as straw (rice straw, wheat straw, sugarcane residue, cotton stalks), waste wood (including branches, leaves, bamboo boards, bamboo branches), and fly ash (containing volcanic ash, silica fume, etc.) can be extensively mixed and incorporated. Among them, waste wood accounts for about 10%, and is used for the construction of hollow panel frameworks; straw accounts for about 85%, and is used for filling the hollow panels and then used as bonding materials after being crushed; fly ash accounts for about 5%, and is used for bonding and the outer layer of the panels. Combined with the traditional wooden structure system's mortise-and-tenon connection form for construction, the load-bearing columns and beams are made of wooden squares, and the building templates are wrapped with glass fiber threads, supplemented by magnesium-based cementitious materials for reinforcement and fixation; the entire building is constructed according to the design drawings, using factory prefabrication and on-site assembly methods. At the joint connection points, the joints are reinforced and connected with magnesium-based cementitious material mortar supplemented by glass fiber mesh strips. After curing, it becomes an integral structure (Fig. 4.47).

Through this manufacturing technology, eco-friendly wall panels and eco-friendly stairs are produced in a single process. They are installed independently and can bear loads of over 5000 kg. They have unique advantages such as shock resistance, fire resistance, water resistance, antifreezing, anticracking, antitermite infestation, antiaging, and anticorrosion. They are easy to install and have extremely high cost-effectiveness. They can be widely applied in both civilian and industrial buildings.

Figure 4.48 Design of diversified magnesium-based low-rise prefabricated buildings.

4.7.1 Characteristics of magnesium-based low-rise prefabricated buildings

4.7.1.1 Diversity in design

Based on the residents' requirements, various architectural categories with diverse construction styles can be designed to meet the needs of different groups such as high-end villas, new rural construction houses, relocation houses, and workers' houses (Fig. 4.48).

4.7.1.2 Factory-based production of prefabricated components

The main building structure and building components can be prefabricated and produced in factories in advance. Under this production method, the construction accuracy of the building structure is guaranteed, the building quality is stable, and the workload of on-site construction workers is reduced.

4.7.1.3 Specialization in construction and installation

For low-rise prefabricated buildings, the construction site only requires a relatively small number of specialized construction personnel. This is the biggest difference between prefabricated buildings and traditional building construction processes. Simple installation procedures, reduced wet work operations, have improved the professional level of construction technology, and the use of high-level mechanical equipment has made the construction method more precise.

4.7.1.4 Integration of building structure

The integration of building structure can realize the prefabrication production of various materials, making the main body of the building structure more unified and sturdier when completed.

4.7.1.5 Scientific management of operation and maintenance

Through computer technology and information technology, the entire process of prefabricated components, from design, production, construction hoisting to later maintenance management, can be made scientific.

4.7.2 Magnesium-based prefabricated wall panel material preparation and molding process

4.7.2.1 Raw materials

Light-burned magnesium oxide, admixture, fly ash, mineral powder, silica fume, rice husk, straw powder, sawdust residue, modifier, glass fiber filaments, bamboo plywood, waste formwork, etc.

4.7.2.2 Reference formula

The reference ratio of the ingredients for magnesium-based prefabricated wall panels is shown in Table 4.12. Taking an external environmental temperature of 25°C–35°C as an example for illustration.

4.7.2.3 Casting process of magnesium composite wall panels (components)

4.7.2.3.1 Mold cleaning

Remove the residues on the inner surface of the mold and apply a release agent.

4.7.2.3.2 Precast mortar

Mix the required mortar according to the production material ratio requirements.

4.7.2.3.3 Panel paneling and forming

Place the precast base material on the surface of the mold, evenly sprinkle glass fiber filaments on the base material, and use a stick to roll and press the glass fiber filaments to ensure they are fully soaked in the mortar (Fig. 4.49).

Table 4.12 Reference ratio of ingredients for magnesium-based prefabricated wall panels (unit: kg).

Light calcined magnesia	Brine (23°Bé)	Fly ash, mineral powder, silica powder, etc.	Rice husks, straw powder, sawdust residue, etc.	Glass fiber filament	Veneer board, waste formwork	GX-4#	GX-12#
100	150	50–60	30	–	–	0.8	1

Note: Depending on the specific properties of the filler materials (such as fineness and moisture content), the proportion and amount of the ingredients can be adjusted appropriately.

Figure 4.49 Laying glass fiber strands.

Figure 4.50 Placing bamboo screens or construction templates.

Then, evenly spread a layer of mortar on the surface, and evenly distribute the premade building templates or bamboo mesh net sheets on the mortar (Fig. 4.50).

Then, cover the surface with mortar to ensure it is evenly covered and soaked. Finally, evenly sprinkle glass fiber filaments on it, and then cover another layer of mortar to ensure the glass fiber filaments are fully soaked (Fig. 4.51).

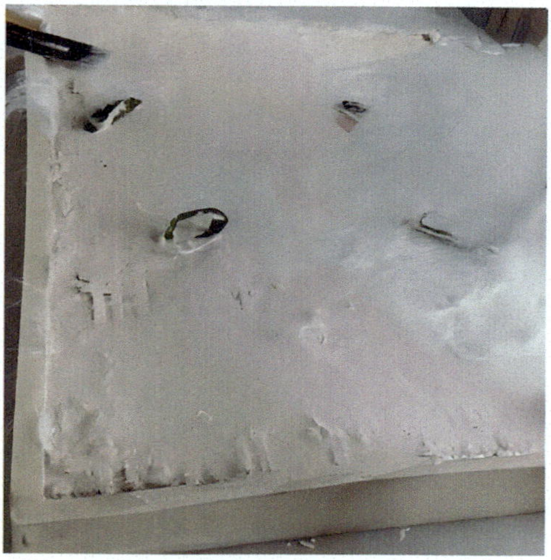

Figure 4.51 Continuing to lay glass fiber filaments.

4.7.2.4 Assembling prefabricated formwork

On the laid board surface, install and fix the long strips of building wood formwork along the periphery of the mold, then stand and fix one at the middle position of the board width. Uniformly stand and lay the formwork on both sides of the board length direction. This forms a middle frame structure with the same thickness as the board. Then evenly spread and cover glass fiber filaments on the board fabric, and evenly pour the material on top. Use brushes and sticks to make the material fully soak the glass fiber filaments. During the process of the material setting and hardening, the base plate and the middle wood formwork frame structure form a unified combination (Fig. 4.52).

4.7.2.5 Assembly of upper and lower board surfaces

4.7.2.5.1 Method one

The prefabricated sheet panels are laid flat on the fixed base plate frame structure, and the embedded joints are formed by soaking the densely-woven glass fiber mesh fabric with magnesium-based cementitious material slurry and connecting the upper and lower panel surfaces (Fig. 4.53).

Figure 4.52 Assembly of prefabricated formwork.

Figure 4.53 Assembly of the top and bottom panels.

4.7.2.5.2 Method two

1. Suspended construction of building wooden formwork method
   Place the cut and sized wooden formwork panels on the frame structure, then apply mortar and combine with glass fiber filaments to form.
2. Hollow frame filled with foam board method
   Fill the hollow part of the frame structure with polystyrene foam boards, then apply mortar and seal the joints, and finally lay the mortar and combine with glass fiber filaments on the foam boards.

Figure 4.54 Foundation construction.

Figure 4.55 Plate verification.

4.7.3 Assembly construction process of ecological house

4.7.3.1 Foundation construction

Carry out foundation construction according to the design drawings. Wait for the foundation to solidify before proceeding to the next step of installation (Fig. 4.54).

4.7.3.2 Board verification

After the ecological boards arrived, the construction team leader conducted on-site counting of the required boards based on the disassembly drawings (Fig. 4.55).

Figure 4.56 Wall assembly.

Figure 4.57 Wall adhesive spot application for fixation.

4.7.3.3 Wall assembly

After the plates have been accurately counted, workers use a crane to place the plates upright on the foundation according to the drawings (Fig. 4.56).

4.7.3.4 Gluing and fixing of the wall

While the ecological panels are placed upright, workers carry out the glue application and curing process (Fig. 4.57).

Figure 4.58 Jointing of the wall.

Figure 4.59 Main body solidification completed.

4.7.3.5 Mortar filling of the wall

After the ecological panels have been spot-glued and cured, workers then commence the mortar filling operation (Fig. 4.58).

4.7.3.6 Completion of the main structure solidification

After the jointing work is completed, wait for the overall solidification of the building and the completion of the main structure construction (Fig. 4.59).

4.7.4 Advantages of magnesium-based lower-layer prefabricated buildings

4.7.4.1 Low construction cost, waste turned into eco-building materials

By using magnesium-based low-rise prefabricated building main plates and components, a large amount of waste such as rice husks, corn stalks, wheat stalks, reeds, bamboo, waste wood boards, construction waste, and slag can be consumed. The cost can be reduced by more than 50%. This is achieved by combining the concept of circular ecology with new technologies to turn waste into valuable eco-building materials. Waste is everywhere, and raw materials are sourced locally, significantly reducing raw material costs. Moreover, the waste has been comprehensively utilized, making it a beneficial project for both the country and the people.

4.7.4.2 Construction speed is fast

The materials required for magnesium-based low-rise prefabricated buildings are preprocessed in factories. At the construction site, only hoisting and assembly are needed. The modular assembly and connection method significantly reduces the wet work process at the construction site. From the moment the project starts on the site, the house can be completed in as fast as 7 days. There is no need for excessive drying or ventilation before entering the decoration stage. The construction speed is fast, the installation speed is even faster, the construction difficulty is low, and it also changes the wet work method on-site, which effectively controls sewage discharge, dust pollution, and noise pollution, achieving more significant energy conservation and environmental protection effects.

4.7.4.3 Building quality is guaranteed

Unlike traditional construction methods with a large amount of on-site work, magnesium-based low-rise prefabricated buildings complete the wall panels, stairs, frames, and other building components in the factory according to the designed specifications and dimensions. Then, they are uniformly transported to the construction site and connected through reasonable connection methods. This way, the complexity of on-site variable situations is avoided to a certain extent, and the possible human work errors in traditional construction methods are effectively avoided. Through this scientific and reasonable method, not only is the building quality guaranteed, but the labor intensity of workers is also reduced, and the construction time can be shortened, thereby improving production efficiency.

4.7.5 Performance test results of magnesium composite building structures

1. Load-bearing capacity test
   This material has a high load-bearing strength, approximately 28 times that of traditional building materials, with extremely strong load-bearing capacity.
2. Fire resistance and flame retardancy test
   The new wall material was tested with a spray gun at a temperature of 1000°C. It did not catch fire at all and its flame retardancy reached grade A.
3. Earthquake resistance test
   Ecological materials can achieve earthquake resistance above level 8. It can resist bending, compression, and tension, with high strength.
4. Waterproof and anticorrosion test
   It has good waterproof sealing performance. Even if it is soaked in water for 10 years, it will not deform, rot, expand, contract, or be damaged.
5. High-temperature boiling test
   After continuous boiling in water for 10 hours, it does not deform, does not foam, does not crack, does not leak water, and is not damaged.
6. Anti-oil stain test
   After being tested with various oil stains such as cooking oil, diesel, and engine oil, the product is easy to clean, does not penetrate, is resistant to oil stains, and remains clean as new after wiping.
7. Compression resistance test
   The compressive strength of this wall panel reaches over 3600 kg/m^2, exceeding that of ordinary traditional wall panels.

4.7.6 Development prospects

The magnesium-based low-rise prefabricated buildings, as the emerging force in the field of prefabricated buildings nowadays, have gained popularity among investors due to their advantages such as low construction costs, unique waste recycling, guaranteed building quality, short construction period, and environmental friendliness. It is believed that they will have a bright future.

4.8 Reinforced steel structure prefabricated homogeneous panel (sulfur-chlorine composite wall panel)

4.8.1 The material characteristics of the homogeneous panels (sulfur-chlorine composite wall panels) used in the prefabricated steel structure construction

The heavy steel structure prefabricated homogeneous panel (sulfur-chlorine composite wall panel) is an advanced building material. Its unique structure and properties give it broad application prospects in modern construction. This wall panel uses a heavy steel structure as its framework and is installed through a prefabrication process, characterized by efficiency, speed, and environmental friendliness (Fig. 4.60).

The main features of the prefabricated homogeneous plate (sulfur-chlorine composite wall panel) made of reinforced steel are homogeneity and high strength. The wall panel is filled with sulfur-chlorine composite material, which ensures that the entire wall panel has uniform density and strength, thereby guaranteeing the stability and durability of the wall. In addition, the sulfur-chlorine composite material also has excellent fire resistance properties, which can effectively enhance the safety of the building.

Compared with traditional wall materials, the heavy steel structure prefabricated homogeneous panel (sulfur-chlorine composite wall panel) has higher construction efficiency and lower cost. Due to the use of

Figure 4.60 Prefabricated steel structure residences.

prefabrication technology, the wall panels can be preassembled in the factory and then directly transported to the construction site for installation, significantly shortening the construction period. At the same time, advanced production processes and technologies are adopted in the manufacturing process of the wall panels, making their cost more reasonable and providing strong support for the sustainable development of the construction industry.

In practical applications, heavy steel structure prefabricated homogeneous panels (sulfur-chlorine composite wall panels) can be widely used in various construction fields, such as residences, commercial buildings, industrial plants, etc. Their excellent physical properties and environmental protection characteristics make them an ideal choice in the modern construction field. As people's requirements for building quality and environmental protection continue to increase, heavy steel structure prefabricated homogeneous panels (sulfur-chlorine composite wall panels) will play a more important role in the future and inject new vitality into the development of the construction industry.

4.8.2 The performance advantages of sulfur-chlorine composite wall panels

1. High strength
   After modification, the compressive strength and flexural strength of the sulfur-chlorine composite wall panel can reach 60 and 9 MPa respectively. This enables it to perform well when bearing the pressure from heavy steel structures.
2. Good air stability and weather resistance
   The sulfur-chlorine composite material is a hard bonding material that can only continue to harden and set in the air, giving it good air stability. At the same time, as the environment becomes drier after the wall panel is cured, it becomes more stable, which enables it to maintain stable performance under various climatic conditions.
3. Mild corrosion to steel
   The sulfur-chlorine composite wall panel uses magnesium sulfate and magnesium chloride as blending agents. While retaining the advantages of high strength and fast curing of the chloromagnesium cementitious material, by combining magnesium sulfate, it neutralizes the incompletely reacted free magnesium chloride in the panel, resulting in low corrosiveness to steel and reducing the risk caused by chloride ion erosion of steel.

4. Environmental protection and energy-saving
   The sulfur-chlorine composite wall panel uses industrial waste materials as the main raw material. Not only does it achieve resource recycling, but its low heat and low corrosive properties also make it outstanding in terms of energy-saving. In addition, the wall panel has good insulation performance, which can effectively reduce energy transfer and loss, and improve the insulation effect of the building.

4.8.3 Application of sulfur-chlorine composite wall panels in prefabricated buildings with heavy steel structures

In prefabricated steel structure buildings, sulfur-oxygen composite wall panels play a crucial role. Due to their lightweight, high strength, insulation, fire resistance and other advantages, sulfur-oxygen composite wall panels are widely used in exterior walls, interior walls, partition walls, and other parts. At the same time, this wall panel can also be closely connected with steel structure components to form an overall stable structural system, improving the overall performance of the building.

4.8.4 Construction process of prefabricated steel structure buildings

The construction process of prefabricated steel structure buildings is a systematic and standardized procedure aimed at ensuring the quality and efficiency of the building. The main steps of this construction process are as follows:

1. Design phase
   a. Conduct preliminary design based on factors such as the building's usage function, geographical location, and environmental requirements.
   b. Design the layout and connection methods of the heavy steel structure frame and the magnesium-sulfur composite wall panels.
   c. Prepare construction detail drawings, including the dimensions of the steel structure, the prefabrication dimensions of the wall panels, and installation nodes.
2. Prefabrication stage
   a. In the factory, heavy steel structures are prefabricated according to the construction details, including cutting, welding, and correction of components such as steel columns and steel beams (Fig. 4.61).

Figure 4.61 Prefabrication of heavy steel structures.

Figure 4.62 Sulfur-chlorine composite wall panel.

   b. Prefabricated sulfur-chlorine composite wall panels are cut and cured according to the position and size of the wall (Fig. 4.62).

3. Transportation phase

   a. Transport the prefabricated heavy steel structure components and sulfur-oxygen composite wall panels to the construction site.

   b. Ensure the integrity and safety of the components and wall panels during transportation.

(A) (B)

Figure 4.63 Foundation construction. (A) Planning layout. (B) Compacting the foundation.

Figure 4.64 Construction of heavy steel structure components.

4. Foundation construction phase
   Carry out foundation construction at the construction site, including foundation treatment and concrete pouring (Fig. 4.63), to ensure the foundation is level and horizontal, preparing for the subsequent installation of the steel structure.
5. Steel structure installation stage
   According to the construction details, the prefabricated heavy steel structure components are lifted to the designated positions (Fig. 4.64). High strength bolts or other connection methods are used to tightly connect the steel columns, steel beams, and other components to form a stable steel structure framework.

Figure 4.65 Hoisting of sulfur-chlorine composite wall panels.

Figure 4.66 Fastening wall panels.

6. Wall panel installation stage
   The prefabricated sulfur-chlorine composite wall panels are lifted and placed at the corresponding positions of the steel structure framework (Fig. 4.65).
   Use special connecting pieces to firmly connect the wall panels with the steel structure components, ensuring the stability and sealing of the wall (Fig. 4.66).
7. Interior and exterior decoration phase
   After the installation of the wall panels, interior and exterior decoration is carried out, including wall treatment, door and window installation, insulation layer construction, etc. (Fig. 4.67).
8. Acceptance phase
   After the completion of the construction, a comprehensive quality inspection and acceptance of the entire building are conducted. Ensure that the building's structure is safe, functions are complete, the appearance is beautiful, and it meets the design requirements and relevant standards.

Figure 4.67 The steel structure prefabricated houses entering the decoration stage.

Through the above construction process, it is possible to efficiently and quickly complete the construction of heavy steel structure prefabricated buildings, while ensuring the quality and performance of the buildings. This construction method not only improves the construction efficiency and reduces costs, but also helps to promote the sustainable development of the construction industry.

4.8.5 The future prospects of sulfur-chlorine composite wall panels

As awareness of environmental protection and energy conservation continues to grow, sulfur-chlorine composite wall panels, as a green and environmentally friendly building material, have very broad market prospects. In the future, with continuous technological advancements and the deepening promotion of applications, sulfur-oxygen composite wall panels will play an increasingly important role in heavy steel structure prefabricated buildings and make greater contributions to the sustainable development of the construction industry.

In conclusion, sulfur-chlorine composite wall panels, as a new type of building material, offer broad application prospects in heavy steel structure prefabricated buildings. Their excellent performance, environmental protection, and energy-saving characteristics, along with their promising market outlook, position them as a significant direction for the future development of the construction industry.

4.9 Lightweight partition panels

With the continuous advancement of building industrialization and prefabricated buildings in our country, the application fields of prefabricated partition panels have become increasingly extensive, presenting a promising future. However, the requirements for partition panels are getting higher. Not only do they need to be lightweight and strong, but also fire-resistant, heat-insulating, modular, and the installation and construction should be more convenient, and they should tend toward integration. Therefore systematic solutions need to be provided for the existing problems of partition panels to make them develop more towards a more humanized direction.

Magnesium-based lightweight partition panels are manufactured using calcined lightweight magnesium as the cementing material, magnesium chloride or magnesium sulfate as additives, along with suitable modifiers, fly ash, sawdust, and other auxiliary materials. These components are mixed into a slurry, and then foaming agents are incorporated and poured into molds for solidification. After curing, they develop into environmentally friendly, energy-saving, lightweight, and high strength wall materials.

4.9.1 Raw materials

Magnesium oxide, magnesium chloride, sawdust, stone powder, fibers, palm strips, etc.

4.9.2 Classification and technical specifications

The classification and technical parameters are shown in Table 4.13.

Lightweight partition panels can be classified into five types based on their structure: hollow strip panels, solid strip panels, composite strip panels, wave-shaped tongue-and-groove partition panels, and insulated homogeneous partition panels.

1. Hollow strip panels

   Prefabricated strip panels with several through holes along the length direction of the panel.

2. Solid strip panels

   Prefabricated strip panels without any holes made of the same material.

3. Composite strip panels

   Prefabricated strip panels composed of two or more different functional materials through compounding or by combining the panel material with the core material.

Table 4.13 Technical parameters of lightweight partition panels.

Project/Type		Indicator (plate thickness)/mm				
		90 (100)	**120**	**150 (160)**	**180**	**210**
Compressive strength/MPa	Cement block panels, composite block panels	≥3.5				
Surface density/(kg/m^3)	Cement block panels	≤90	≤110	≤ 130	–	≤180
	Composite block panels	≤90	≤110	≤ 130	≤150	≤160
Water content/%		≤12/10/8				
Tensile strength of the composite strip board panel perpendicular to the board surface/MPa		≥0.2				

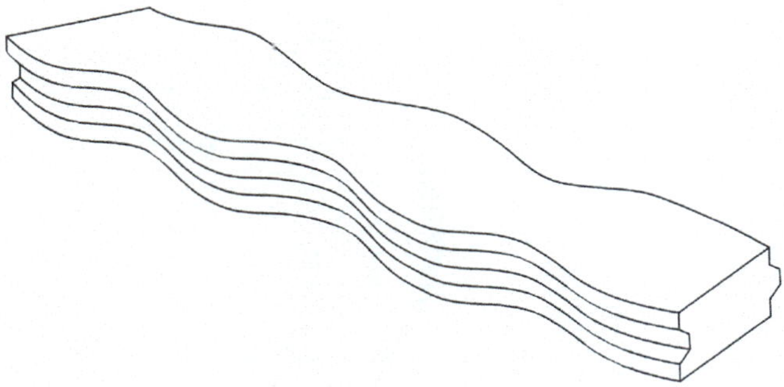

Figure 4.68 Wave-shaped tongue-and-groove partition panel.

4. Wave-shaped tongue-and-groove partition panel
Due to the influence of temperature, the partition panel undergoes thermal expansion and contraction, and the tongue-and-groove area of the panel will generate internal stress, resulting in cracking and unevenness of the exterior decorative layer. To address this issue, MJT, based on years of practical experience, designed a wave-shaped tongue-and-groove partition panel (patent number: CN2012206901614). The groove (mortise) of this tongue-and-groove (mortise) is a wave-shaped curved surface (Fig. 4.68), and the corresponding convex groove (mortise head) on the other side is also a wave-shaped curved surface. Compared to a flat surface, this increases the contact area and reduces the internal stress. Moreover, the mortise and mortise head are not completely fitted but leave a certain amount of space. During installation, elastic sealing glue (or polyurethane foam) is filled into the reserved pores. When the partition panel expands due to heat, the curved surfaces of the mortise and mortise head only come into contact at a few points, and most of the internal stress is counteracted by the elastic sealing glue (or polyurethane foam). The few contact points, due to the characteristics of the curved surface structure, mutually cancel out internal stress F_1 and F_2 (Fig. 4.69), thereby evenly distributing the stress in the vertical direction of the panel surface, and eliminating the concentration of stress. The wave-shaped tongue-and-groove effectively solves the problem of stress concentration in the partition panel causing deformation, cracking, and bulging of the decorative layer.
Regarding the cracking or cracks of prefabricated partition panels, not only the forming process needs to be addressed, but also the adaptability and flexibility of the sealing and decorative surface materials, flexible connection materials, lifting and installation equipment, and on-site installation at the construction site should be considered.

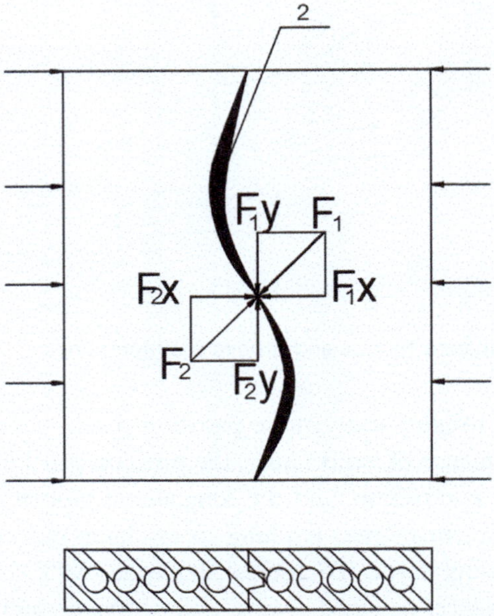

Figure 4.69 Force analysis diagram of the two-wave-shaped flange opening.

Figure 4.70 Insulation homogeneous partition panel.

5. Insulation homogeneous partition panels
   Insulation homogeneous partition panels (Fig. 4.70) integrate the extruded board and homogeneous board into one from the original partition panels, and apply a layer of bonding mortar on the outside of the homogeneous board to become a new type of partition panel with better insulation effect and more convenient installation.

4.9.3 Production process

The production processes and materials of various partition boards vary, but they can be roughly divided into the following four parts: material and

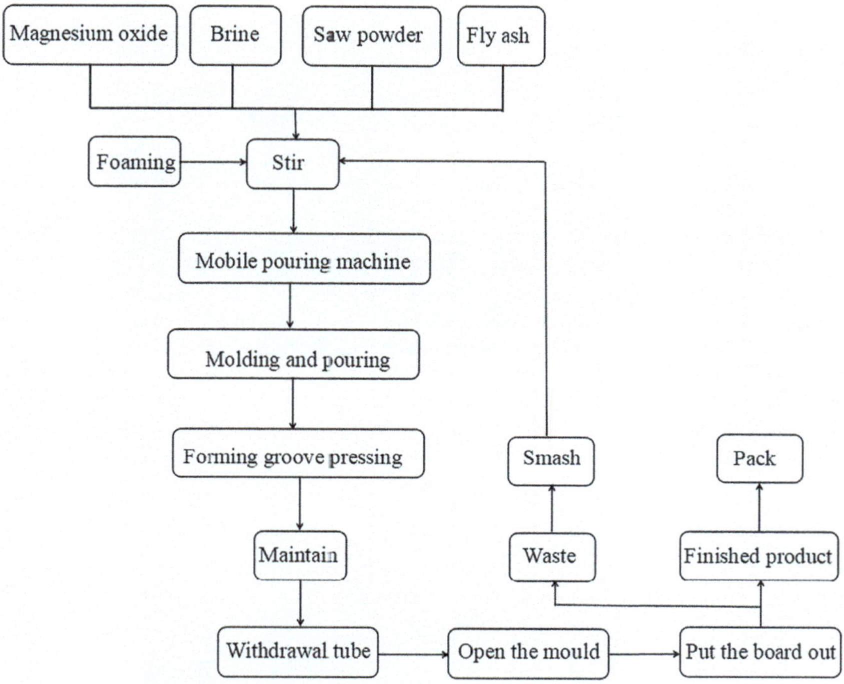

Figure 4.71 Production process flow.

equipment preparation, mixing, pouring and molding, and demolding, packaging, and sorting.

The horizontal core-pulling method is a traditional production method for magnesite partition boards. After assembling the molds, the foaming material slurry is poured in, the mold is formed, and then the core is removed. The specific production process flow is shown in Fig. 4.71.

1. Preparatory work. Clean the mold thoroughly, apply the release agent, and assemble it; place the core tube in the mold beforehand, and adjust the specific gravity of the brine or magnesium sulfate solution (Fig. 4.72).
2. Mix all the raw materials evenly. Through the introduction of the foaming process, a large amount of foam is added to it to improve the fluidity of the slurry and reduce its density.
3. Pour the evenly mixed slurry into the mold through the pouring machine. After the mold is filled with the slurry, manually smooth the upper surface (Fig. 4.73).

Figure 4.72 Assembly mold.

Figure 4.73 Casting slurry.

4. After pouring, the mold undergoes a 6- to 8-hour reaction and curing process. Then, the core tube can be withdrawn (Fig. 4.74), the mold is opened, and the sheet is removed and packed (Fig. 4.75).

4.9.4 Product advantages and market analysis

4.9.4.1 Environmental protection

The panels are prefabricated in the factory, reducing wet work on-site and minimizing dust and wastewater pollution. In the inorganic cementitious materials used in production, a certain amount of waste materials can be mixed in to reduce pollution while also lowering costs.

Figure 4.74 Core extraction.

Figure 4.75 Remove the sheet material.

4.9.4.2 Lightweight

By adding foaming agents, polystyrene granules, etc., the density of the board is significantly reduced, effectively reducing the self-weight of the building. At the same time, it can reduce the economic investment in the foundation, facilitating construction, and transportation.

4.9.4.3 Sound insulation

The wall of the lightweight partition board can achieve a sound insulation of 40–50 dB, ensuring that the product can be used in buildings such as offices, commercial buildings, entertainment venues, and residences.

4.9.4.4 Good seismic resistance

The elastic modulus is low, and it absorbs shock energy quickly. The seismic resistance level can reach 8.

4.9.4.5 Thermal insulation and heat preservation

Introducing sealed air holes into the board can significantly enhance the thermal insulation performance of the material. Additionally, the addition of polystyrene particles reduces heat conduction and lowers the thermal conductivity.

4.9.4.6 Easy to construct, high efficiency

The lightweight partition board products can be drilled, sawed, and nailed. The board surface is flat, with mother and daughter slots, which facilitate installation and construction.

The country strongly advocates environmental protection and energy conservation. The lightweight partition board is a perfect example of environmental protection and energy conservation. It transfers the on-site work to the factory for prefabrication, then transports it to the construction site for direct use. This reduces on-site operations and avoids water and dust pollution. The materials are lightweight and high strength, with standard specifications, and can have mother and daughter slots, reducing transportation costs. It is easy to construct, effectively reducing labor costs. The partition board reduces the thermal conductivity of materials through various processes, reduces heat transfer, and meets the national energy-saving standards.

4.9.5 Common problems and solutions

4.9.5.1 Cracks

The occurrence of cracks in partition wall panels is caused by various factors and can be summarized into the following two points: (1) Raw material control, the quality of raw materials varies greatly. If the calcium content is high, the poor stability will lead to cracking of the panels; (2) Insufficient curing time and conditions. During curing, necessary temperature and humidity environments should be provided to ensure adequate reactions and meet the strength requirements. In addition, sufficient curing time is also needed. Wet panels that have not fully reacted and are directly installed on the wall will also cause cracks.

4.9.5.2 Collapse mold

During production, foaming process is often employed. After foaming, mold collapse is inevitable. The main reason is that the foam is insufficient to support the volume of the slurry. The causes can be analyzed from multiple aspects. Firstly, there may be problems with the foam itself, such as thin bubble walls and short supporting time. Before the slurry is set, the foam disappears. At this point, it is necessary to consider whether the foaming agent and foaming liquid are mixed evenly, whether there is sedimentation, whether the water dilution ratio is correct, whether the steel wire balls of the foaming machine are rusted, and whether the foam produced is uniform, dense, and fine. Apart from the problem of the foaming agent itself, other raw materials also need to be considered. For example, whether the sawdust added is too dry, causing water absorption and bubble breaking; whether the added fly ash and other materials contain impurities, which cause the foam to be unstable, etc.

4.9.5.3 Uneven interface

During on-site construction, it was discovered that the interface of the sheet material was uneven. This was due to the influence of thermal expansion and contraction, where the size change at the interface caused the unevenness. Therefore the designed sheet materials now have special interfaces. For instance, the wave-shaped interface designed by MJT can solve the problem of unevenness at the thermal expansion and contraction joints.

4.10 Purification plate

4.10.1 Definition of purification plate

Purified corrugated steel sheet, commonly known as purified board or clean board, is a composite board made by using corrugated steel sheet or stainless steel sheet as the surface material, and rock wool, paper honeycomb, glass fiber board, etc. as the core material, and bonded together. This type of board is mainly used in clean rooms, partitions, and ceilings, and has the advantages of dust prevention, antibacterial, corrosion resistance, rust prevention, and antistatic. Due to these characteristics, purified boards have been widely applied in purification engineering fields with high requirements such as medical, electronics, food, biopharmaceuticals, aerospace, and precision instruments (Fig. 4.76).

Figure 4.76 Purification board wall panel.

4.10.2 Classification of purification board core materials

The core material is an important component of the purification board, directly influencing the quality and performance of the purification board. In the market, there are various types of purification board core materials, including rock wool board, silica rock board, glass fiber reinforced manual board, sulfur-magnesium hollow purification board, etc.

Here are some common purification board core materials and their characteristics.

4.10.2.1 Mechanized foam purification board

The core material is made of flame-retardant EPS foam board, the surface layer is hot-dip galvanized coated sheet and high strength adhesive, and the purification board is formed by cold pressing bonding (Fig. 4.77).

4.10.2.2 Flame-retardant paper honeycomb purification board

The core material of the paper honeycomb is made of flame-retardant paper, and the top and bottom panels are composite steel sheet or stainless steel sheet, as shown in Fig. 4.78. The paper honeycomb core material only carbonizes when exposed to fire, does not burn, and does not release toxic substances. The flame-retardant paper honeycomb purification board has high rigidity, high strength, strong load-bearing capacity, and good heat insulation and sound insulation effects.

Figure 4.77 Mechanism foam purification plate.

Figure 4.78 Flame-retardant paper honeycomb-type purification board.

4.10.2.3 Rock wool purification board

The rock wool purification board is a "sandwich" structure board composed of a color steel profile as the surface layer, structural rock wool as the core material, and bonded with special adhesives, as shown in Fig. 4.79. It is a clean board with extremely strong fireproof effect. It can be sealed on all four sides. Reinforcing ribs can be added in the middle of the board to make the surface smoother and have higher compressive strength.

Figure 4.79 Rock wool purification board.

Figure 4.80 Hollow glass-magnesium composite panel.

4.10.2.4 Hollow glass-magnesium composite board

The hollow glass-magnesium composite board (purification board) has a colored coated steel sheet as the surface layer, a hollow glass-magnesium plate as the core material, and is bonded together in a continuous molding machine under pressure using heat-curing adhesive, as shown in Fig. 4.80. The glass-magnesium plate is a new type of noncombustible decorative material made by a ternary system of magnesium oxide, admixture and water, supplemented by a modifier, reinforced by a medium alkaline glass fiber mesh, and filled with lightweight materials. It has the characteristics of fire resistance, no odor, no toxicity, no freezing, no decay, no cracking, high strength, and light weight.

Figure 4.81 Silica rock color steel sandwich panel.

4.10.2.5 Silica rock color steel sandwich panel

The silica rock color steel sandwich panel (purification panel) uses silica rock board as the core material, and galvanized or painted board as the surface layer. The two layers are bonded with high strength and are formed through a high-speed continuous automated forming machine by heating and cold pressing. It has the characteristics of thermal insulation and easy installation. It is a new type of fireproof panel with stronger fire resistance among similar (sandwich panel series), as shown in Fig. 4.81.

4.10.2.6 Magnesium sulfate color steel sandwich panel

The magnesium sulfate color steel sandwich panel (purification panel) uses magnesium sulfate as the core material, and galvanized or painted board as the surface layer, as shown in Fig. 4.82. The water repellency rate of the magnesium sulfate color steel sandwich panel is greater than 98%. It does not deform when encountering water and has good heat insulation and sound insulation effects.

4.10.3 Forming process of purification board core material

4.10.3.1 Hollow glass-magnesium composite board

The hollow glass-magnesium composite board is mostly produced by manual molding. The glass-magnesium board manufacturer cuts the

Figure 4.82 Magnesium sulfate steel composite panel.

3000 × 1220 × 5 mm glass-magnesium boards produced into 4-cm-wide strips according to the customer's requirements, and then sends the boards and the strips to the purification board factory for secondary processing.

The purification board factory cuts the already coated with film panels according to the size of the finished purification board specified by the customer. Coating the film is to reduce scratches on the surface of the purification board during later production, transportation, and installation, in order to achieve the maximum aesthetic effect.

The cut panels are then manually coated with foaming glue. The advantage of manual gluing is that the glue is applied more evenly, resulting in better adhesion. Moreover, since most purification boards are non-standard boards with varying size specifications, manual operation offers greater flexibility. Currently, the commonly used glue by purification board manufacturers is "Yihang · Shark," which has good bonding performance but is relatively expensive. Furthermore, since this glue is of the foaming type, there is an issue of residual glue in the pipeline when applying glue by equipment. If the glue in the pipeline foams up during production gaps or pauses, it can block the entire pipeline, leading to waste.

After the steel panels coated with polymer adhesive are applied, they are positioned atop a full 5-mm-thick glass-magnesium sheet. Subsequently, glass-magnesium strips cut to a width of 4 cm are placed at regular intervals across the top surface to serve as the primary frame. Concurrently, the secondary frame is installed. Both the secondary frame and the primary frame collaborate to provide structural support. Typically, the secondary frame is fabricated by extruding steel strips, which are then cut to the specified dimensions and secured at the four corners using specialized plastic fasteners.

After the main frame and the frame are installed, the same steps are repeated. Glue is applied, and the entire glass-magnesium sheet and the upper poly-coated steel panel are bonded. Finally, the bonded purification panels are neatly stacked. After stacking to a certain height, cold pressing and gluing are carried out. The finished product must be compressed for 48 hours before it can be shipped.

Now, to enhance the fire resistance and sound insulation effect of the hollow glass-magnesium composite board, the hollow glass-magnesium board is often combined with rock wool. Rock wool strips are filled between the main frame to form a new type of manual glass-magnesium rock wool purification board.

4.10.3.2 Silica rock board

The silica rock board (Fig. 4.83) is a new type of purification board core material that emerged in recent years. It is based on polystyrene foam board as the core material. The fireproof slurry prepared by mixing magnesium oxide and magnesium sulfate is modified and then penetrated into the interior of the polystyrene foam board through a vacuum negative pressure process to form a fireproof board.

Figure 4.83 Silica rock plate.

The silica rock board was first applied by Yifeng Clean Technology Jiangsu Co., Ltd., which used a ship-shaped vibration extrusion molding equipment. However, the production efficiency was low and a lot of slurry was wasted. Later, a vacuum adsorption type equipment designed by Jinan Meiyatu Machinery and Equipment Co., Ltd. (Patent Number: CN2015203400913) was adopted. The daily output of standard boards can reach 500 m^3. This successfully solved the problems of thick plates not being able to be absorbed completely and low efficiency for thin plates. Once it was put into the market, it received unanimous praise from various production enterprises. With the promotion and popularization of the silica rock board and its sales across the country, it has occupied a significant share in the industry.

4.10.3.3 Hollow magnesium sulfate board

The hollow magnesium sulfate board (Fig. 4.84) serves as an alternative product to traditional rock wool board, hollow glass fiber board, and silica rock board. It is made by physically foaming and reducing the density of magnesium-based cementitious materials, then combining with polystyrene foam particles to enhance the insulation and heat preservation performance. After being rolled by a flat panel production line to become a 2.5-cm-thick corrugated board, it undergoes natural curing and solidification. Subsequently, it is processed through sandblasting, gluing, and cutting to produce a 5-cm-thick board. Its comprehensive performance is excellent, and it is currently highly praised as the new generation of purification boards in the market.

Figure 4.84 Hollow magnesium sulfate plate.

4.10.4 Common problems and analysis of several types of purification board core materials

4.10.4.1 Hollow glass-magnesium board

The core material of the hollow glass-magnesium board is mainly made from glass-magnesium board. Due to the traditional production process, it is very prone to the phenomenon of re-humidification in high humidity environments, which is commonly referred to as the appearance of water droplets on the glass-magnesium board. The main component of the curing agent used in producing glass-magnesium boards is $MgCl_2$, and Cl^- has strong hygroscopicity. When the glass-magnesium board products get wet, there will be sticky damp stains on the surface of the products. These damp stains increase with the increase of humidity or moisture, and accumulate to form "water droplets."

Through the analysis of the components of the water droplets on the surface of the glass-magnesium board, the main components are shown in Table 4.14.

According to the national standard GB/T 15957, the main corrosive substances in the atmosphere are sulfur dioxide and chlorides. The concentration of Cl^- in the water droplets on the surface of the glass-magnesia board reaches 40.13 g/L. If it remains in a high-humidity environment for a long time, a large number of water droplets will form on the surface of the glass-magnesia board, and the Cl^- containing water droplets will continuously erode the color steel panel inside the glass-magnesia purification board, resulting in large-scale corrosion phenomena.

When the clean room is in a certain temperature and humidity environment for a long time, the Cl^- in the water droplets on the surface of the glass-magnesia board begins to corrode the backside of the color steel panel that is bonded to the glass-magnesia board first. Over time, the corrosion phenomenon will extend from the backside of the color steel panel to the outer surface, ultimately presenting a corrosion phenomenon on the outer surface of the purification board.

Table 4.14 Main components of water droplets on the surface of glass-magnesium board.

Component	Cl^-	Mg^{2+}	K^+	Na^+
Concentration/(g/L)	40.13	13.73	12.24	4.16

When rusting problems occur in ordinary hollow glass-magnesia purification boards, they have already been installed and constructed, and it is relatively impossible to repair them. Therefore during the production process, strict requirements must be imposed, qualified raw materials must be selected, and reasonable production ratios must be formulated. The following is a simple analysis of several prone-to-problem areas.

4.10.4.1.1 Light-burned magnesium oxide powder

The quality indicators of light-burned magnesium oxide usually include the content of active magnesium oxide, loss on ignition, free calcium content, and fineness. The content of active magnesium oxide is strictly controlled within a certain range. An excessively high or low content will have negative impacts.

When the activity is too low (generally below 50%), the solidification process is slow, and there is a large amount of brine surplus. A large amount of Mg^{2+} and Cl^- exist in the form of free in the product, causing the product to absorb moisture and reoxidize.

When the activity is too high (more than 70% or above), during hydration, a large amount of $Mg(OH)_2$ colloids is formed in a short period of time, which will increase the liquid phase alkalinity and prevent the normal reaction of magnesium oxide and magnesium chloride materials. The magnesium chloride will exist in the form of $MgCl_2{\cdot}6H_2O$, thereby exacerbating the absorption of moisture and reoxidation of the product.

During winter production at low temperatures, it is often necessary to increase the loss on ignition to accelerate the slurry reaction. However, when magnesium oxide with a loss on ignition exceeding 10% is mixed with brine, the brine quickly penetrates into the interior of the magnesium oxide. The amount of brine required is much higher than that for magnesium oxide with an appropriate loss on ignition, resulting in a significant amount of free chloride magnesium, which causes the product to absorb moisture and reoxidize.

If the loss on ignition is too low, usually below 4% or less, the internal crystalline structure of this magnesium oxide powder is relatively dense. When mixed with brine, the solution is less likely to penetrate into the structure, resulting in a slower curing reaction rate, a longer setting and hardening time, and incomplete reaction. There will also be a large amount of free magnesium chloride in the product, leading to moisture absorption and reliquefaction of the product.

The standard requires that the fineness of magnesium oxide should be greater than or equal to 200 mesh. If the fineness is not up to standard, it will also affect the chemical reaction products and hardening speed of the product, causing the chemical reaction of the product to be insufficient and resulting in excessive magnesium chloride, which leads to the product absorbing moisture and regenerating brine.

4.10.4.1.2 The moisture absorption and reoxidation caused by the noncompliance of magnesium chloride

Due to the influence of many factors such as the origin, raw materials, production technology, and production process of magnesium chloride, the quality varies greatly. Some products have the phenomenon of excessive KCl and NaCl content. KCl and NaCl are harmful components in industrial magnesium chloride. They have strong hygroscopicity and do not participate in chemical reactions. They exist in the products in a free form, increasing the probability of moisture absorption and re-bleeding of the products.

4.10.4.1.3 Moisture absorption and reoxidation caused by unreasonable production formula

When the concentration of brine is too high (exceeding 30°Bé) or too low (lowering below 23°Bé, and in extreme cases, dropping below 18°Bé), a large amount of MgO and $MgCl_2$ cannot undergo normal hydration reactions. The excess components exist in a free form. Currently, with an increasing number of manufacturers of glass fiber–reinforced plastic boards and increasingly fierce competition, each manufacturer engages in a price war to seize the market. They blindly increase the amount of brine, thereby adding more sawdust and sanding waste, which in turn reduces production costs. However, this method results in the products containing a large amount of free moisture. This moisture dissolves a large amount of magnesium chloride. When the moisture inside the product migrates to the surface and reaches the surface of the product, the moisture evaporates into the air, causing the product surface to condense and form small water droplets or even large water droplets and crystals, resulting in severe moisture absorption and reoxidation.

4.10.4.1.4 Improper maintenance measures

The maintenance of gypsum board products must meet three conditions: temperature, humidity, and time. These three conditions are mutually dependent and cannot be lacking. During the hardening process of gypsum board, there is a certain humidity requirement. If the humidity in the maintenance room is too low and the temperature is too high, the moisture on the surface of the product will be carried away by the large amount of internal reaction heat, causing the surface of the product to fail to harden in time. A large amount of $MgCl_2$ that has not participated in the reaction exists in a free form in the interior and surface of the product, thereby causing moisture absorption and readsorption.

The maintenance time is a necessary condition for the hardening of all magnesium-based products. Without sufficient time, it is impossible to generate sufficient strength. If the magnesium oxide and magnesium chloride in the product have not fully undergone chemical reactions, a large amount of free chloride magnesium will exist, which also poses a potential risk of moisture absorption and readsorption for the product.

4.10.4.2 Silica rock board

1. During the production process, silica rock board cannot be fully penetrated.
   Silica rock board is made by using the principle of negative pressure vacuum to allow inorganic fireproof slurry to penetrate into the gaps between the foam particles of polystyrene foam board. If the gaps are too small, the resistance to slurry penetration is high, and the phenomenon of not being able to penetrate easily occurs. The particle gaps can be controlled by adjusting the air pressure and injection time during the injection molding process of polystyrene foam board, that is, adjusting the "degree of maturity."
   Traditional production equipment has a simple structure, rough workmanship, and poor sealing performance. The conveyor belt is directly connected to the baffle and the vacuum suction box. With the passage of time, the wear is severe, and the pores become larger, resulting in air leakage and insufficient vacuum suction force. Currently, various regions have successively implemented energy-saving standards of 65% or 75%, and the boards have become thicker, making the problems of insufficient suction force and inability to penetrate more prominent. It is recommended to install rubber strips between the conveyor belt and the baffle, the vacuum box, to improve sealing performance, reduce friction-induced belt wear, and extend the service life.
2. Dripping of material after penetration of silica rock panels.
   When the gaps between the particles of polystyrene foam board are too large, the resistance after penetration is small. Due to the effect of

gravity, the slurry flows and drips before it begins to set, causing inconsistent density on the top and bottom of the panel, resulting in a density difference and leading to an uneven surface, which affects usage. Therefore the base plate should "set as much as possible" while ensuring complete penetration.

Dripping of material is also related to the quality of light-burned magnesium oxide. If the powder has poor activity, the slurry will have poor viscosity and a slow curing rate. After the slurry enters the base plate, it drips and flows out under the effect of gravity.

Nowadays, the competition for silica rock panels on the market is fierce. To retain a reasonable profit margin, it can only be achieved by blindly increasing the water volume and reducing the density. The larger the water volume, the lower the viscosity of the slurry, and the more obvious the dripping phenomenon. Therefore by using a curing agent in combination, the reaction rate of the slurry can be adjusted, the initial setting rate can be accelerated, and the bonding strength between the slurry and the foam particles can be increased, thereby eliminating the dripping of the slurry.

3. The strength of the silica rock plate decreases, it becomes powdery and undergoes crystallization.

The board may begin to crumble and lose strength over time after storage and use, primarily due to the degradation of internal inorganic materials. This is often attributed to the use of lightweight calcined magnesia that fails to meet required quality standards. During the curing process, the unique stable "5 · 1 · 7" phase of sulfur oxy-magnesium cement is not formed. Moreover, the modification is not adequate. Over time, exposed to wind and sun, its chemical stability deteriorates, leading to powdering.

Surface powdering and the formation of crystalline substances are primarily caused by improper mixing ratios, a blind reduction in production costs, and the use of an excessively large water-cement ratio. In particular, some manufacturers, attempting to cover up the defect of their equipment's insufficient suction capacity, promote the claim of "buying the equipment comes with the formula." However, the mixing ratio is seriously imbalanced, with excessive moisture content. Lightweight calcined magnesia fails to fully react with brine (or magnesium sulfate solution) to form the crystalline phase and instead directly reacts with water to form magnesium hydroxide. As moisture evaporates, magnesium hydroxide migrates to the surface of the board, resulting in powdering. In severe cases, this affects bonding strength later and leads to board detachment.

The crystallization occurs because there is an excess of magnesium chloride and magnesium sulfate, and the reaction is insufficient. In a low-temperature environment, it becomes supersaturated and forms frosty crystals. Currently, the methods for producing sulfuric magnesium permeable plates mostly involve dissolving magnesium sulfate on-site. However, when magnesium sulfate is dissolved, it absorbs a large amount of heat, resulting in a low solution temperature, poor dissolution effect, uneven dispersion, and low-temperature is not conducive to the reaction and curing of the sulfuric magnesium system, causing the board to have low strength. Additionally, magnesium sulfate will crystallize and precipitate later. It is recommended to prepare the solution in advance, store it for later use, provide time for the magnesium sulfate to fully dissolve and heat up, and if possible, manufacturers can establish a curing room to maintain a certain temperature, allowing for more thorough curing reactions of the system.

4.10.4.3 Hollow sulfuric magnesium plate

4.10.4.3.1 Moisture absorption and reoxidation

Although hollow magnesium-sulfur oxide board is officially named as such, in reality, over 80%–90% of enterprises continue to use magnesium chloride as the binder for its production. Worse still, some opt to purchase unspecified local chemical water instead of brine prepared from halite. These unspecified chemical waters often have a relatively low effective magnesium chloride content and are also rich in certain other impurities. Even when production is strictly conducted according to scientific ratios, some residual magnesium oxide and magnesium chloride inevitably remain. These residual magnesium oxide and magnesium chloride exist in a free state within the board and, under suitable temperature and humidity conditions, absorb moisture from the air and undergo secondary reactions, leading to reoxidation and cracking issues.

4.10.4.3.2 Low strength

Many manufacturers, in an attempt to maximize profits, blindly increase the foaming ratio and the amount of sanding waste. A reasonable amount of sanding waste can, to a certain extent, help adjust the reaction speed and increase the viscosity of the slurry. However, excessive amounts will slow down the reaction speed and reduce strength.

At the same time, it is often believed that within the hollow sulfur-magnesium oxide board system, the cost of foaming is the lowest, since more than 90% of the diluted foaming liquid is water. The most apparent outcome of excessive foaming is that more boards can be produced, achieving the greatest benefit with minimal cost. However, this frequently leads to losses. Excessive foaming introduces a large amount of water, reduces the concentration of the brine, slows the curing speed of the board, and decreases its strength. In hollow sulfur-magnesium oxide boards, the most ideal ratio is for the foaming slurry and polystyrene foam particles to each account for half. This not only preserves the product's insulation and soundproofing performance but also avoids excessive loss of product strength.

4.11 Floor heating module

4.11.1 Advantages of underfloor heating

Firstly, the most prominent advantage of the floor heating system is the rationality and uniformity of its heating gradient. The heating effect provided by the floor heating system makes people feel more comfortable. The uniform heat dissipation across the entire floor of the room significantly reduces the relative temperature difference, and the surface temperature of the floor inside the room is higher than the temperature of the user's respiratory system. This is particularly in line with the theory of traditional Chinese medicine that "heat originates from the head and cold enters from the feet," and "warm the body first by warming the feet"—this is precisely the key to the success of floor heating. Additionally, the floor heating system has a good conditioning and care effect on the reflex zone at the bottom of the feet, thereby having a positive impact on the internal functions of the body.

Secondly, the floor heating system is beneficial for the maintenance and improvement of the indoor environment. For traditional water-based heating radiators, the outlet water temperature is mostly above 70°C, which easily generates dust clusters. This is why the furniture and walls near the radiator tend to become black and dirty, and are difficult to clean. However, the floor heating completely avoids such problems, creating a fresh, clean, healthy, and comfortable environment.

In addition, underfloor heating can effectively save floor space in the living area and improve the rational layout of the home. Careful users can easily notice that traditional radiators are only about 10 cm thick, but they cause several square meters of living space to be wasted or affect the normal use of the room. By using underfloor heating, the impact on the living area and the placement of furniture can be minimized to the greatest extent.

4.11.2 The current status of floor heating modules

4.11.2.1 European market

In the 20th century, Europe had already begun to use modules for floor heating installation (Fig. 4.85). Due to the standardized and tool-based construction concept, Europe has fully upgraded to modular installation, and the heating market started earlier and is more mature.

4.11.2.2 Domestic market

After the floor heating modules were introduced from Europe, many domestic manufacturers began to produce them. However, due to limitations such as technology, production capacity, and production lines, the current situation of the domestic floor heating module market is characterized by significant price differences, few users, a wide variety of types, inconsistent quality, and no unified construction standards. Currently, most of the floor heating modules in China use extruded plastic boards as the base material, and there are also floor heating modules using silicate cement as the bonding material (Fig. 4.86).

Among them, the structure of the extruded board floor heating module is to process the extruded board into module size and process the groove for the water pipe. On the end of the extruded board where the groove is made, a layer of aluminum foil sheet is adhered. This form of extruded

Figure 4.85 Houses in the European market that adopt floor heating modules.

(A) (B)

Figure 4.86 Common floor heating modules in China. (A) Extruded plastic board floor heating module. (B) Silicate cement floor heating module.

Figure 4.87 Magnesium heat-insulating module.

board has good insulation effect and low thermal conductivity, but its biggest drawback is that it is not fire-resistant. The fire resistance grade is B, and the safety is poor.

For the floor heating module with silicate cement as the base material, it uses silicate cement as the cementing material and adds foam particles to reduce density. The obtained floor heating module in this way has a large mass, although it is fire-resistant, the insulation effect is poor and the heat loss is large.

To address the aforementioned issues, MJT has innovatively designed a floor heating module made by adding foam to a magnesium-based cementitious material as the base material. This floor heating module is a prefabricated trench-type one (Fig. 4.87). Magnesium-based cementitious materials are lighter, stronger, more environmentally friendly, and pollution-free compared to silicate cement; they have higher strength, better fire resistance (with a fire rating of A grade), higher safety, and can be reused with graphene heat transfer sheets, resulting in good heat conduction and low heat loss; the use of foaming to reduce density also has the characteristics of heat insulation and sound absorption.

4.11.3 Definition, composition, and manufacturing process of magnesium thermal insulation modules

4.11.3.1 Definition and composition

The overall geothermal heating system mainly consists of four components: the heat source system, the piping system, the control system, and the insulation system. These components form a complete geothermal heating system, each playing an important role in ensuring that the indoor temperature is suitable (Fig. 4.88).

The insulation system is set up to maintain the temperature balance inside the room and prevent heat loss. The floor heating blocks are the main components of the insulation system, used to enhance the overall energy efficiency and insulation effect of the floor heating system.

The magnesium floor heating module is made by using light-burned magnesium oxide, admixtures (magnesium chloride or magnesium sulfate solution), modifiers, and supplementary materials such as fly ash, mixed in a reasonable proportion and sequentially added to the mixer for uniform

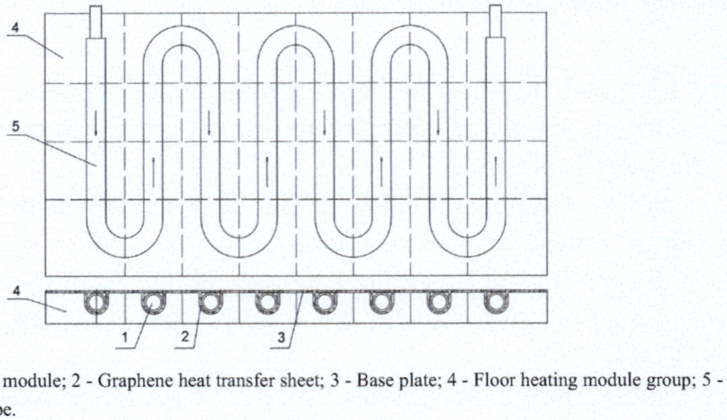

Figure 4.88 Structure of magnesium floor heating module. 1—Floor heating module; 2—Graphene heat transfer sheet; 3—Base plate; 4—Floor heating module group; 5—Floor heating pipe.

stirring. Then, foam is produced using a foaming device and added to the slurry for further uniform stirring. It is poured into the module mold and solidified to form a module structure. After treatment on the surface, graphene heat transfer sheets are embedded on the side of the groove for water pipes, and then a new type of floor heating module (patent name: A floor heating structure; patent number: CN202320820876.5) is obtained.

4.11.3.2 Manufacturing process

Raw materials: Light-burned magnesium oxide, admixture (brine or magnesium sulfate), foaming agent, modifier, fly ash, short fibers, etc.

The design ratio as per requirements is shown in Table 4.15.

Weigh the various materials, mix the calcined magnesium oxide with the admixture in sequence, add the modifier and short fiber, stir evenly, then add the fly ash, and foam it in the foaming machine until it is evenly mixed in the mixer.

1. Pour the mixed slurry into the module mold.
2. Cure and shape, then remove the mold and repair.
3. Embed the graphene heat transfer sheet into the module groove.

4.11.4 Heating system and future development of buildings

As the world becomes increasingly urgent in demanding carbon peak and carbon neutrality, how to save energy and reduce carbon emissions has become a common consensus among governments of various countries and regions. MJT has designed an intelligent fireproof and thermal insulation floor heating system with zero carbon emissions. This system consists of solar power generation and heating insulation panels, energy conversion units, intelligent control units, floor heating module units, and exterior wall fireproof and insulation units. It combines photovoltaic solar panels with exterior wall insulation panels into a single panel. According to the photovoltaic effect principle, it converts solar radiation energy into electrical energy, and the solar water heater converts light energy into heat energy. Through the intelligent control unit, it controls whether the water in the water storage tank is heated (or other household appliances are turned on or off), and then delivers the heated water to the pipe of the floor heating module unit. Through the pipe circulation for heat dissipation, the indoor temperature reaches the set value. The intelligent control unit determines whether to continue heating. In summer, the intelligent control unit can control the start and stop of the air conditioning cooling mode.

Table 4.15 Design ratio of magnesium thermal insulation module (unit: kg).

Light-burned magnesium oxide	Sulfate magnesium solution	Modified			Fly ash	Short fibers	Foaming agent
		GX-15#	GX-13#	GX-16#			
100	120	1	1	1	20	0.5	0.5

Additionally, the floor heating module unit and the exterior wall insulation unit both have fireproof and insulation properties. The entire system adopts a fireproof and energy-saving building system to achieve zero carbon emissions and intelligent control (patent name: A Safe Floor Heating Module; patent number: CN202320820872.7).

4.12 Door core panel

4.12.1 The concept and function of fireproof doors

Fireproof doors are typically doors installed in enclosed areas such as stairways, elevator lobbies, cable shafts, and smoke exhaust ducts (Fig. 4.89). They are essential equipment in fire prevention work, providing fire resistance stability, heat insulation, and the ability to prevent the spread and diffusion of fire, ensuring the evacuation of personnel and facilitating firefighting operations.

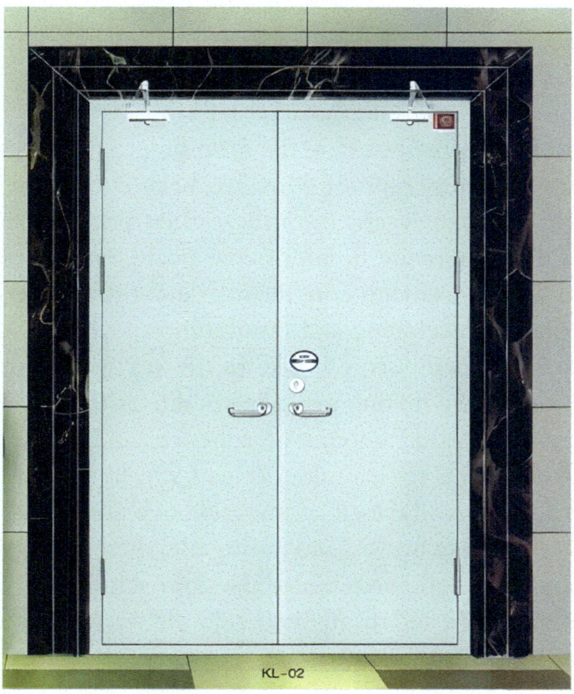

Figure 4.89 Fireproof door.

4.12.1.1 Composition of fireproof doors

Fireproof doors consist of door frames, panels (filled with insulation materials inside), fireproof hardware components, and electromagnetic door catches. The door panels are connected by hinges to form a structure. The door is equipped with door closers and fire locks. For double-paned doors, additional features such as concealed latches (installed on the fixed panel side) and sequential devices are added to prevent the overlapping of the door panels. Fire detectors and fire linkage control systems are also linked to the normally open fireproof doors.

4.12.1.2 Classification of fireproof doors

1. Classification by material.
   Fireproof doors can be classified by material into wooden fireproof doors, steel fireproof doors, steel-wood composite fireproof doors, and fireproof doors made of other materials.
2. Classification by fire resistance performance.
   Fireproof doors are classified based on their fire resistance limit into three types: Class A fireproof door (fire resistance limit ≥1.5 hours), Class B fireproof door (fire resistance limit ≥1 hour), and Class C fireproof door (fire resistance limit ≥0.5 hours).
3. Types of fireproof door core boards.
   The fireproof door core board, as the main fireproof and heat insulation filling material of the fireproof door, has been continuously upgraded and replaced in recent years: from the earliest perlite board, silica gel cotton board to magnesite fireproof core board, and then to gypsum-based core board, cement core board, and sulfur-magnesium core board, constantly developing and innovating.
   a. The expanded perlite core board for fire-resistant doorsets is a type of board made by mixing expanded perlite as the main component with an appropriate amount of adhesive, and is used to fill the interior of the fire-resistant door (Fig. 4.90). In the early days, water glass was commonly used as the adhesive for these perlite fire-resistant door core boards, and during use, there was a phenomenon of alkali return and corrosion of the door skin. Later, a mixture of magnesium oxide and magnesium chloride was used as the adhesive, and expanded perlite was mixed to form the board. However, during the operation, scientific proportioning was not paid attention to, and a loose production process was adopted, adding water when

Figure 4.90 Expanded perlite fireproof door core board.

the mixture was thick and adding powder when it was thin. This often led to a large amount of magnesium chloride remaining, moisture absorption and re-alkalization, and corrosion of the door panel. To fully utilize the fire-resistant and lightweight characteristics of perlite, the water-repellent treated expanded perlite was added as an auxiliary material to the foaming slurry of magnesium-based cementitious materials, achieving the effect of reducing the density of the board and increasing the overall fire resistance limit.

b. The expanded vermiculite core board for fire-resistant doorsets is a type of board made mainly from expanded vermiculite, with adhesives and other modifiers added and pressed into shape, having noncombustible fire-resistant and thermal insulation properties, and used to fill the interior of the fire-resistant door (Fig. 4.91).

c. The foamed concrete core board of fire-resistant doorset (foamed concrete core board for fire-resistant door) is produced by physically or chemically introducing bubbles into a slurry composed of cement, admixtures, additives, and water, through mixing, pouring, molding, and curing. It serves as a sheet material used to fill the interior of the fire-resistant door (Fig. 4.92). However, during use, it has been observed that the foam concrete core board of fire-resistant doorsets often has an excessively high density and is brittle, leading to significant damage during cutting and transportation.

Figure 4.91 Expanded stone fireproof door core board.

Figure 4.92 Foam concrete fireproof door core panel.

d. Magnesite core board of fire-resistant doorset is a type of board used to fill the interior of fire-resistant doors. It is made by mixing light-burned magnesium oxide, magnesium chloride, or magnesium sulfate (or a mixture of magnesium chloride and magnesium sulfate) with water as the cementing material. Through the addition of reinforcing materials, modifiers, and foaming agents, it is processed through foaming, molding, and curing (Fig. 4.93).

e. Graphite fire-resistant door core board (GEP core board of fire-resistant doorset) is made from silicon and calcium-based mineral raw materials, bonding modifiers, and other inorganic binding

Figure 4.93 Magnesite fireproof door core board.

Figure 4.94 Graphite fireproof door core board.

materials. It is combined with expanded graphite polystyrene foam particles (also known as expanded polystyrene particles), supplemented by appropriate foaming, hydrophobic and other additives (without magnesium chloride or chloromagnesium oxide). It is formed through processes such as adding water for mixing and stirring, molding in molds or equipment, curing, and processing, and is a kind of insulation board with good fire resistance properties used in building energy-saving projects (Fig. 4.94).

Graphite homogeneous board, as an emerging type of fire-resistant and insulation building material in recent years, has been widely used in the core of wooden fire-resistant doors. It has initiated the "filling process upgrade 4.0" revolution in the wooden fire-resistant door industry.

4.12.2 Production process of magnesium-based fireproof door core board

When manufacturing magnesium fireproof door core panels, the raw materials are first mixed, then physically foaming agents are added and stirred evenly. Depending on the production form, the process can be either roller pressing or pouring molding. After the core panels are formed and cured, they need to undergo flattening, sanding, cutting, and engraving treatments. The specific production process flow is shown in Fig. 4.95.

4.12.3 Performance of magnesium fireproof door core panels

4.12.3.1 Fire resistance performance

The main performance of fireproof door core panels is their fire resistance. During a fire, fireproof door core panels can withstand high temperatures and prevent the spread of fire, providing valuable time for personnel evacuation. According to different national and regional standards, the fire

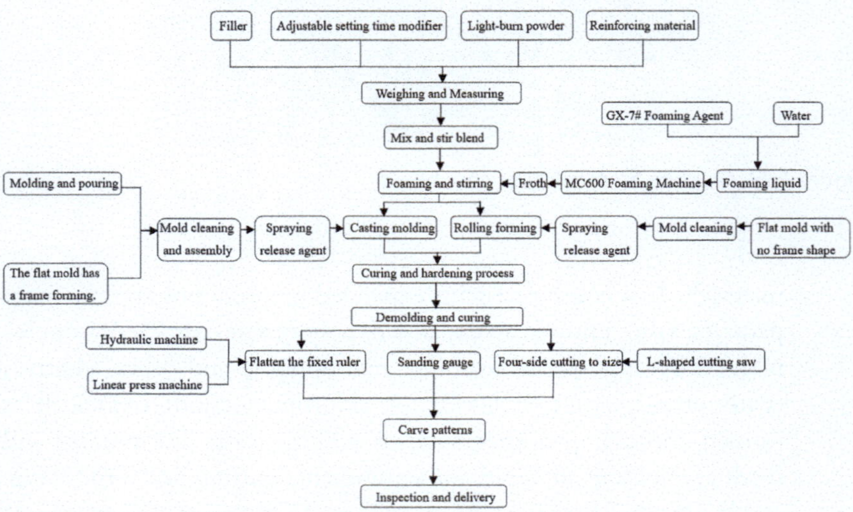

Figure 4.95 Production process flow of magnesium-based fireproof door core board.

resistance time requirements for fireproof door core panels vary within a certain range. For example, the Chinese national standard requires that the fire resistance time of grade A fireproof doors should be no less than 1.5 hours.

4.12.3.2 Heat insulation performance

Fireproof door core panels have excellent heat insulation performance, which can effectively prevent heat transfer and reduce the impact of temperature on the door panel during a fire, thereby protecting the structure and integrity of the door.

4.12.3.3 Strength and stability

Fireproof door core panels have been specially treated and have high strength and stability, capable of withstanding the normal opening and closing of the door and impacts, and are not prone to deformation or damage.

4.12.3.4 Environmental performance

The main material of the fireproof door core board is inorganic minerals, which are harmless to the human body and do not release toxic gases. Moreover, the production process of this fireproof door core board is also environmentally friendly and meets the requirements of modern green buildings.

4.13 Magnesium-steel fire-resistant composite duct panel

The fire suppression and smoke exhaust pipeline project mainly consists of two parts: fire suppression ducts and smoke exhaust ducts. According to the material selection for the high-pressure system, the smoke exhaust system and smoke prevention system will be activated in case of a fire, which could not only promptly remove the smoke and heat in the room, but also supply a large amount of fresh air to the interior (Fig. 4.96). The fire suppression and smoke exhaust pipeline insulation and fire-resistant materials should have excellent fire resistance, corrosion resistance, sealing, durability, etc., and should not produce toxic gases during combustion, and no large amounts of smoke should be generated.

Magnesium steel fire-resistant composite duct panel is officially known as sulfur-magnesium inorganic cement composite color steel surface fire suppression and smoke exhaust duct. It is a specialized panel for fire suppression and smoke exhaust, consisting of a single or double-sided

Figure 4.96 Fire extinguishing and ventilation pipes.

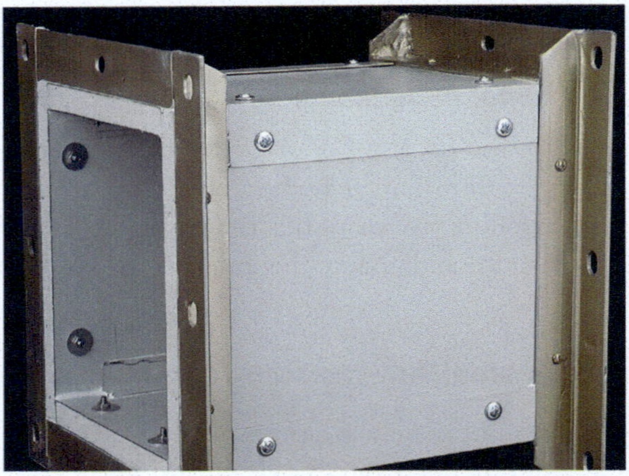

Figure 4.97 Magnesium steel fire-resistant composite duct panel structure.

color steel surface layer and a sulfur-magnesium core layer (Fig. 4.97). This product is an environmentally friendly new type of fire suppression and smoke exhaust solution. Compared to other types of smoke exhaust ducts, it features fire resistance, waterproofing, antibending, heat insulation, sound insulation, nontoxicity, light weight, and good appearance. The construction process is straightforward and production is convenient. Currently, magnesium steel fire-resistant composite duct panels have been successfully applied in major projects such as hospitals, schools, underground parking lots, airports, subways, high-speed railway stations, commercial complexes, and international exhibition centers.

Figure 4.98 Magnesium sulfate steel composite panel.

The core material of the structure, magnesium oxide board, is a ternary system composed of magnesium oxide, magnesium sulfate, and water. Through proper proportioning and modification, with the upper and lower layers of glass fiber mesh fabric as surface-reinforcing materials, the middle layer is sandwiched with composite mineral admixtures, polyethylene particles or perlite, reinforcing fibers, and supplemented with slight foaming, and then formed by mechanical roller pressing.

4.13.1 Product characteristics

The base material used for the magnesium steel fire-resistant composite duct panel is the sulfur-oxygen-magnesium color steel composite board. The sulfur-oxygen-magnesium color steel composite board is made by combining sulfur-oxygen-magnesium sheets with color steel through a fire-resistant bonding process, which well meets the fire resistance and fire endurance requirements of fire prevention and smoke exhaust ducts (Fig. 4.98).

4.13.1.1 Excellent fire resistance performance

The magnesium steel fire-resistant composite duct panel is composed of core material and surface layer. The core material uses sulfur-oxygen-magnesium insulation sheets (noncombustible A1 grade), and the surface layer is generally color steel. The two are bonded together through fire-resistant adhesives, making it have excellent fire resistance. Tests have proved that it can withstand temperatures exceeding 1200°C. The fire endurance limit of 8 mm magnesium steel fire-resistant composite duct panel is greater than or equal to 1 hour, that of 14 mm panel is greater than or equal to 2 hours, and that of 18 mm panel is greater than or equal to 3 hours. The test results of the National Fireproof Building Materials Quality Supervision and Inspection Center show that the magnesium steel fire-resistant composite duct panel meets the fire resistance standards GB 8624–2006, GB/T 20285–2006, and GB/T 5464–2010.

4.13.1.2 Excellent thermal insulation and heat preservation properties

The core material of the magnesium-steel fire-resistant composite duct board uses hydrophobic perlite and fire-resistant foam particles as fillers, achieving a certain level of insulation effect. The fire resistance limit test shows that when the external temperature reaches above 1000°C, the internal temperature of the duct remains below 180°C.

4.13.1.3 Corrosion-resistant, acid-proof, nontoxic

The magnesium-steel fire-resistant composite duct board core material is composed of sulfur-magnesium cementitious material combined with inorganic mineral materials. It is corrosion-resistant, acid-proof, and nontoxic, ensuring that the duct does not deform or deteriorate in a humid basement environment.

4.13.1.4 Superiority

The combination of color steel and magnesium oxide board has solved the current problems of the fire exhaust system. The traditional galvanized iron sheet ducts cannot meet the fire resistance limit of 2 hours for the fire exhaust system, and the galvanized steel sheet ducts have no sound insulation performance, so a sound absorber must be added. In addition, the galvanized iron sheet pipe has a large heat conductivity and no insulation performance, and an insulation layer needs to be wrapped separately. It is not suitable for use in environments with high humidity, as it is prone to corrosion and rusting in such conditions. After the magnesium oxide core material is combined with the galvanized iron sheet, it can perfectly solve the problems of heat insulation, sound absorption, and fire prevention.

4.13.1.5 Sound insulation and sound absorption performance

The core material of the sulfur-magnesium board is composed of a porous crystal structure, which can absorb the noise in the duct and reduce resonance.

4.13.1.6 Flexural strength

The sulfur-magnesium board core material is combined with color steel, possessing both the toughness of color steel and the rigidity of the core material. Its perfect combination demonstrates an extremely strong flexural strength.

4.13.2 Product-related parameters and applications

4.13.2.1 Product parameters of magnesium oxide board core material

Product specifications: 1200 × 2440 (3000) × 8, 10, 14 or 20 mm.

4.13.2.2 Technical parameters of magnesium steel fire-resistant composite duct board

Technical parameters of magnesium steel fire-resistant composite duct board are shown in Table 4.16.

Table 4.16 Technical parameters of magnesium steel fire-resistant composite duct board.

Name	Technical requirements	Product parameters
Density/(kg/m^3)	0.8–1.0	0.85
Flexural strength/MPa	≥ 10	21.1
Anti-halogen leakage property	No water droplets, no condensation	No water droplets, no condensation
Chloride ion content/%	≤ 10	0.2
Softening coefficient/%	Soaked for 7 days, with a softening coefficient of ≥ 85	88
Fire performance	Grade A1	Grade A1

4.13.3 Production and manufacturing process of magnesium sulfate air duct panels

4.13.3.1 Production process flow

The production process flow is shown in Fig. 4.99.

The reference slurry ratio for magnesium sulfide duct sheet is shown in Table 4.17.

4.13.3.2 Maintenance process

During the curing stage of the magnesium sulfate air duct panels, two curing rooms are required, namely the first-level curing room and the second-level curing room.

1. The first-level curing room is equipped with heating and dehumidification devices as well as temperature and humidity meters. The purpose is to allow the panels to be stored on the curing frame for 8–12 hours, ensuring that they reach a certain strength and moisture content (Fig. 4.100) during the curing process. When removing the molds, the bottom surface of the panels does not stick to the molds, the removal process is smooth, and the surface is clean without any residue. The roller shutter door of the curing room is in a closed state. After about 4 hours, the temperature inside the curing room can reach above 50°C. Once this temperature is reached, it should be maintained for 1 hour. During this period, the hardening speed and structural strength of the panels will rapidly increase. The steam inside the curing room is close to saturation, and the humidity meter shows a humidity of nearly 100%. Then, the ceiling dehumidification device of the curing room is started. After approximately 20 minutes, the temperature on the temperature and humidity meters will drop to around 40°C, and the

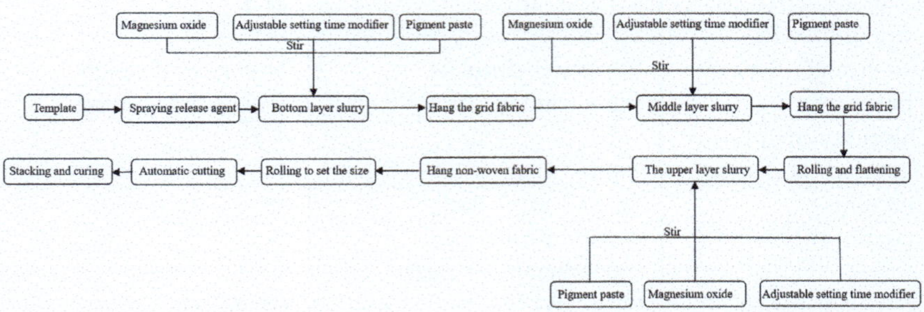

Figure 4.99 Production process flow of magnesium sulfate air duct panels.

Table 4.17 Reference slurry ratio.

Light-burned magnesium oxide/kg	Magnesium sulfate solution/(26°Bé)	Talcum powder/kg	Crushed material or sanding powder/kg	Closed-cell perlite/kg	EPS particles/kg	Modifier/kg	Foaming agent/kg
250	300	100	50–60	12	6	2	0.5

Figure 4.100 The laid-on boards in the curing.

humidity will drop to around 70%. Then, the dehumidification device is turned off, and the curing room remains in a sealed state. After 3 hours, this operation is repeated once. This cycle is repeated to achieve the purpose of using heating measures to obtain better structural strength for the product during the first stage of curing, and to reduce the time for the second stage of curing by eliminating the moisture in the product.

2. The secondary curing room adopts sealed curing measures and is ventilated regularly. After the boards are demolded, they are stacked in groups of five, with flat wooden strips evenly spaced in the middle (Fig. 4.101). The roller shutter door of the curing room is in a closed state. There is a temperature and humidity meter inside, and a vertical exhaust fan is placed on the ground. During the storage of sulfur-magnesium boards, due to the continuous hydration reaction, the structural strength keeps increasing. Although the heat generated by the boards is not as intense as in the early stage, it is still rapidly releasing heat. When the temperature in the curing room rises to 50°C and the humidity reaches 90%, the exhaust fan is turned on to expel the moisture from the curing room. The boards are generally stored in the secondary curing room for 3 days. By the third day, the temperature generated by the heat release from the stack can be reduced to within 40°C. This indicates that the rapid hydration reaction in the early stage has ended, and the structural strength has reached about 80% of the final strength, with the moisture content approaching 13%.

Figure 4.101 Interval between boards and wooden strips.

Figure 4.102 Storage of magnesium oxide board products—corner of the workshop.

4.13.4 Cutting of magnesium sulfate insulation air duct panels

After 3D curing, the magnesium sulfide board can be transferred to the factory for storage and curing in batches. Usually, it can be cut after being stored for 7–10 days at a workshop environment temperature ranging from 10°C to 38°C (Fig. 4.102). The edges of the cut boards are neat and without burrs. By adjusting the spacing between the upper and lower rollers on the production line, boards of different thicknesses can be produced, such as 8 , 10 , and 14 mm; by changing the size of the template, boards of different lengths can be produced, such as 2440 and 3000 mm.

4.13.5 The manufacturing process of magnesium-steel fire-resistant composite ducts

Magnesium-steel fire-resistant composite duct is a type of fire-resistant duct made of steel plates and magnesium oxide sheets. It mainly consists of the following production steps.

4.13.5.1 Material preparation

1. Steel surface treatment.
   The steel surface of galvanized steel plates is treated, including cleaning, rust removal, and painting, to ensure a smooth, flat, and rust-free surface. The steel plates are used to enhance the load-bearing capacity of the duct.
2. Magnesium oxide sheet treatment.
   The sheets are treated, including drying (controlling the moisture content of the sheets), sheet cutting, and dust removal. The surface of the magnesium oxide sheets after cutting must be smooth and must not have obvious scratches or scars. After processing, the magnesium oxide sheets are drilled, bent, and pressed. The magnesium oxide sheets are used to improve the fire resistance and insulation performance.

4.13.5.2 Bonding and assembly

The magnesium oxide sheets and steel plates are bonded with special glue. During bonding, attention should be paid to the applicable temperature of the glue and the bonding pressure. The bonded magnesium steel plates are assembled one by one according to the design plan, connected with angle iron, and fixed with connecting parts such as rivets and bolts. During assembly, attention should be paid to gap treatment and size adjustment.

4.13.5.3 Postprocessing

1. Sealing
   The edges of the duct need to be treated. The edges should be sealed and interfaces reserved, which will facilitate the connection during later installation (Fig. 4.103).
2. Fire protection
   Perform fire protection treatment on the assembled ducts, including applying fireproof coatings and covering with fireproof materials, to ensure their fire resistance performance.

Figure 4.103 Mouth closure.

3. Welding
   Carry out welding treatment on the assembled ducts to ensure their structural stability and good sealing performance.
4. Inspection
   Conduct quality inspections on the fabricated magnesium-steel fire-resistant composite ducts to ensure they meet the design requirements and relevant standards.
   The above is the manufacturing process of magnesium-steel fire-resistant composite ducts. Magnesium-steel fire-resistant composite ducts are stable in performance, safe and reliable ducts, and are widely used in the industrial field. During the manufacturing process, quality control must be strictly ensured to ensure that the produced products can meet the standard requirements.

4.13.6 Common problems and solutions

4.13.6.1 The sheet material lacks strength and has a high brittleness

4.13.6.1.1 Reasons

1. The quality control of raw materials such as magnesium oxide, magnesium sulfate heptahydrate, and glass fiber mesh fabric was not strict. Problems such as low activity content or excessive or insufficient loss during the calcination process occurred with magnesium oxide. The magnesium sulfate was of agricultural grade or made from waste acid, resulting in poor quality. The glass fiber mesh fabric had low weight, thin warp and weft fibers, and large mesh size, with limited surface reinforcement effect.

2. The formulation and ingredients were unreasonable. The sole pursuit of low cost led to the addition of too much crushed materials, highly water-absorbent perlite or glass microspheres, ultra-low weight polystyrene particles, and excessive foaming foam to reduce density and fillers.
3. Appropriate modifiers for sulfuric acid-magnesium oxide modification were not selected, and single chemical raw materials or unscientific additives failed to achieve good modification functions.
4. The curing room during the curing stage did not pay enough attention or did not know how to do it properly.

4.13.6.1.2 Solutions

1. There should be scientific means for raw material testing. Select qualified magnesium oxide and magnesium sulfate. Unlike chloride magnesium sheet materials, magnesium sulfate sheet materials have higher requirements for the quality of magnesium oxide. Good magnesium oxide and magnesium sulfate are the prerequisite conditions for ensuring product quality.
2. The proportioning should be scientifically reasonable. The amounts of each material should be finely adjusted to a stable ratio to avoid the inconsistent consistency of the slurry produced in each batch, which would result in uneven products.
3. The addition of the modifier should have the functions of enhancing the structural strength and toughness of the product; delaying early setting, achieving balanced and uniform hardening, improving the water resistance of the product and preventing shrinkage and cracking, and optimizing the crystal structure. The birth of the modifier is the result of a large number of experiments in the laboratory, data analysis, combined with the production situation of the products on-site, and continuous optimization and improvement. It should be scientific and reasonable.
4. Attention should be paid to the scientific and reasonable layout of the curing rooms. It is recommended to set up two curing rooms, namely the first-level curing room and the second-level curing room. Through the heating and dehumidification methods of the curing rooms, the strength of the board can be significantly improved, and the moisture content of the product can be effectively controlled.

Figure 4.104 Surface scale layer of magnesium sulfate plate.

4.13.6.2 Surface alkali and frost on the board

4.13.6.2.1 Causes

The reaction between magnesium oxide and magnesium sulfate solution is insufficient, resulting in the simultaneous residual and excess of magnesium oxide and magnesium sulfate, which will form a frost layer on the surface of the board (Fig. 4.104). In severe cases, there will also be hair-like phenomena on the board surface.

4.13.6.2.2 Solution

Apart from what was mentioned in the solution to Problem 1, select qualified raw materials, mix ratios, formulas, conduct modification treatment, and maintenance in the curing room; MJT, through long-term on-site guidance to customers, has found that the temperature and concentration of the magnesium sulfate solution, as well as the presence of undissolved magnesium sulfate crystal particles remaining in the water solution during use, all can cause incomplete curing reaction of the magnesium oxide-magnesium sulfate slurry, the precipitation of magnesium sulfate crystals, and the occurrence of efflorescence and frosting. Now, based on the successful cases after the improvement, this will be explained.

To prepare the magnesium sulfate solution, three tanks are needed. The first tank is used for mixing water and dissolving the magnesium sulfate. The concentration of the magnesium sulfate solution in this tank should be higher than the required production concentration. The freshly dissolved solution has a temperature of 1°C, close to the freezing point. The magnesium sulfate solution should be settled and stored for later use; this tank can serve as a sedimentation tank. The second tank pumps the supernatant from the precipitated magnesium sulfate solution into it, then adds water and mixes thoroughly to adjust to the required production concentration. The third tank can be replaced with a semitransparent plastic tank. The solution prepared in the second tank is pumped into the plastic tank, which should be located in the workshop and preferably close to the production line. Then, compressed air is introduced at the bottom of the plastic tank, and a constant-temperature airflow is circulated and bubbled in the solution, transitioning the solution from a cold state to a state close to the workshop temperature. This not only warms the magnesium sulfate solution but also makes the mixed solution more uniform. Through these three treatment steps, magnesium oxide can better combine with the magnesium sulfate solution, avoiding issues of efflorescence and frosting in the magnesium oxide-magnesium sulfate products during later stages.

4.14 Magnesium sulfate fireproof board

The development of innovative building materials plays a vital role in the building materials industry, representing an emerging sector. Sulfur-magnesium fireproof panels are environmentally friendly, emitting no formaldehyde. They are lightweight yet exhibit high strength, possess Class A fire resistance, and demonstrate resistance to corrosion and moisture. With excellent processing properties and a long service life, they meet the dual requirements of architectural decoration and fire safety. Consequently, their application is becoming increasingly widespread in the construction and decoration sectors (Fig. 4.105).

Sulfur-magnesium fireproof board is a ternary system composed of magnesium oxide, magnesium sulfate, and water. Through proper proportioning and modification, the flexibility is adjusted by using fiber threads. The board surface is reinforced with fiber mesh fabric laid down, and composite plant fibers and mineral admixtures are added internally. The board is formed by mechanical rolling. Its unique physical and chemical properties enable the fireproof board to have fire resistance, water resistance, light weight, high strength, bending resistance and load-bearing capacity, and high strength. On this basis, it can be cut with grooves, combined in a card-like manner, have holes drilled and sawn, and be nailed or hammered freely. The processing is fast and it fully meets the functional requirements of deep processing and various industrial products.

Figure 4.105 Tray made by splicing sulfur-magnesium fireproof boards.

4.14.1 Production and manufacturing process of magnesium sulfate fireproof board

4.14.1.1 Raw materials

The raw materials include calcined magnesium oxide, magnesium sulfate, talc powder, heavy calcium powder, ground material or sandblasted powder, sawdust, fiber threads, modifiers, foaming agents, etc.

4.14.1.2 Reference for ingredient composition

At a lower production environment temperature (8°C–15°C), the reference ingredient ratio is shown in Table 4.18.

4.14.1.3 Basic forming process

The sulfur-magnesium fireproof board is produced using a flat mold assembly line equipment. The equipment typically includes three sets of mixing systems. The first set is used to mix and prepare the surface layer slurry. Depending on board requirements, pigments can be added to the surface layer slurry to produce decorative boards in colors such as iron red, yellow, and dark gray. The second set is for main material preparation, serving as the core of the mixing process. The last set is used for applying slurry to the back of the board and is generally not used in standard production. It is primarily employed for sanding fireproof boards to create smooth surfaces and for producing double-sided coated decorative boards.

Table 4.18 Ingredient ratio of slurry (unit: kg).

Light calcined magnesia	Magnesium sulfate solution 28°Bé	Talcum powder (required for the sole layer alone)	Crushed material or sandblasting powder	Heavy calcium powder	Saw powder	Cellosilk	Modifier		Foaming agent
							A: Magnesium sulfate modifier	B: Hemostatic and solidifying powder	
300	390	100	90–100	120	90–100	0.5	2.4	6~9	0.5

1. Forming process.
   The thickness of the board can be adjusted by changing the distance between the upper and lower rollers on the production line. For example, 5 , 8 , 10 , 12 mm, etc. can be produced. By changing the size of the template, boards of different specifications can be produced, such as 1220 × 2440 , 1220 × 3000 mm, etc.
2. Demolding and curing process, cutting process of the sulfur-magnesium fireproof board, see Section 4.13 of the magnesium steel fire-resistant composite duct board described above.

4.14.2 Application of magnesium sulfate fireproof board

Based on the parameters established during the advanced processing stage, following extensive physical and chemical performance testing, as well as a series of process coating experiments, the substrate can be utilized to produce the following products.

4.14.2.1 Magnesium fireproof decorative board

By drying the magnesium oxide board and performing double-sided sanding treatment, it is possible to hot-press and composite with melamine resin to create thousands of different decorative boards (Fig. 4.106).

Trisodium cyanamide veneered magnesium board is a type of magnesium decorative panel formed by immersing paper of different colors or textures in trisodium cyanamide resin adhesive, drying it to a certain degree

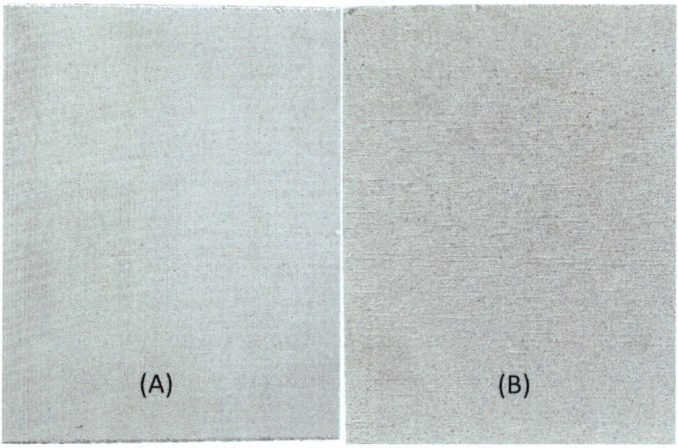

Figure 4.106 Magnesium oxide board. (A) Bright surface. (B) Sanded surface.

Figure 4.107 Magnesium decorative panel.

of curing, laying it on the sulfurized magnesium sand-polished board, and then subjecting it to heat pressing (Fig. 4.107). It has bright colors, high hardness, good wear resistance and heat resistance, and at the same time possesses excellent functions that natural wood cannot have. It is often used in large shopping malls, hotels, villas and home decoration and renovation.

4.14.2.2 Home decoration field (cabinets, wardrobes)

The requirements for the materials in the home decoration field are to be environmentally friendly and have zero formaldehyde emission. This material is made by bonding a double-sided magnesium fireproof board with a wooden ecological board as the base, and then hot-pressing a fireproof skin on both sides of the double-sided fireproof board (Fig. 4.108). It has the advantages of fire resistance, antipermeation, easy cleaning, realistic effect, smooth surface, environmental friendliness, etc.

In the assembly and manufacturing process, the double-sided magnesium fireproof board with a wooden ecological board is bonded by the method of hot pressing glue, using sulfur-magnesium cement as the bonding material, through modification treatment, and then adding water-soluble resin for stirring and compounding. It is combined into a fireproof composite board through the method of high-pressure heat fixation. Finally, the fireproof composite board is hot-pressed and bonded with a fireproof skin on both sides.

Figure 4.108 Interior decoration wardrobe panels.

Fireproof leather is a surface decoration material made by pressing together phenolic resin glue, imported kraft paper, and decorative color paper under high-temperature and high pressure. Its surface features various textures and colors, capable of imitating wood grain or stone patterns. It comes in a wide variety of styles and is currently one of the most popular surface decoration materials.

4.14.2.3 Artificial wood flooring

Magnesium-based boards can be processed to make artificial wood flooring. It consists of magnesium-based boards, a decorative layer, a protective layer, and a connection system (Fig. 4.109). The magnesium-based boards are made of sandblasted boards. The decorative layer is the outer surface of the artificial wood flooring, printed with wood grain or stone patterns, and then combined with the magnesium sandblasted boards through heat pressing technology. The protective layer is a transparent plastic coating that protects the floor surface from scratches or wear. The connection system refers to making grooves on the side of the magnesium sandblasted board and then using a snap-fit combination for two boards (Fig. 4.110).

Figure 4.109 Magnesium imitation wood flooring.

Figure 4.110 Magnesium-based sandblasted board snap-fit assembly.

The advantages of imitation wood flooring lie in that it has a very similar appearance to traditional wood flooring, but it is more durable, easy to maintain, and less prone to moisture-induced deformation.

4.14.2.4 Magnesium sulfate container floor

The magnesium sulfate container floor is a new type of fire-resistant, waterproof, high strength, pollution-free floor material specifically used for movable houses, rental apartments, and temporary sheds. The product is made of magnesium oxide, high-quality magnesium sulfate, neutral glass

Figure 4.111 Magnesium sulfate glazed container floor.

fiber mesh fabric, plant fibers, noncombustible lightweight closed-cell perlite, mineral powder, board crushing materials, modifiers, with slight foaming, produced through scientific proportioning and made according to scientific and mature maintenance processes.

During the production process of magnesium oxide container floor, the surface slurry is treated through modification, and water-soluble resins, which are used to enhance brightness and form a film, are deliberately added. As a result, the surface of the board becomes oilier and shinier. For decorative panels that incorporate color slurry internally, they can display long-lasting bright colors without fading (Fig. 4.111).

For decorative panels that incorporate color slurry internally, they can be used directly without painting, which is simple to install and has good processing performance. The magnesium oxide floor is A-level fire-resistant, water-resistant, moisture-resistant, has good surface hardness, high wear resistance, and strong load-bearing capacity (when placed suspended, 1.5 m below the center has no support, and it will not break when bearing a load of 150 kg). It does not breed bacteria and is environmentally friendly and nontoxic. Decorative panels with color slurry incorporated internally can be used directly without painting, and the construction is simple and the processing performance is good.

In application fields, magnesium-based panels can be used for hot-pressed melamine, pressure-laid fireproof skins, surface slurry incorporated internally for decoration, and can also be produced through polyethylene

Figure 4.112 Magnesium sulfate floor strip (width 30 cm).

terephthalate (PET) composite, PVC coating process, to manufacture fireproof furniture, sound-absorbing panels, MGO decorative floors, prefabricated building materials, electrical equipment heat layer boards, etc. (Fig. 4.112).

4.14.3 Performance advantage

4.14.3.1 Fireproof, moisture-proof, and freeze-thaw resistant properties

The sulfur-magnesium fireproof board has excellent fire resistance. It is a noncombustible material. The duration of continuous flame combustion is zero. It does not burn at 800°C and has no flame at 1200°C, reaching the fireproof and noncombustible level A1. The partition system composed of the sulfur-magnesium fireproof board and excellent keels has a fire resistance limit of more than 3 hours.

The sulfur-magnesium cement 5 · 1 · 7 dissolves in water at a rate of 0.034 g/100 g water (gypsum is 0.205; aluminosilicate cement is 0.029; silicate cement CSH (C/S = 1.5) is 0.084). Its water resistance is far superior to gypsum and is equivalent to cement products. In dry and cold as well as humid weather, the performance of the sulfur-magnesium fireproof board remains stable without being affected by condensed water droplets and humid air. Even if it is soaked in water for several days and then naturally air-dried, it will not deform or become soft, and can be used normally. It has been tested by the National Building Materials Testing Center and shows no water droplets or moisture reentrainment.

The sulfur-magnesium fireproof board also performs exceptionally well in freeze-thaw resistance. The retention rate of strength after 100 freeze-thaw cycles is greater than 70%, and there are no cracks, powdering, or peeling on the surface.

4.14.3.2 High strength

Although the sulfur-magnesium fireproof board is lightweight, it has a tight structure and good stability, without deformation. It has the toughness similar to that of wood and performs exceptionally well in terms of impact resistance, compressive strength, tensile strength, and fracture resistance. Its bending strength is greater than 8 MPa, compressive strength can reach 90 MPa, impact resistance is greater than 34 MPa, and the flexural strength is up to 322 kgf/cm^2 (vertical) and 216 kgf/cm^2 (horizontal). The dry shrinkage rate is as low as 0.3%, and the wet expansion rate is 0.6%.

4.14.3.3 Excellent building thermal performance

The sulfur-magnesium fireproof board has excellent heat storage capacity. The heat storage coefficient S = 2.20 $W/(m^2{\cdot}K)$ (metal iron is 112.1 $W/(m^2{\cdot}K)$, wood is 4.9 $W/(m^2{\cdot}K)$). It is an energy-saving material for regulating room temperature. The thermal coefficient of the sulfur-magnesium fireproof board is 0.15–0.29 $W/(m{\cdot}K)$ (concrete is 1.28 $W/(m{\cdot}K)$; building bricks are 0.69 $W/(m{\cdot}K)$; glass is 1.09 $W/(m{\cdot}K)$), magnesium products are materials with low heat conduction, and have excellent insulation properties; it is superior to the thermal capacity performance of cement and gypsum products, with a thermal capacity value of 1000 $J/(kg{\cdot}K)$; therefore its insulation performance is far superior to cement and gypsum products.

4.14.3.4 Environmental and health benefits

Magnesium sulfate fireproof board contains no asbestos, formaldehyde, benzene, or radioactive elements. When exposed to fire, it produces no smoke, is nontoxic, and has no odor. The materials used in its production are natural mineral powder and plant fibers. At the same time, it can consume industrial waste such as waste residue, fly ash, stone powder, slag, furnace slag, industrial garbage, and agricultural and forestry processing by-products like wood chips and plant straw. It is an excellent material for the comprehensive utilization of waste. The production process is naturally cured, consumes little energy, produces no pollutants, is energy-saving, and environmentally friendly. When used, the board surface will not lose powder. Its unique natural fine pore structure can regulate indoor temperature, making home and office environments more comfortable.

4.14.3.5 Pest and mold resistance

Sulfur-magnesium fireproof board has the functions of preventing mold, bacteria, insects, and termites. It complies with the antimold standards of European and American countries.

4.14.3.6 Cost-effective

Sulfur-magnesium fireproof board has stable and reliable quality, high cost-effectiveness, light weight and high strength, excellent processing performance, superior toughness, and is not prone to breakage. It is also easy to install.

4.15 High-pressure extruded thermosetting eco-board

The problem of straw burning in rural areas of our country has existed for a long time. Straw burning has caused serious disturbances to people's lives and the normal operation of the economy, and has become a major chronic problem in society. In response to the frequent straw burning incidents in various regions, the government has introduced antiburning measures, but the results have been rather limited.

To reduce environmental pollution and protect the ecological environment, the utilization of agricultural residues is particularly important. Magnesium-based plant fiber ecological panels are a kind of green and environmentally friendly new building decoration panels. Their structural composition uses modified magnesium-sulfate cement as the binding material, and a large amount of composite crushed agricultural residues, straw, sawdust, etc. are added, and through special forming equipment, they are heated, pressurized, and extruded (Fig. 4.113). Through this technological process, the ecological panels have high density, good structural strength, excellent dimensional stability, pest resistance, zero formaldehyde release, good processing performance, and the waste can be reused, protecting the ecological environment. This panel will be a good project that the country strongly advocates for the benefit of the country and the people.

Figure 4.113 Magnesium-based plant fiber eco-board.

The high-pressure extrusion thermosetting method refers to the process of making the components in the billet closely contact each other and promoting the solidification and polymerization of the magnesium-based cementing material under the condition of pressure and heating, so as to achieve the required properties of the material. The elements of hot pressing are the hot pressing temperature, the unit pressure, and the pressing time. The cementing material of magnesium oxide is fully premixed with the crushed materials of the artificial board (wood fibers or waste materials such as straw, rice straw, etc.) and then added with a quantitative amount of magnesium sulfate solution and MJT series modifiers. After stirring and mixing, it is further mixed in the kneader to form plastic material (plastic index $10 \leq Ip \leq 17$), which is evenly fed into the extruder through the feeding system and compressed into the desired thickness of the board under the strong compression of the 2000 t press at a temperature of 100°C. For example, when the plastic material is spread at a thickness of 5 cm, it becomes a 1 cm thin board after extrusion. Through continuous stacking and uniform pressing, ecological boards can be batch-produced. If decorative embossed boards are arranged in the extrusion device, decorative boards with a sense of three-dimensionality can be obtained.

4.15.1 Production process of magnesium-based ecological panels

The main components of the magnesium-based ecological board are over 90% plant fibers such as straw, rice straw, and sawdust. These plant fibers need to be dried after being crushed to ensure a moisture content of less than 8%. The sulfur-magnesium cementitious material is mixed with the crushed materials (such as wood fibers or waste like straw and rice straw) to form a semidry state. Then, it is compressed into a high-density board by a heat-pressing composite device, with a density ranging from 1300 to 1400 kg/m^3.

1. Mix the cementitious material magnesium oxide with the crushed materials of the artificial board (wood fibers or waste such as straw and rice straw) in a dry powder mixing machine for thorough premixing, ensuring uniform mixing.
2. Prepare a magnesium sulfate solution, and gradually add the measured GX series modifiers to the magnesium sulfate solution, stirring evenly.
3. The mixed magnesium oxide and plant fiber composite material is conveyed through the conveyor belt into the kneader, the equipment is started to rotate, and then magnesium sulfate mother liquor is added for thorough mixing and kneading to form plastic material.
4. The mixed plastic material is uniformly transported by the feeding system to the flat plate assembly line for even distribution and discharge. The thickness of the discharged plates is controlled by the combing circular rollers, usually around 5 cm.
5. The plastic material after spreading and combing is conveyed by the flat mold assembly line to the hot pressing machine.
6. In the high-pressure extrusion hot curing process, the plastic material entering the frame is pressed into a dense plate layer by layer and heated, usually the plastic material with a thickness of about 5 cm can be pressed to a thickness of 1 cm (Fig. 4.114).
7. The hot-pressing determination stage takes about 15–20 minutes. Then, the conveying equipment transports the plates to the storage area for cooling and dehumidification.
8. After drying, the plates are stacked and stored. After 2–3 days, they can be cut and packaged, as shown in Fig. 4.115.

4.15.2 Key elements of high-pressure extrusion thermosetting process

1. Generally, the higher the hot-pressing temperature, the faster the curing and consolidation speed of the magnesium-based cementitious material, and the corresponding shortening of the pressing time, resulting in higher production efficiency and further reduction of production costs. The analysis results of the DSC/TG curve for the magnesium oxide

Figure 4.114 A 5-cm-thick sheet of material.

Figure 4.115 Packaged laminated 1-cm-thick sheet.

4.1.7 phase show that at 131°C, the 4.1.7 phase will lose four crystalline water and form a weaker 4.1.3 phase. When using magnesium oxide as the cementitious material for magnesium ecological boards, to balance strength, fire resistance performance, and production efficiency, the heating temperature should be controlled at 90°C–100°C.

2. The unit pressure is closely related to the density of the board. The greater the unit pressure, the greater the density of the board, and the shorter the time to reach the specified thickness. Generally, the unit pressure can be 15–22 kg/cm^2. For example, when producing boards with a size of 1300 × 3000 mm, a press machine of no less than 1000 t should be selected. Producers can determine the production based on the pressing ratio and board density according to their production requirements.
3. The heat pressing time has a significant impact on the quality of the sheet (such as sheet thickness, density, surface quality, and consolidation strength). The appropriate pressing time corresponds to the appropriate degree of solidification of the magnesium oxide and the proper evaporation of moisture. When the sheet is subjected to high-temperature pressing, its moisture content is generally within (10 ± 2) %. Moreover, when the sheet leaves the press machine, it should have sufficient strength, and the sheet body should not be delaminated or cracked. Insufficient pressing time will prevent the magnesium-based cementitious material from fully solidifying, thereby reducing the strength of the sheet. If the moisture content of the sheet is too high, during the depressurization process, stratification and explosion phenomena may occur. And during the atmospheric equilibrium process of the produced sheet, it is prone to warping deformation due to the large amount of water leakage.

4.15.3 Characteristics and advantages of magnesium-based ecological panels

4.15.3.1 Low cost, turning waste into treasure

In the raw material configuration of the hot-pressed magnesium eco-board, over 90% are plant fibers such as straw, rice straw, and sawdust. Comprehensive utilization of industrial and agricultural waste makes it a green and environmentally friendly product. The raw materials are sourced locally and can be used after rough processing. While fully utilizing the waste, the cost of the board is significantly reduced.

4.15.3.2 Fast forming speed, short postcuring period

During the forming process of this board, the heat pressing method is adopted, which significantly accelerates the forming speed of the hardened body. After being loaded onto the upper heat pressing device and then placed on the lower frame, it becomes an integrated board that can be lifted, taking only 15–20 minutes. This is completely different from the traditional magnesium-based boards that require about 12 hours of curing time to remove the mold. This effectively improves production efficiency. Moreover, the boards produced under this process have a low moisture content, typically around 13%. The surface of the board emits heat, and when in contact with the air, it is more conducive to eliminating the excess moisture inside. After 2–3 days of internal curing and storage in the factory, it can be cut and packaged, and then ready for shipment.

4.15.3.3 Flat and dense

The board formed under strong pressure is densely compressed and mixed together. The surface flatness is high, the bending resistance is good, and the density is high, truly achieving the visual effect of imitation wood.

The high-pressure extrusion thermosetting process for producing magnesium ecological boards is a scientific, reasonable, and efficient modern technology. By using scientific means to recycle and utilize waste materials and turn them into resources, it not only creates economic value but also reduces environmental protection burdens. It benefits the country and the people and is worthy of vigorous promotion and application.

4.16 Tunnel fire protection panel

Tunnels and underground transportation facilities are important means of transportation. Tunnel fires are one of the main sources of disasters that threaten human lives. Therefore to effectively protect tunnels from fires, higher requirements must be set for the fire protection design of tunnels. Comprehensive safety measures must be considered, and all measures must be organically complementary and coordinated.

Tunnel fire protection panels (blast panels) usually refer to new and environmentally friendly panels used for fire protection of tunnel concrete structures (Fig. 4.116). The main purpose is to protect the tunnel's concrete structure during a fire, thereby preventing the concrete from cracking and the tunnel structure from being damaged in the event of a tunnel fire. In China, various types of top fire protection panels are used, with the main ones being glass fiber board and calcium silicate board. This book only introduces the tunnel fire protection panels made of glass fiber board. Among them, glass fiber board is also known as magnesium board, which is mainly composed of magnesium oxide and additives (such as brine or

Figure 4.116 Tunnel with fire protection panels installed.

magnesium sulfate solution, etc.), along with quartz sand, plant fibers, and a certain proportion of GX series modifiers, mixed into a slurry, and then poured with the slurry and laid with grid glass fiber cloth (and nonwoven fabric, etc.) while being poured, after cutting, curing, and solidification to obtain.

Its features include high strength, anticorrosion, weather resistance, environmental friendliness, long service life, no pollution, no termite infestation, and fire resistance level of A1 grade, etc. It can meet the needs of different types of buildings and is implemented according to the standard "Tunnel Fire Protection Board" GB 28376–2012.

4.16.1 Classification

The fire protection plates for tunnels are fixedly installed on the concrete surfaces of highways and urban traffic tunnels, and they can enhance the fire resistance limit of the tunnel structure.

4.16.1.1 According to the material type

1. A single tunnel fire protection panel, which is composed of a single homogeneous material (Fig. 4.117).
2. A composite tunnel fire protection panel, which is made by combining two or more materials (including decorative panels) (Fig. 4.118).

Figure 4.117 Single tunnel fire protection panel.

Figure 4.118 Composite tunnel fire protection panel.

4.16.1.2 According to the protected area

1. Tunnel top fire protection panel.
2. Tunnel side wall fire protection panel.

4.16.1.3 According to the temperature rise curve of the fire resistance test

1. Class BZ: Tunnel fire protection panels that undergo heating and measurement following the standard temperature rise curve specified in GB/T 9978.1 (applicable to Class IV tunnels with fire resistance of 2 hours).

2. Class HC: Tunnel fire protection panels that undergo heating and measurement following the HC temperature rise curve specified in GA/T 714–2007 (applicable to Class III tunnels with fire resistance of 2 hours).
3. Class RABT: Tunnel fire protection panels that undergo heating and measurement following the RABT temperature rise curve specified in GA/T 714–2007 (applicable to Class I and Class II tunnels with fire resistance of 3.83 hours and 3.3 hours, respectively).

4.16.2 Development prospects

The tunnel fire protection board technology has been continuously evolving to enhance the safety and other functions of the tunnels. This section introduces several of the latest products.

The tunnel fire protection decorative board adds decorative functions to the traditional tunnel fire protection board. The rigid base material and the flexible finish are perfectly combined, offering better mechanical and structural properties. Commonly, it includes inorganic precoated boards and steel-magnesium boards.

1. Inorganic precoated board, also known as "tunnel decorative board," is made from magnesium-based cementitious material boards, processed through surface treatment, fluorocarbon coating, etc., into a decorative board that can be directly installed without surface treatment. The product has a service life of over 25 years and is suitable for tunnel enclosure and other decorations (Fig. 4.119).

Figure 4.119 Inorganic precoated board.

Figure 4.120 Steel magnesium plate.

2. Steel-magnesium board is made from magnesium-based cementitious material boards, combining high-weather-resistant fluorocarbon-coated or inorganic magnetic-coated high-performance aluminum plates with it, and produced through processes such as high-temperature pressing and static pressure curing (Fig. 4.120). The product has stable chemical properties and a service life of over 20 years. It is mainly used in urban rail transit platforms and urban tunnels.

4.16.3 Common problems and solutions

4.16.3.1 Warping and deformation

The common fireproof boards have large widths and small thicknesses, resulting in significant absolute deformation during construction. There are cracks at the joints of the boards and warping and concave bulging phenomena on the large surfaces.

Solution: After the fireproof boards are formed, they must be placed in an environment with certain temperature, humidity, and sufficient curing time. Based on the actual production environment temperature and under the premise of ensuring the quality of raw materials is qualified, an appropriate amount of GX series modifiers should be added to improve the performance.

4.16.3.2 Brittle failure

Due to poor production quality, some manufacturers produce fireproof boards that have excellent toughness when they are newly manufactured and can roll the entire thin board into a tube. However, as the storage time increases, the boards gradually become brittle and prone to fracture.

During the handling process, certain theoretical knowledge should be mastered and analyzed from the following aspects.

1. Check whether the raw materials used meet the requirements: If the activity of magnesium oxide is low, it will affect the strength of the board; or if the content of calcium oxide is excessive, it will cause the board to be brittle. If the added glass fiber cloth is of high alkalinity, the initial strength will be good, but the strength will decline significantly in the later stage. Whether the filler material sawdust is too little, and whether the added quartz sand and other materials are excessive, these are all issues that need to be considered.
2. Whether the ratio of magnesium oxide and the admixture is scientific: If the molar ratio is too large, there will be too much magnesium oxide, resulting in good hardness of the board, but high brittleness.
3. Whether a matching modifier is added: Adding GX-12# modifier to the slurry can increase toughness, which is also a very important step.
4. Whether the curing process can meet the requirements: The curing process can be referred to in the section on curing process in this book.

4.17 Built-in insulation system in the main module

4.17.1 Definition

Place the magnesium permeable (or homogeneous) insulation board inside the outer formwork. Use the cast-in-place concrete exterior wall as the base layer. The inner surface of the magnesium permeable (or homogeneous) insulation board is combined with the concrete after it is formed. Then, on the outer side of the magnesium permeable (or homogeneous) insulation board, apply the insulation mortar leveling layer, the facing mortar composite glass fiber mesh anticrack layer, and the decorative layer to form an exterior wall external insulation system. This system is referred to as the magnesium large mold built-in insulation system.

4.17.2 Magnesium-based large mold built-in insulation system

The performance requirements for the built-in insulation system of the magnesium large mold are shown in Table 4.19.

Table 4.19 Performance requirements for built-in insulation system of large mold.

Project		Organization	Performance indicators		Test method
			Apply decorative coating		
Weather resistance	Appearance		There are no visible cracks, no powdering, hollowing out or peeling phenomena		JGJ 144
	The tensile bond strength of the facing material to the insulation layer	MPa	≥0.15	The damaged area should be located within the insulation layer.	
Impact resistance	For layers 2 and above		3 J grade		
	For the first floor		10 J grade		
Freeze-thaw resistance performance			After 30 freeze-thaw cycles in cold regions, there is no hollowing, detachment, or water seepage cracks in the system; after 50 freeze-thaw cycles in extremely cold regions, there is no hollowing, detachment, or water seepage cracks in the system.		
		MPa	The tensile bond strength between the protective layer and the insulation layer is not less than 0.1 MPa, and the damaged part should be located in the insulation layer.		
Heat resistance			The thermal resistance of the composite wall meets the design requirements.		
Water impermeability of the facing layer			2-h water impermeability		
Water vapor permeability resistance of the protective layer			Complies with the design requirements		
Absorption capacity			After being immersed in water for 1 h, the water absorption of both the system with only the facing layer and the system with all protective layers shall not exceed or equal to 500 g/m^2		

Figure 4.121 Homogeneous type large-scale built-in insulation board.

Table 4.20 Allowable deviation of specification size of insulation board (unit: mm).

Length	Thickness	Height	Flatness	Diagonal length
± 1.0	± 1.0	± 1.0	± 1.0	± 1.0

4.17.3 Types of insulation board materials

The magnesium-type large mold is equipped with an insulation system. The insulation panels used are molded polystyrene boards (EPS), extruded polystyrene boards (XPS), modified polyurethane rigid foam boards (PIR), graphite polystyrene boards, thermosetting composite polystyrene insulation boards of permeable type and pressed type (Fig. 4.121), etc., as the construction layers of the insulation materials.

4.17.4 Insulation board parameter requirements

1. The compressive strength of the insulation board should be no less than 0.20 MPa, and the tensile strength perpendicular to the board surface should be no less than 0.15 MPa. The combustion performance of the insulation board should comply with the requirements of "Classification of Combustion Performance of Building Materials and Products" GB 8624–2012, which should not be lower than B1 grade. The density, thermal conductivity, water absorption rate, and dimensional stability should meet the requirements of relevant standards.
2. The allowable deviation of the specification size of the insulation board should comply with the provisions in Table 4.20.

3. The specifications and dimensions of the dovetail grooves and tongue-and-groove joints of the insulation board and fire isolation belt are as stipulated in Table 4.21. The schematic diagram of the insulation board is shown in Fig. 4.122.
   When using magnesium permeable boards or homogeneous boards as the internal insulation core material of the large mold, the boards are usually processed into standard plates of 1200 × 600 mm. After curing and shaping, they are then grooved using a double-edge six-sided slotting machine (Fig. 4.123).
4. The fire isolation belt should be made of inorganic fire-resistant materials, and the performance indicators are specified in Table 4.22.
5. The performance of the anchor bolts should comply with the provisions of the current industry standard "Anchor Bolts for Exterior Wall Insulation" JG/T 366. The standard value of the tensile bearing capacity of the anchor bolts should be greater than or equal to 0.60 kN, and the standard value of the tensile pull-out force of the anchor disc should be greater than or equal to 0.50 kN. The standard value of the tensile pull-out force of the special connection bridge anchor disc for positioning should be greater than or equal to 0.50 kN. The area of the anchor disc should not be less than 0.0025 m^2, and other performance should comply with the requirements of relevant standards. The special connection bridge for positioning is shown in Fig. 4.124.

4.17.5 Important notes

1. The thickness of the external facing mortar of the magnesium-based large mold's built-in insulation system should not exceed 15 mm.
2. When the built-in insulation system of the large mold uses insulation materials with a combustion performance grade of B, the thickness of the first-level protective layer should not be less than 15 mm, the thickness of other protective layers should not be less than 5 mm and should not exceed 6 mm, and horizontal fire isolation belts should be set at each layer within the built-in insulation system of the large mold. The design and construction of the fire isolation belts should comply with the current national standards "Code for Fire Protection Design of Buildings" GB 50016 and "Technical Code for Fireproof Isolation Belts of Building Exterior Insulation" JGJ 289.
3. The actual thickness of the insulation board is the sum of the design thickness and the depth of the single side dovetail groove.

Table 4.21 Specifications and dimensions of the dovetail grooves and tongue-and-groove joints of the insulation board and fire isolation belt (unit: mm).

Chamfered groove			Buttress groove			
Width	Depth	Spacing	Upper opening width	Lower opening width	Depth	Spacing
15	15	20	10	15	10	≤50

Note: The flange openings should be double flange openings; the two dovetail grooves should be arranged at opposite sides and centered.

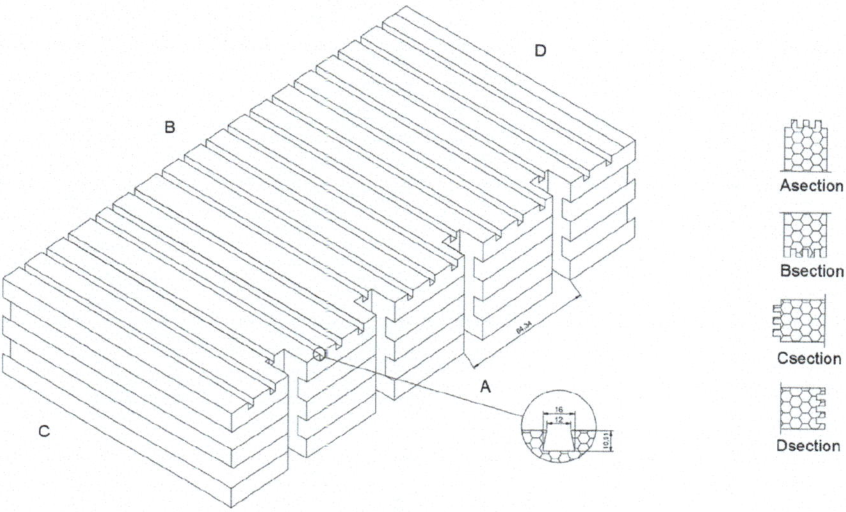

Figure 4.122 Insulation board schematic diagram.

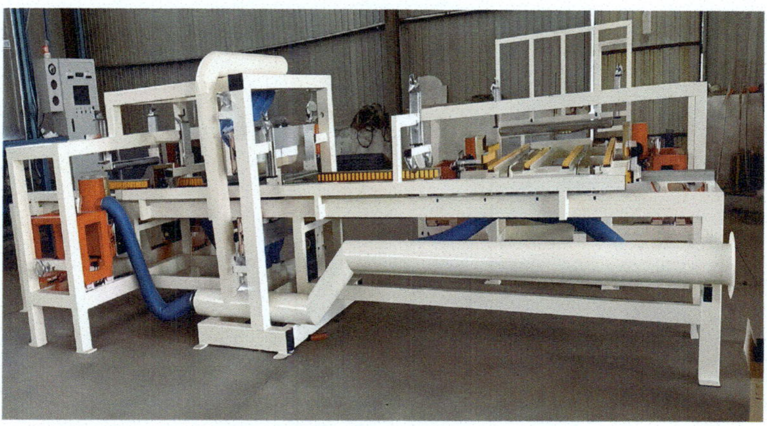

Figure 4.123 Double-seam six-sided slotting machine.

4. The decorative layer of the magnesium-based large mold's built-in insulation system should preferably use light-weight materials such as light-colored coatings and decorative mortars.

4.17.6 System construction design and technical requirements

1. The construction diagram of the magnesium large mold's built-in insulation system is shown in Fig. 4.125.

Table 4.22 Performance indicators of fire isolation belt.

Serial number	Item	Unit	Indicator	Testing method
1	Apparent density	kg/m^2	≥90	GB/T 5486
2	Compressive strength	MPa	≥0.15	
3	Thermal conductivity	W/(m ·K)	≤0.046	GB/T 10294
4	Tensile strength perpendicular to the surface	MPa	≥0.10	GB/T 30804
5	Flame resistance	–	Grade A	GB 8624

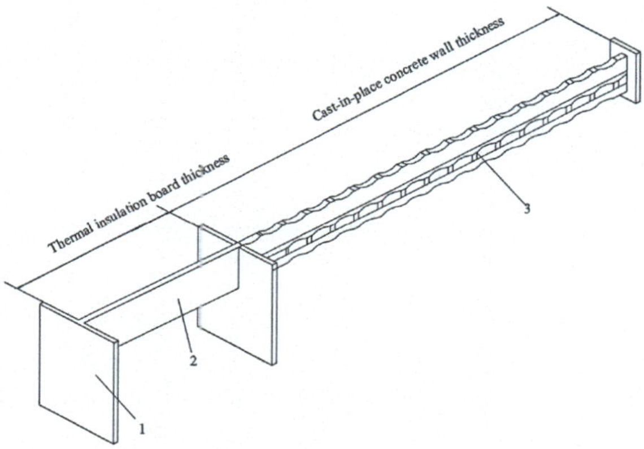

1 - Anchor plate; 2 - Connecting insert; 3 - Connecting rod.

Figure 4.124 Diagram of the special connection bridge for limit positioning. 1—Anchor plate; 2—Connecting insert; 3—Connecting rod.

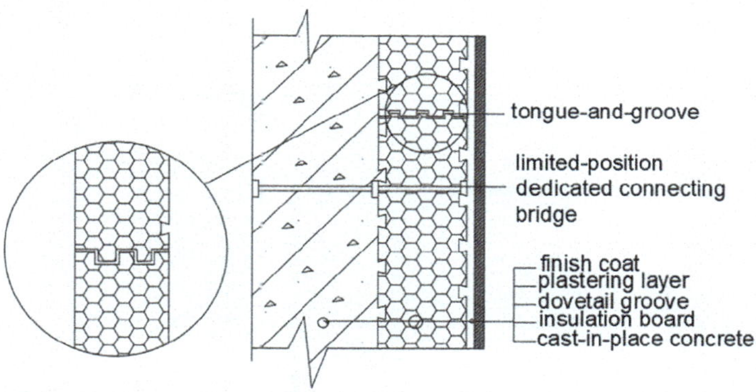

Figure 4.125 Construction diagram of the insulation system integrated in the magnesium large mold.

2. The insulation board design and installation should start from the corner area on the sunny side and be laid horizontally toward the opposite corner. The corner construction designs are shown in Figs. 4.126 and 4.127.

4.17.7 Installation construction method for built-in insulation panels in the main structure

4.17.7.1 Installation and construction

The installation and construction of insulation boards embedded in large formwork are conducted using the method of external wall insulation with

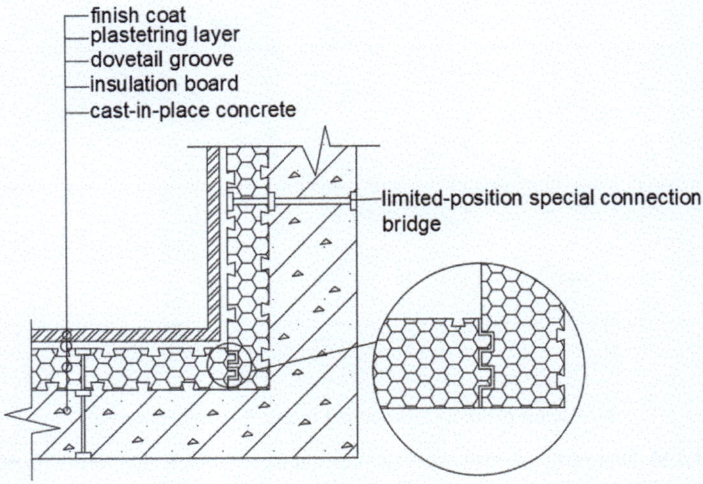

Figure 4.126 Diagram of the concealed corner area of the magnesium large mold's internal insulation system.

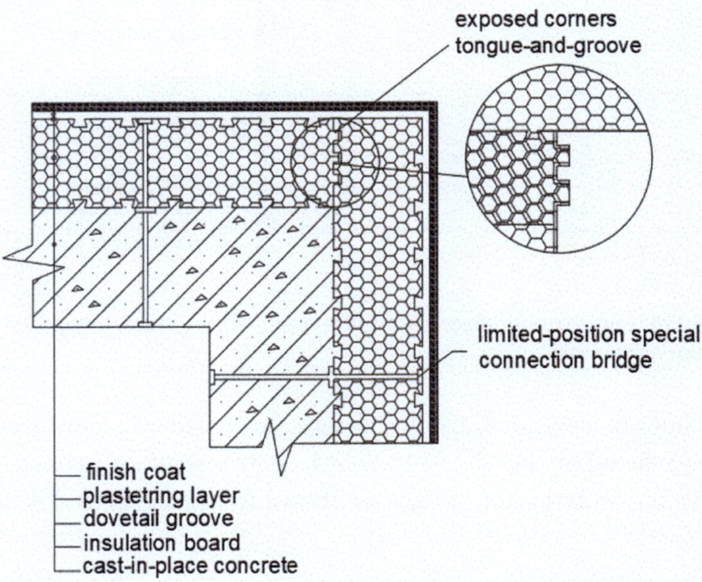

Figure 4.127 Diagram of the built-in insulation system at the corner of the magnesium large mold.

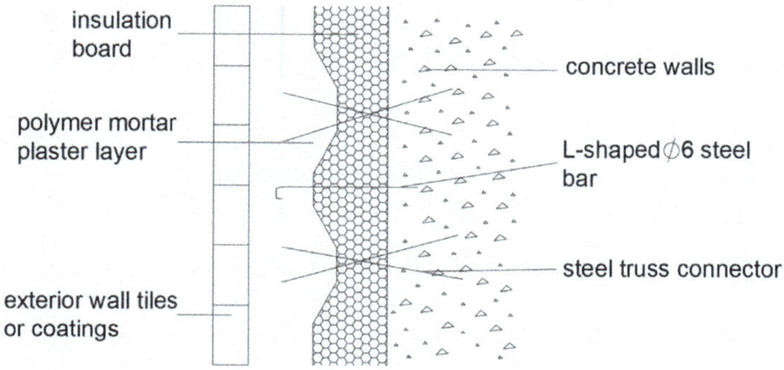

Figure 4.128 Installation diagram of built-in insulation board in the main mold.

extruded insulation boards embedded in the formwork. The insulation board is placed on the inner side of the exterior wall formwork and poured simultaneously with concrete. The concrete wall and the backside of the insulation board are organically combined. Polymer mortar is applied to the exterior side of the insulation board, and then ceramic tiles are adhered (or the surface is decorated with paint) on the outer side of the polymer mortar (Fig. 4.128).

4.17.7.2 Construction features

1. The external insulation board and the concrete of the wall are poured simultaneously, which facilitates the construction process, enhances work efficiency, and saves costs.
2. The connection between the external insulation board and the wall is more secure.
3. The external insulation board and the wall form an organic whole, which is beneficial for extending the service life of the external insulation of the wall.

4.17.7.3 Application scope

It is applicable to the cast-in-place concrete shear wall structure system with exterior finishes of tiles or coatings.

4.17.7.4 Basic principle

Install the insulation board on the outside of the wall reinforcement, insert L-shaped ϕ6 steel bars (or nylon anchor bolts, with the embedded length in the wall not less than 50 mm) on the board, and bind them with the wall reinforcement. This serves both as temporary fixation and a connection measure between the insulation board and the wall. Then install the large formwork. After the concrete is poured, the insulation layer is organically combined with the wall.

4.17.7.5 Construction process and key operations

4.17.7.5.1 Construction process

The construction process is shown in Fig. 4.129.

4.17.7.5.2 Key operating points

1. After the wall reinforcement is inspected and found to be qualified, the insulation board installation can be carried out.
   According to the required wall thickness as per the design, draw the wall thickness lines on the floor surface to determine the thickness of the exterior wall. At the same time, bind mortar pads (not plastic pad clips) on the outside of the exterior wall reinforcement (not using plastic pads), with no less than three pads per square meter of insulation board.
2. Installation of the insulation board
   The front side of the extruded polystyrene insulation board has a trapezoidal groove and is equipped with a single sheet steel wire mesh frame connected through the extruded polystyrene board (also known as the belly wire, using galvanized steel wire) to form a three-dimensional space. First, install the insulation boards at the corners, then install the insulation boards between the wall panels: When installing the insulation boards, the height difference between the boards should be bonded with special adhesive, and the vertical joints between the boards should be tied with fire-welded wires between the steel wire mesh, with a spacing of less than or equal to 150 mm. After the insulation board is in place, pass an L-shaped ϕ6 steel wire through the insulation board (using an electric soldering iron to make the hole), extend into the wall for a length of no less than 100 mm (the steel wire should be treated for rust prevention), and bind it firmly with the wall

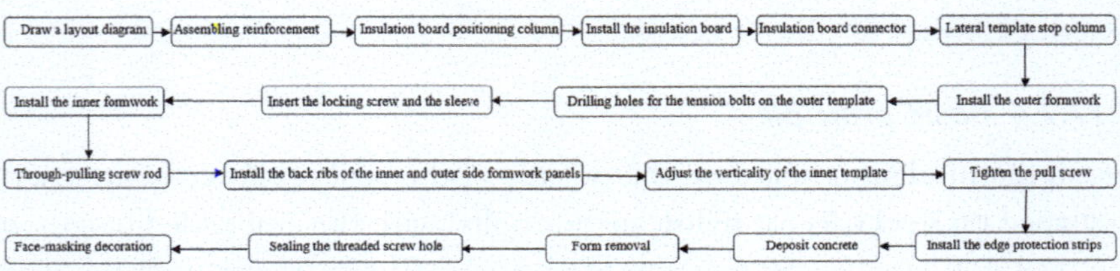

Figure 4.129 Construction process of the large formwork.

reinforcement using fire-welded wires. The spacing of the L-shaped ϕ6 steel wire is 900 mm. When installing the insulation board to the window and door openings, according to the size of the window and door openings and the node construction requirements, cut the insulation board for assembly, and the low-carbon steel wire mesh outside the insulation board should be disconnected at each floor height and not connected to each other.

3. Template installation
Use a large template, determine the size and quantity of the template according to the thickness of the insulation board, install the template at the position where the exterior wall lines are marked, and start installing the next layer of template when the strength of the bottom layer concrete is not less than 7.5 MPa. Before installing the exterior wall template on the outside, it is necessary to take reliable positioning measures at the root of the cast-in-place concrete wall or on the outside of the insulation board to prevent the template from pressing against the insulation board. Place the triangular platform frame using the bolt holes of the lower layer exterior wall before installing the exterior wall template on the outside. Before installing the exterior wall template on the outside, it is necessary to take reliable positioning measures at the root of the cast-in-place concrete wall or on the outside of the insulation board to prevent the template from pressing against the insulation board. The connection must be tight and firm to prevent misalignment and leakage of concrete.

4. Concrete pouring
The slump of the cast-in-place concrete should be greater than or equal to 150 mm. Before pouring the wall concrete, measures must be taken to cover the top surface of the insulation board. A slot protection cover with a shape like "II" should be placed, with a width equal to the thickness of the insulation board plus the thickness of the template. The concrete should be poured in layers, with a height controlled at 500 mm. The concrete feeding points should be scattered and continuously poured, with an interval of no more than 2 hours. The vibration distance of the vibration rod should generally be less than 500 mm. The duration of each vibration point should be based on the appearance of floating concrete and no further sinking, and the vibration rod must not be placed close to the insulation board. When pouring concrete at the opening, the concrete should be fed simultaneously on both sides of the opening to ensure that the pouring height on both sides is roughly the same. The vibration rod should be placed more than 300 mm away from the edge of the opening.

5. Removal of formwork
The removal of formwork can only be carried out when the concrete strength of the wall reaches no less than 1.0 MPa at normal temperature, or no less than 7.5 MPa during winter construction and reaching 30% of the designed strength value of the concrete. During the removal process, the compressive strength of the test blocks under the same conditions should be taken as the reference. The exterior formwork of the exterior wall should be removed first, followed by the interior formwork of the exterior wall. After the removal of the wall penetrator, the holes in the concrete wall should be filled with dry-hard mortar, and the holes in the insulation board area should be blocked with insulation materials (extruded polystyrene board or polystyrene particle slurry). The horizontal steel wires on the insulation board surface after removal must be aligned with the groove, and the distance from the bottom of the groove should be greater than or equal to 8 mm.

4.17.7.6 Quality control

1. Familiarize with the relevant drawings and data provided by all parties, and refer to the relevant construction techniques.
2. Understand the material properties, master the construction essentials, and clarify the construction sequence.
3. Conduct technical briefings to the construction personnel based on the material and technical quality standards.
4. The insulation board shall be extruded board, and its material performance shall comply with various performance indicators of Extruded Polystyrene Foam for Thermal Insulation (GB/T10801.2–2002). The groove lines of the extruded board should be formed using model molds, with accurate dimensions and uniform spacing.
5. The performance indicators of the polystyrene granule insulation mortar used for external insulation on the window side, local leveling, and hole blocking should comply with the requirements of the industry standard "Adhesive Polystyrene Granule External Wall External Insulation System" JC/T 158–2004.
6. To prevent the corrosion of the steel wire mesh, improve the flame retardancy of the insulation board, protect the extruded polystyrene board from the influence of sunlight and rain, and ensure a good bond between the surface of the insulation board and the external leveling mortar, the interface agent should be sprayed on the surface of the insulation board and the steel wire mesh frame. The interface agent for

the extruded polystyrene board on the mesh system should have a strong adhesion to the steel wire, and should not fall off after being repeatedly bent at 90°C for five times. Its technical performance should comply with the requirements of the industry standard "Concrete Interface Treatment Agent" JC/T 907–2002.

7. The solubility of the extruded board adhesive in the extruded board should be less than or equal to 0.5 mm, and the bonding tensile strength between the extruded boards should be greater than or equal to 0.1 MPa.
8. The quality requirements for the steel wire mesh frame of the insulation board are shown in Table 4.23.
9. The transverse steel wires should be aligned with the center of the groove; the adhesion between the interface agent, the steel wires, and the extruded board is strong; the coating is uniform and consistent; there should be no exposed base; and the thickness is not less than 1 mm. The extruded board deforms less than 10% under a pressure of 60 kg/m^2.
10. The inclined inserted steel wires (transverse wires) should be more than 100 per square meter, and should not exceed 200. The performance indicators of the low-carbon steel wires used for the steel wire mesh frame and the inclined inserted transverse wires are shown in Table 4.24.
11. For the already constructed insulated walls, no holes should be randomly drilled. Strengthen the protection of the finished products.

4.17.7.7 Safety precautions

1. Construction workers must wear safety helmets and fasten the chin straps when entering the site. Drinking alcohol is strictly prohibited. Smoking and fighting are not allowed on the site.
2. During hoisting, a dedicated person should be assigned to direct and use unified signals.
3. During the construction process, it is strictly prohibited to drill or make holes on the insulation board at will.
4. The erection and use of all scaffolds must be inspected by the safety officer, and can only be used after passing the acceptance.
5. When performing high-altitude work, the insulation board and tools should be placed neatly to prevent falling objects from injuring people.
6. Adequate lighting measures must be provided during night construction.

Table 4.23 Quality requirements for the steel wire mesh frame of the insulation board.

Serial number	Project	Quality requirements
1	Appearance	The front surface of the insulation board has trapezoidal concave-convex grooves, with a distance of 100 mm between the grooves. The surface of the board and the steel wire are uniformly sprayed with interface agent.
2	Welding point strength	The tensile strength is greater than or equal to 330 N, and there is no overburning phenomenon.
3	Welding point quality	The number of missing or detached welds on the mesh does not exceed 8‰ of the total number of welds, and they should not be concentrated in one place. The number of continuous detachment welds should not exceed 2 points. The welds in the 200 mm section at the board end are not allowed to be detached or have weak connections. The detached or missing welds of the inclined reinforcement bars should not exceed 3%.
4	Steel wire head selection	The overhang of the mesh edge is less than or equal to 6 mm, the overhang of the inserted wire is less than or equal to 5 mm, and the overhang of the penetrating reinforcement bar is greater than or equal to 30 mm.
5	Butt joint of extruded board	For long boards less than or equal to 3000 mm, the butt joint of the extruded board should not exceed two places, and the joint area should be firmly bonded with polyurethane glue.
6	Quality	The weight per square meter is less than or equal to 4 kg/m^2.

Table 4.24 Performance indicators of low-carbon steel wire.

Diameter/mm	Tensile strength/(N/mm^2)	Cold bending test—repeated bending 180 degrees	Purpose
2.0 ± 0.05	≥550	≥6 times	Steel wire mesh frame
2.5 ± 0.05	≥550	≥6 times	Galvanized inclined inserted core wire

7. The insulation board should be stored away from fire sources, and the construction site should prevent electric welding and gas welding from approaching the insulation board.
8. Operations must be carried out in strict accordance with the "Safety Technical Regulations for Construction and Installation Engineering."

4.18 Magnesium self-insulating composite blocks

Those concrete small blocks that are produced by compounding lightweight aggregates and/or filling insulation materials in the cavities, and which have the insulation function for the constructed walls, are referred to as self-insulating blocks. Magnesium self-insulating composite blocks refer to the blocks made by using magnesium cement as the binder material, solid waste as the filler, adding modifiers, coarse and fine aggregates, and insulation materials, mixing the slurry, and then pouring and molding after achieving uniform mixing (Fig. 4.130).

According to the different heat transfer coefficients required for the exterior walls of buildings in the north and south, the thickness can be increased or reduced to achieve factory production. Buildings constructed

Figure 4.130 Magnesium self-insulating composite blocks.

with such blocks can not only meet the specified building energy-saving standards, but also overcome the defects such as wall cracking, detachment of the external insulation layer, and poor durability of the insulation layer that exist in the current external wall insulation. Self-insulating composite blocks are suitable for both winter insulation in the north and summer heat insulation in the south, and have wide regional adaptability.

Magnesium self-insulating blocks are a lightweight and high strength wall material, and have been used in low-rise buildings, multistory concrete small-block buildings, and reinforced small-block buildings. Magnesium self-insulating blocks have the characteristics of integrating insulation and load-bearing, and the insulation materials have the same lifespan as the building. They have broad market prospects.

At present, Xinjiang Uygur Autonomous Region has introduced policies or standards to support the production and application of self-insulating external blocks, as follows:

Building materials industry standard: JCT 407–2013 Self-insulating Concrete Composite Blocks.

Xinjiang standard: XJJ 109–2019 Application Technical Standard for Self-insulating Blocks.

The main performance indicators for self-insulating blocks are specified in Table 4.25.

4.18.1 Product advantages

1. The process is simple and the equipment investment is low. This product has a simple process, is easy to produce and implement, and is conducive to popularization and promotion.
2. The thermal insulation performance is good, with a thermal conductivity lower than 0.12 W/(m·K).
3. Lightweight and high strength. The magnesium self-insulation blocks with a density of 650 kg/m^3 have an compressive strength higher than 5 MPa. Replacing the silicate cement with magnesium clay results in higher strength, faster setting, and easier molding.

4.18.2 Production process

4.18.2.1 Raw materials

Light-burned magnesium oxide, magnesium chloride, EPS particles, modifiers, foaming agents, etc.

Table 4.25 Main performance requirements of self-insulating masonry blocks.

Project		Unit	Performance indicators	Testing method
Dry density moisture content		kg/m^3	≤900	GB/T 4111
Moisture content		%	≤ 10	
Absorption rate		%	≤ 18	
Dry shrinkage value		mm/m	≤0.65	
Frost-resistance	Quality loss	%	≤5	
	Loss of strength	%	≤25	
Compressive strength		MPa	≥5.0	
Softening coefficient		–	≥0.85	
Carbonization coefficient		–	≥0.85	
Radioactivity		–	Should comply with the provisions of GB 6566	GB 6566
Thermal conductivity coefficient		W/(m·K)	Meet the design requirements	GB/T 13475

4.18.2.2 Reference ratio

The reference ratio for magnesium-based homogeneous self-insulation blocks is shown in Table 4.26.

4.18.2.3 Production process flow

The production process flow is shown in Fig. 4.131.

4.18.2.3.1 Raw material processing

1. Solid waste treatment: If the industrial waste residue is in the form of blocks, it should be crushed to the required size for the process and then sieved for use.
2. Preparation of the brine solution: To precipitate the magnesium chloride impurities in advance, dissolve the halite sheet at least 24 hours in advance, prepare the brine solution, let it stand and clarify, and calibrate the osmotic degree of the brine solution. Before use, discard the precipitate and only use the clear liquid, and re-calibrate the osmotic degree.
3. Preparation of magnesium sulfate solution: Due to the heat absorption when magnesium sulfate dissolves in water, dissolve the magnesium sulfate at least 24 hours in advance, prepare the magnesium sulfate solution, and calibrate the osmotic degree of the solution.

4.18.2.3.2 Mixing of raw materials and pulping

Use a microcomputer-controlled batching system to measure the amount of one-time input according to the production formula and the size of the mixer. The mixer speed should be controlled at 90–100 r/min, and the mixing time should be no less than 5 minutes.

The sequence of mixing by adding materials and pulping is as follows.

1. Add the measured magnesium chloride/magnesium sulfate solution, then add the modifier (some modifiers need to be added to the slurry) and anticracking short fibers to the solution, and stir evenly.
2. Add calcined magnesia, fly ash, and waste materials, mix them with the solution thoroughly to form a uniform slurry, add the modifier, and stir evenly.
3. Add polypropylene particles to the slurry and mix them with the slurry by combining forward and reverse rotation of the mixer to make the slurry and the particles mix evenly, so as to achieve the purpose of completely covering the polypropylene particles and having no gaps in the accumulation.
4. Start the foaming machine, inject a certain amount of foam into the particle slurry, stir evenly, and make the foaming particle slurry.

Table 4.26 Reference composition ratio of magnesium-based homogeneous self-insulating masonry blocks.

Block categories	Light-burned magnesium oxide/kg	Magnesium sulfate / (30°Bé/kg)	Brine / ((27°Bé)/kg)	Fly ash/kg	EPS particles/m^3	Modifier	Foaming agent
Chloromagnesium Oxide Homogeneous Self-Insulation blocks	100	0	80–85	20	0.3–0.4	GX-1# 0.5%–1% GX-4# 6‰~ 8‰	GX-7# Foaming Agent diluted with water at a ratio of 1: 80
Sulfur-chlorine composite homogeneous self-insulation blocks	100	20	85–90	35–40	0.3–0.4	GX-1# 0.5%–1% GX-4# 6‰~ 8‰	GX-7# Foaming Agent diluted with water at a ratio of 1: 80

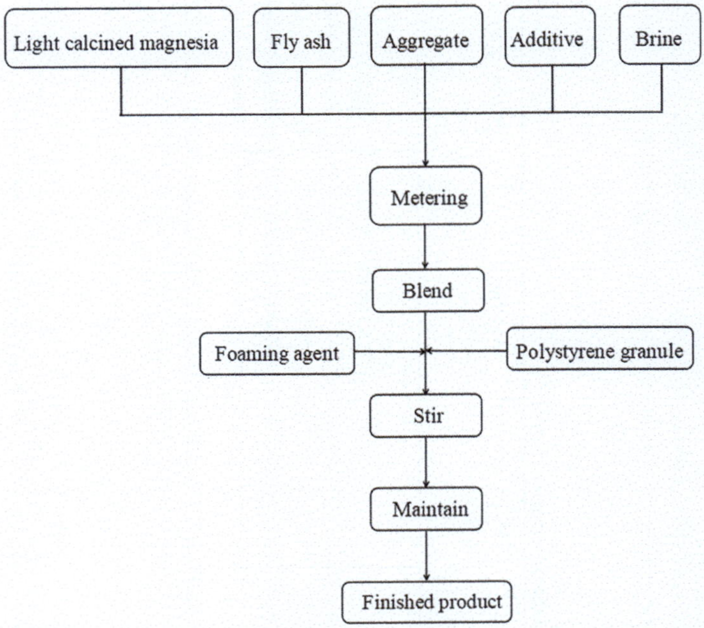

Figure 4.131 Production process flow chart of magnesium-based self-insulating masonry blocks.

4.18.2.3.3 Pouring and forming

The pouring process is divided into mobile pouring and fixed-point pouring. Generally, people choose fixed-point pouring, as it can reduce the number of pouring vehicles and requires the addition of mobile tracks. The mixed foam particle slurry is placed in the mold box, and workers level the surface of the mold and then move it to the designated position.

4.18.2.3.4 Initial curing and hardening

The initial curing stage mainly involves the increase in the strength of the product. The curing temperature should not be too high, as excessively high temperatures can cause excessive heat release from the mold, leading to the cracking of the mold. The optimal temperature for the initial curing stage is between 20°C and 30°C, and the mold can be removed and cut after 2–3 hours. Special attention should be paid to the temperature changes at the center of the mold, as magnesium products start to release heat, and it is difficult for the heat to spread outward, easily causing rapid internal heating and excessive thermal stress concentration, resulting in cracks. Generally, production is carried out in the afternoon of the same day, and the mold can be removed and cut in the morning of the next day. During the cutting process, the condition should be checked to ensure that the product does not stick to the mold and the structural strength is sufficient for lifting and transportation.

4.18.2.3.5 Mold cutting

When the initial curing reaches the cutting strength, the mold can be cut.

4.18.2.3.6 Later curing

The main purpose of post maintenance is to promote the hydration reaction and complete the normal growth of crystal whiskers. The optimal temperature for post maintenance is above 20°C, and the optimal humidity should be maintained at around 60%. During the maintenance period, outdoor storage is prohibited to prevent direct sunlight and rainwater erosion. The post curing period is 7 days. After 7 days of natural drying, it can be delivered in 28 days.

4.18.2.4 Production technical key points

1. The board has a high bulk density, or a low bulk density but insufficient strength.
    a. In terms of raw materials, scientific analysis and testing are required to ensure the quality of the raw materials is qualified.
    b. When mixing the ingredients, the formula ratio should be scientifically reasonable, allowing magnesium oxide to fully participate in the hydration reaction, and the dosage of magnesium oxide should not exceed the dosage of magnesium sulfate solution.
    c. Pay attention to modification. After the magnesium oxide composite cement is modified by the modifier, the product strength and its physical and chemical properties are improved.
    d. Repeatedly conduct wet curing to ensure the smooth progress of the later curing reaction.
2. The magnesium-based self-insulating blocks have a high-water absorption rate. How to improve water resistance.

    Add GX-13# waterproof agent to improve the waterproof and hydrophobic performance of the magnesium-based self-insulating blocks, and at the same time, it is beneficial for on-site construction of the board and prolongs the service life.

3. Poor weather resistance, and the strength decreases over time.
   The main reason is that there are unreacted free chlorine ions in the chloromagnesium self-insulating blocks, and under the action of CO_2 and H_2O in the air, the crystal structure of chloromagnesium will decompose, the board loses strength, softens, and the service life is shortened.
   Solution:
   a. Scientific and reasonable proportioning: Based on the quality changes of raw materials, dynamic proportioning should be carried out, and the concentration and dosage of brine should be appropriately reduced.
   b. Add modifiers: By adding antihalogen agents, the generation of free chlorine can be inhibited, and the phenomenon of back-bleeding can be reduced.
4. Reduce thermal conductivity.
   With the continuous improvement of energy-saving requirements in China, traditional insulation materials are no longer sufficient. Therefore some insulation materials, such as homogeneous self-insulating blocks, need to continuously reduce their thermal conductivity. The traditional solution involves introducing low density and low-conductivity materials like SiO_2 aerogel, which is composed of nanoparticles, has high porosity, a high specific surface area, low density, and low thermal conductivity. However, its compatibility with inorganic cementitious materials is poor, its price is relatively high, and its application in the field of energy conservation is limited. On the other hand, EPS particles are produced from EPS beads as raw material, expanded and foamed to form a closed-cell foam plastic. They exhibit characteristics such as low density, high porosity, low water absorption rate, and low cost. Currently, EPS particles have been used as lightweight insulating aggregates in the field of energy conservation. However, the porosity of the pores between EPS particles is related to the size of the EPS particles. To increase the overall closed porosity of homogeneous panels, MJT proposed a particle distribution scheme.

4.19 Functional porous magnesium products

Porous magnesium materials are advanced materials with a unique porous structure. They have numerous connected or closed pores inside the body. Porous magnesium materials possess characteristics such as low volume density, high porosity, high specific surface area, and high thermal

Figure 4.132 Porous magnesium materials.

insulation performance. Moreover, they have the properties of high-temperature resistance, high pressure resistance, resistance to acid and alkali and organic medium corrosion, good mechanical properties, controllable pore structure, long service life, and good chemical stability. They can be applied in various media for precise filtration and separation, high-pressure gas exhaust and sound absorption, gas distribution, and have broad application prospects in environmental governance, engineering sound absorption and noise reduction, and wastewater treatment (Fig. 4.132).

4.19.1 Production process

Depending on the different products of related applications, porous magnesium materials can be processed and manufactured into various shapes and with different functions. This section lists the preparation methods for porous structures. Currently, the available preparation methods mainly include two types.

1. Prepare foam by using the physical foaming method. Mix the foaming agent with water to form a foaming liquid, then use a foaming machine to produce foam. The foam is fully mixed with the magnesium slurry, molded, hardened, demolded, cured, and processed to obtain the desired shape products.

2. Prepare pores by using the chemical foaming method (hydrogen peroxide, stabilizer). Add a set amount of hydrogen peroxide and stabilizer to the modified magnesium slurry. In the constant temperature medium of the slurry, the magnesium slurry will expand in volume, containing a porous structure. The density, pore size, and distribution of the product can be adjusted according to the consistency of the slurry, the temperature of the slurry, the amount of hydrogen peroxide added, the rotation speed of the mixer, and the time.

Through these two foaming methods, various pore structures such as small air holes, honeycomb pores, uneven flat circular pores, and connected pores can be produced, meeting the application requirements of different products.

4.19.2 Product characteristics

4.19.2.1 High porosity

Through the regulation of the formulation and process technology, the key feature is the presence of a large number of uniform and controllable pores. Pores can be classified as open pores (those in the magnesium-based matrix that are in direct contact with the atmosphere) and closed pores (those in the magnesium-based matrix that are not in direct contact with the atmosphere). Open pores have functions such as filtering, absorbing, adsorbing, and eliminating echoes; while closed pores are conducive to blocking the transmission of heat, sound, and liquids to solid particles.

4.19.2.2 High strength

After modification, the Mohs hardness of magnesia-based cementitious materials ranges from 6 to 7. The compressive strength of the pure paste specimens can reach up to 130 MPa, and the density is 600 kg/m^3. The compressive strength of the foamed products is greater than or equal to 5 MPa.

4.19.2.3 Stable physical and chemical properties

The porous magnesium-based materials can resist acid and alkali corrosion, and can withstand high temperatures and high pressures. They are nontoxic, harmless, and have stable properties. They are a kind of green and environmentally friendly functional material.

4.19.3 Product application

4.19.3.1 Used as a catalyst carrier

Porous magnesium materials possess excellent adsorption capacity and reactivity. In environmental engineering, porous magnesium materials are widely applied in aspects such as air purification and water treatment. After being coated with catalysts, the reaction fluid passes through the magnesium pores, significantly enhancing the conversion efficiency and reaction rate. Due to the characteristics of this material, such as high specific surface area, good thermal stability, wear resistance, nontoxicity, and low density, it has been widely used as a carrier for automotive exhaust catalytic purifiers.

4.19.3.2 For sound absorption and noise reduction devices

As a sound-absorbing material, porous magnesium materials mainly utilize their diffusion function, that is, by dispersing the pressure caused by sound waves through the porous structure, to achieve the purpose of sound absorption. As a sound-absorbing material, porous magnesium materials require small pore diameters (20–150 μm), high porosity (above 60%), and high mechanical strength. Porous magnesium materials can be used as fireproof isolation belts and applied in high-fire-resistant areas such as high-rise buildings, tunnels, and subways, as well as in places with high sound insulation requirements such as television transmission centers and cinemas.

4.19.3.3 For filtration and separation devices

The pore size of porous magnesium materials can be adjusted. The filtration device composed of flat plates, spheres, or tubular products of porous magnesium materials has the characteristics of large filtration area and high filtration efficiency. It is widely used in wastewater purification, oil separation filtration, organic solutions, acid-base solutions, other viscous liquids, compressed air, coke oven gas, steam, methane, ethylene, etc. separation. Due to the excellent properties of porous magnesium materials, such as high-temperature resistance, wear resistance, chemical corrosion resistance, and high mechanical strength, they show unique advantages in corrosive liquids, high-temperature fluids, molten metals, etc.

4.19.3.3.1 Wastewater treatment

Using porous magnesium materials for filtration of industrial wastewater and domestic sewage can remove suspended solids and fine particles in water. It has become an important development direction for wastewater treatment and purification, suitable for treating various polluted wastewater, with high efficiency and low cost.

4.19.3.3.2 Corrosive liquid filtration

The acid and alkali corrosion resistance of porous magnesium materials makes it show unique advantages in filtering acidic, alkaline, etc. corrosive liquids or gases.

4.19.3.3.3 Molten metal filtration

The filtration by porous magnesium materials can remove most inclusions and gases in molten metals, improving the strength and other intrinsic qualities of the metal materials. It is particularly important in electronic components, wires, and precision casting metals.

4.19.3.3.4 High-temperature gas filtration

The dust removal of high-temperature flue gas and the purification of high-temperature gas such as coal gas all require the use of high-temperature-resistant porous magnesium materials.

4.19.3.3.5 Filtering radioactive substances

Nuclear power plants and other facilities produce a large amount of radioactive waste. After combustion, it can become chemically stable solid powder. Porous magnesium materials can solidify it and store it conveniently and economically.

4.19.3.4 For heat insulation and preservation

Due to their high porosity and low thermal conductivity coefficient, porous magnesium materials exhibit high thermal resistance and low volumetric heat capacity. The lining of traditional kilns and high-temperature electric furnaces is commonly made from porous magnesium materials. To enhance their thermal insulation performance, the internal gas can be evacuated. Advanced porous magnesium materials can even be applied for insulating spacecraft hulls and missile heads. The world's best thermal insulation materials are precisely this type of porous structured product.

4.19.3.5 For grinding tools

Porous magnesium materials can be used as grinding balls, grinding wheels, grinding foot stones, grinding stones, etc., which have grinding functions.

Figure 4.133 Magnesium-based grinding balls.

1. Grinding balls are a common type of grinding medium used to finely grind materials for abrasives. They are widely applied in fields such as grinding, mixing, dispersion, and ball milling (Fig. 4.133). Their shape is mostly spherical and are typically made of various materials, such as ceramics, glass, steel, and natural stone. Different fields require the selection of different materials and types of grinding balls to meet manufacturing needs. Compared with traditional grinding techniques, grinding balls have obvious advantages such as high efficiency, precision, low noise, and low energy consumption.
   Magnesium-based cementitious materials, after modification, have excellent bonding and internal admixture composite properties. The products made from them have excellent mechanical structural strength. Through compression testing experiments, the maximum strength can reach up to 130 MPa. Such excellent inorganic cementitious materials can be added with abrasives (such as silicon carbide powder, brown corundum powder, etc.) during the batching process to enhance the structural strength and wear resistance of magnesium grinding balls. Compared with nonmetallic materials such as ceramics, glass, and natural stone, magnesium grinding balls have their own unique advantages.
2. Grinding wheels, also known as consolidated tools, are the most widely used and utilized type of grinding tools. When in use, they rotate at high speed, capable of performing rough grinding, semifinish grinding, and finish grinding on the outer circumference, inner circle, plane, and various profiles of metal or nonmetal workpieces, as well as slotting and cutting operations.

Figure 4.134 Magnesium-based grinding wheel.

Magnesia grinding wheels (Fig. 4.134) are composed of modified magnesia cement and abrasive materials in a certain proportion. They inherit the advantages of high strength and hardness of magnesia cement, while retaining the excellent wear resistance of the abrasive materials. Magnesia grinding wheels are classified into ordinary abrasive magnesia grinding wheels and super-hard abrasive magnesia grinding wheels based on the type of abrasive used. The former is made with common abrasives such as corundum and silicon carbide, while the latter is made with super-hard abrasives such as diamond and cubic boron nitride. The abrasives that have been developed and utilized in the market include black corundum, black silicon carbide (coarse sand, fine powder), zirconia beads, white corundum (coarse sand, fine powder), brown corundum (coarse sand, fine powder), and corundum.

3. The foot rubbers are made of magnesite materials and come in various sizes and shapes (Fig. 4.135). These foot rubbers have the functions of promoting blood circulation in the feet and maintaining human health. At the same time, they can remove the old (hard) skin on the feet, provide skin care and beauty, and contain dozens of minerals that are beneficial to the human body. They are excellent health care products for middle-aged and elderly people.

   Artificial foot rollers are made from magnesium-based cementitious materials through chemical foaming. The pores are controllable, and the color and weight can be adjusted. The raw materials are abundant in

Figure 4.135 The stumbling block.

Figure 4.136 Yangjiang eight sons magnesium-based grinding stone for knives.

reserves, the production cost is low, and they inherently contain soluble salts, which have a bactericidal effect.

4. Grinding stones are tools used for sharpening knives, making the blade more refined and sharp (Fig. 4.136). For example, they are used to sharpen kitchen knives, swords, axes, etc. The magnesite material can bond composite coarse and fine abrasive materials to create different grinding surfaces. The coarse abrasive surface smooths out the gaps in the knife, while the fine abrasive surface performs a meticulous grinding. Some water can be sprinkled on the finished blade, making it shiny and sharp quickly.

Take the magnesia grinding wheel as an example. Its production process is usually manual and mold forming. Customized sand wheel molds (commonly made of cast iron, with good temperature resistance, wear resistance, and dimensional stability) should be prepared according to the requirements. Before use, evenly apply a release agent on the inner wall and upper edge of the mold. In terms of mixing and preparing the casting slurry, first mix the magnesia slurry according to the set formula proportion, then add modifiers and water-soluble emulsions to the slurry. After thorough mixing, add silica powder and abrasives (brown corundum, silicon carbide) successively until fully mixed and homogeneous. Pour the prepared slurry into the mold, then place it on a vibration platform for vibration and leveling to eliminate residual bubbles. Place it in an environment with a temperature of 10°C or above for natural curing for 8–12 hours before demolding. After demolding, place it on the curing rack for 7–15 days of natural curing before stacking and packaging. Refer to the ratio in Table 4.27.

4.19.3.6 For air distribution device

The porous magnesium material can also be used for the mixing of gas-liquid and gas-powder phases, which is commonly referred to as gas dispersion and gas distribution. Through the gas dispersion effect of the porous magnesium material, the contact surface between the two phases is enlarged, accelerating the reaction. The porous magnesium material gas distribution device used in the activated sludge method for treating urban wastewater is relatively successful. Not only does it have a good gas distribution effect, but also has a long service life. By blowing gas into the powder using the porous material, the powder can be in a loose and fluidized state, achieving rapid and uniform heat transfer, accelerating the reaction speed, and preventing powder caking. It is suitable for powder transportation, heating, drying, and cooling, and has a good application prospect in the production and transportation of powders such as cement, lime, and alumina powder.

4.20 Other new products

4.20.1 Magnesium-based negative ion eco-board

With the increasing awareness of the environment and health among people, as well as the continuous breakthroughs in technology, a new type of environmentally friendly material—negative ion ecological board—has emerged. This board is based on the understanding of negative oxygen ions. At the end of the 19th century, German physicist Dr. Philip Lionad first

Table 4.27 Reference formula for materials (unit: kg).

Light-burned magnesium oxide	Magnesium chloride solution	Silica powder	Black corundum (fine powder)	Silicon carbide (coarse sand)	GX-1# or GX-8#	GX-4#	Water-soluble resin emulsion
100	110–120	15	25	40	0.8	0.6	2

Note: ① In this list, light-burned magnesium oxide is 90 powder, with an activity content of no less than 65%; ② Magnesium chloride solution is prepared from superior grade brine sheets (> 45% content) or high-purity magnesium chloride (> 98.5% content). The osmotic degree of the brine is (30–32°Bé); ③ Depending on the difference in production environment temperature, choose GX-1# or GX-8# to adjust the solidification speed.

expounded the efficacy of negative oxygen ions on human health in an academic manner, making remarkable contributions to negative oxygen ions.

Firstly, what is a negative ion? In fact, negative ions are everywhere in our lives. We know that the main components of air are nitrogen, oxygen, and carbon dioxide, among which only oxygen and carbon dioxide have affinity for electrons. Negative ions are formed when these gas molecules combine with free electrons.

The oxygen content in the air is 700 times that of carbon dioxide. Therefore the majority of the negative ions generated in the air are oxygen negative oxygen ions. For example, natural discharge phenomena, ultraviolet radiation from the atmosphere, waterfall impact, forest trees discharge, and some natural minerals release, etc. can all cause the surrounding air to ionize and form negative ions.

The negative ion ecological board is generally made by ultra-microizing natural minerals that can produce negative ions, and then forming micro-micron or nano-level ultra-micro powders. After that, it is mixed, impregnated, coated, etc., with other organic or inorganic materials in a certain proportion to be prepared. The treated natural minerals can release high concentrations of air negative ions (more than 5000 per cubic meter), without side effects, and meet environmental protection requirements. What is discussed in this section is the negative ion ecological board made using this natural mineral, magnesium oxide.

The negative ion eco-board has the functions of eliminating pollution, purifying air, settling dust, and disinfecting. After being tested by authoritative institutions, the negative ion eco-board has a purification rate of 84.5% for formaldehyde, and an antibacterial rate of up to 99.9% against *Escherichia coli* and *Staphylococcus aureus*, thus safeguarding people's health. The "Negative Ion Functional Building Interior Decoration Materials" JC/T 2040–2022, which is under the jurisdiction of the Standardization Technical Committee for Environmentally Friendly and Health-Beneficial Building Materials in the Building Materials Industry, has been officially released and will come into effect on April 1, 2023.

4.20.1.1 Function principle of negative ion eco-board

1. The dust purification and health care functions of negative oxygen ions.
   a. The property of negative oxygen ions having a negative charge has an extremely strong capturing ability, which can actively capture pollution substances such as dust, formaldehyde, and second-hand smoke in the air, and degrade these pollutants, thereby purifying the air (Fig. 4.137).

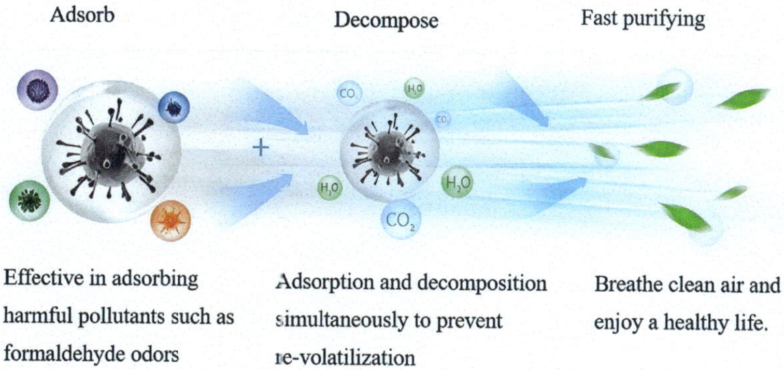

Figure 4.137 Principle of negative ion air purification.

b. Due to its negative charge, the activity of negative oxygen ions is greatly enhanced, and their migration distance becomes longer. Therefore negative oxygen ions can bring more oxygen into the human respiratory system, cardiovascular system, cerebral cortex, and nervous system, exerting a series of health benefits such as enhancing lung function (for example, medical device oxygen machines), improving sleep quality, and lowering blood pressure.

2. The antibacterial effect of negative ion ecological panels.
Negative ion ecological panels are prepared using ultrafine powder (micron or nano) magnesium oxide as the main material. The antibacterial mechanism of ultrafine powder magnesium oxide has two types: oxidative damage by reactive oxygen species (ROS) and mechanical damage by adsorption.
   a. Oxidative damage by ROS.
   The mechanism of oxidative damage is that the oxygen vacancies on the surface of ultrafine powder magnesium oxide can catalyze the reduction reaction of dissolved oxygen in water to produce superoxide anion O^{2-}. Since O^{2-} has strong oxidizing properties, it can destroy the protein peptide chains of bacterial cell membranes, thereby quickly eliminating bacteria.
   b. Mechanical damage by adsorption.
   The surface of ultrafine magnesium oxide powder contains a large number of crystalline-limiting hydroxyl groups, free hydroxyl groups, and various ionic and other active sites. These can all serve as adsorption and surface reaction sites. The mechanical damage mechanism of adsorption is a supplement to the oxidative damage

mechanism of active oxygen. It can not only explain the problem that magnesium oxide still has good antibacterial properties without the presence of ROS, but also the adsorption effect of particles can cause cell membrane damage. MJT further verified the mechanism that the smaller the particle size of ultrafine powder magnesium oxide, the higher the antibacterial performance. Therefore the antibacterial performance of magnesium oxide can be improved by reducing the particle size of magnesium oxide, increasing the specific surface area, and enhancing the adsorption effect.

4.20.1.2 Application and development

The ecological board made from specially formulated magnesium oxide and processed by special techniques will become a new type of functional inorganic material. Especially in the field of antibacterial materials closely related to human survival and health, it shows unique advantages, such as long-lasting antibacterial activity, low cost, noncolor change, and no biological toxicity. It will be widely applied in hospital, restaurant, shopping mall, home interior decoration, and other scenarios in the future, and will be increasingly favored.

4.20.2 Inorganic quartz stone series products

As the country's requirements for environmental pollution and carbon emissions continue to increase, the tile and stone industry, due to its inevitable "high energy consumption," "high emissions," and "resource dependence" flaws, urgently needs to undergo industrial upgrading and replacement. Inorganic quartz stone (Fig. 4.138) has no need for firing in the production process compared to natural stones and tiles. It has significant advantages such as green production process, no emissions, low energy consumption, diverse resources, and recyclability, which solves the pain points of the tile and natural stone industries and meets environmental protection requirements and policy directions. The performance comparison of several materials is shown in Table 4.28.

Reference can be made to the execution standards "Processing, Decoration and Construction Quality Acceptance Specifications for Artificial Stone" JC/T 2300–2014 and "Artificial Stone Building Panels" GB/T 41919–2022.

MJT is further exploring new applications of fully solid waste-based cementitious materials and is working together with friends from home and

Figure 4.138 Inorganic quartz stone.

abroad to contribute to reducing emissions and waste in society and creating an environmentally friendly ecological environment suitable for human survival.

4.20.3 Spray-foamed magnesium material

In underground engineering, water conservancy projects, and rock and soil construction, the spraying concrete technique is commonly used. During the construction process, a spraying machine is typically employed to spray concrete onto the exterior surface of the engineering structure. Under the influence of the accelerator, the concrete rapidly solidifies, providing certain protective functions for the engineering structure. The spraying concrete construction is fast and efficient, and is widely used in tunnel projects.

In actual production, some steel supports used in tunnels and mines are difficult to resist the erosion of temperature and humidity, and require certain protective treatment. Due to the large internal space of tunnels and mines, manual painting is inconvenient. Therefore the spraying concrete technique is generally adopted to cover the steel supports and ensure their durability. In fact, the strength of these steel support protective layers does not need to be very high; as long as they can provide coverage and isolation functions is sufficient. Moreover, the materials for the spraying concrete technique are numerous and the cost is high, and there is redundancy in its strength. Now, a process of spraying construction using foamed concrete has begun to appear on the market.

Table 4.28 Comparison of properties of several common stones.

Project	Tiles	Natural stone	Inorganic quartz stone	Comprehensive comparison
Production process	Firing process (1100°C–1250°C)	–	Nonburning waste recycling mixing polymerization forming process technology	Revolutionary green and environmentally friendly production process
Energy conservation	1 square meter of ceramic tile consumes 8 kg of standard coal (The national output of ceramic tiles is approximately 10 billion square meters, consuming 8 million tons of standard coal)	High pollution	5.44 $kW \cdot h/m^2$ Converted to standard coal is approximately 2.2 kg/m^2 (Producing 100 billion square meters of inorganic quartz stone results in an approximate consumption of 220 million tons of standard coal)	Can reduce the consumption of 580 million tons of standard coal compared to ceramic tiles
Emission reduction	High pollution, high emissions (The consumption of 8 million tons of standard coal will produce 208 million tons of carbon dioxide)	High pollution	During the production process, there is no pollution or emissions	Can reduce the emission of 150 million tons of carbon dioxide compared to ceramic tiles
Recycling	Limited to a single type of clay	Limited to the mining of natural stones	Various natural hard rock particles, such as quartz sand, glass particles, ceramic tile particles, and waste tailings, can be used as raw materials	Compared to ceramic tiles and stone raw materials, which do not have restrictions on the use of specific natural resources and do not consume specific resources, renewable resources can be utilized
Recycling	Cannot be recycled	Cannot be recycled	Can be recycled for a second time	100% utilization of raw materials, and 100% recyclable utilization of inorganic quartz stone, achieving 100% circularity

Figure 4.139 Spray application of foamed magnesia material.

Foamed concrete, as the protective layer for steel supports, is lightweight, low cost, sound-absorbing, and has good heat insulation properties. It can withstand temperature changes in underground engineering and will not experience cracking of the protective layer due to stress. In fact, apart from foamed concrete, magnesium sulfate foaming slurry is also suitable for use as a protective layer. Besides having many of the advantages of foamed concrete, the extremely small amount of free sulfate ions contained in the magnesium sulfate slurry can cause the steel support surface to form a passivation layer, thereby protecting the steel support from corrosion (Fig. 4.139).

4.20.3.1 Process

In the construction of sprayed foamed magnesium materials, the density is usually reduced by using the physical foaming method. During the construction process, the peristaltic pump of the spraying machine separately extracts the magnesium slurry and the preprepared foaming liquid. Under the action of the air source, the foaming liquid generates foam at the foam generation device of the spraying machine, and then mixes with the magnesium slurry. The mixed slurry is in front of the nozzle, where it is fully mixed by the internal structure design of the spraying machine and then sprayed out at the nozzle with the air pressure. In addition, in some processes, the magnesium slurry and the foam generated by physical foaming will be premixed, and then sprayed directly using the spraying machine. This process requires high stability of the foam and certain requirements for equipment pressure control and spraying air pressure control.

4.20.3.2 Construction steps

1. Preparation work, check the water, electricity, and gas pipelines, check if the equipment is in good condition, preprocess the construction interface, remove floating dust, impurities, etc., and lay waste padding materials below the spraying point to receive the backfill waste materials.
2. Mixing work, dilute the foaming liquid to an appropriate concentration, mix the magnesium sulfate solution, magnesium oxide, additives, waterproof emulsion, etc. in proportion evenly and keep them ready.
3. Spraying work, turn on the power supply and air source, use the spraying machine to separately extract the magnesium slurry and the foaming liquid for spraying work. During spraying, it should be done from bottom to top and as perpendicular as possible to the interface.
4. Completion work, first stop the material supply, use clean water to rinse the equipment, then turn off the power supply and air source, and pay attention to cleaning the rebound slurry when leaving the site, and keep samples for curing and tracking the strength changes.
5. Follow-up work: When the spraying thickness fails to meet the construction requirements, a respraying operation can be carried out after the sprayed slurry has solidified.

4.20.3.3 Reference ratio

1. The foaming agent can be diluted according to the requirements specified in the product manual. When using GX-7# foaming agent, it can be diluted at a ratio of 1: 40 to 1: 20.
2. The magnesium sulfate solution can be prepared at a degree of 28–30 Brix. The raw material ratio can be referred to the appendix.
3. When using lightweight calcined magnesia with an activity content of 60%, the ingredients can be proportioned at a ratio of magnesia: magnesium sulfate solution = 1: 1 to 1.2, and 1% of GX-15# sulfur-magnesium oxide modifier can be added. According to the thickness requirements, 7%–10% of silica fume or fly ash can be added for adjustment, or antisagging additives, short fibers, etc. can also be added for adjustment.
4. According to the construction requirements, an appropriate amount of waterproof emulsion can be added to improve the water resistance of the protective layer. A certain amount of setting accelerators can be added, such as setting powder, GX-8# setting antisalt agent, seed crystals, etc., to increase the solidification speed of the protective layer and facilitate the implementation of subsequent work.

(A) (B)

Figure 4.140 Self-leveling. (A) Inorganic material self-leveling decoration. (B) Organic material self-leveling decoration.

4.20.4 Foamed magnesium cement self-leveling material

In building decoration, the application of self-leveling technology is becoming increasingly widespread. There have been two types of self-leveling processes: cement self-leveling and epoxy self-leveling. However, with the improvement of living standards and the continuous development of technology, magnesium cement foaming self-leveling has emerged to replace the traditional cement self-leveling. Magnesium cement, after adding an appropriate amount of foaming agent, becomes lightweight and strong, featuring insulation, heat insulation, fire resistance, and environmental friendliness. It can also be combined with floor heating pipes to form a new type of floor heating insulation form.

Currently, self-leveling is classified according to materials as inorganic self-leveling (magnesium cement, silicate cement, etc.) and organic self-leveling (epoxy resin, etc.) (Fig. 4.140). Common cement self-leveling is inorganic, using a certain grade of cement as the base material, single-component, and water is added during construction. Epoxy self-leveling is organic, using epoxy resin as the adhesive material and adding pigments, fillers, and additives to form, usually requiring the addition of a curing agent during use, and no other auxiliary materials are added during construction. Epoxy self-leveling mainly uses epoxy resin as the main material, mixed with curing agent, diluent, solvent, dispersant, defoamer, and certain fillers. Epoxy self-leveling consists of two parts: concrete base layer and epoxy self-leveling floor coating. These two types of self-leveling materials have certain limitations. MJT has developed foaming magnesium cement self-leveling, and a brief introduction will be given here.

Foamed magnesium cement self-leveling is a new method that uses magnesium-based inorganic cementitious materials. Under the guidance of a scientific formula, a certain amount of additives and fillers are added, and then it is evenly mixed. The preprepared foaming solution is foamed using a foaming machine and added to the slurry according to the requirements. The foaming slurry is then pushed to the designated floor (or roof) using a pumping system. After natural leveling, it is manually smoothed to achieve a flat and naturally cured surface.

4.20.4.1 Characteristics, applications, and composition of magnesium foamed self-leveling

4.20.4.1.1 Characteristics

1. Magnesium foamed self-leveling has excellent insulation, heat preservation, sound insulation, and lightweight properties. Compared with calcium silicate self-leveling, it has higher strength. Under the same strength, less foamed magnesium cement can be used without occupying space. Magnesium materials are also environmentally friendly, which is superior to other traditional insulation materials.
2. Using foamed magnesium cement as a ground insulation and heat preservation material greatly improves the bonding and adhesion performance between the insulation layer and the floor base surface, reducing the occurrence of delamination, cracking, and hollowing (Fig. 4.141).

Figure 4.141 Foamed magnesium cement self-leveling material.

3. The compressive strength of foamed magnesium cement is higher than that of wood boards, having a bearing and load-bearing advantage. Moreover, magnesium materials are nontoxic and fire-resistant, providing better safety.
4. The service life of foamed magnesium cement self-leveling is longer than that of standalone epoxy self-leveling. The service life of foamed magnesium cement self-leveling is over 10 years, while that of epoxy self-leveling is generally 3 years.
5. The foamed magnesium cement self-leveling material can be pumped and transported, significantly improving the construction efficiency.
6. It comes in various styles, allowing users to choose the appropriate color and shaping effect according to their needs.

4.20.4.1.2 Application scope

The foamed magnesium cement self-leveling material has a wide range of applications. It can be used for the floors of high-rise residences, as well as industrial plants, workshops, warehouses, commercial showrooms, exhibition halls, gymnasiums, hospitals, offices, etc. It can also be used for the heating and insulation layer framework structure of walls in homes and villas, as well as roof insulation layers. The magnesium foamed cement self-leveling material can also be used as the decorative surface layer or wear-resistant base layer for various insulation layers.

4.20.4.1.3 Structural composition

The foamed magnesium cement self-leveling material is divided into a cushion layer self-leveling and a surface layer self-leveling.

1. The cushion layer self-leveling is used under wooden floors, plastic floors, and carpets.
2. The surface layer self-leveling is directly used as the floor and can also serve as the base ground for floor paint. The technical requirements are relatively high (Fig. 4.142).

4.20.4.2 Construction techniques

During the construction process, controlling the water content of the foamed magnesium cement is an important measure to prevent cement precipitation and surface cracking; a reasonable transportation method can effectively reduce the fragmentation of bubbles, thereby reducing the moisture content; choosing appropriate admixtures, high-performance foaming agents, and certain auxiliary materials can provide a reliable

Figure 4.142 The base surface of the floor paint.

Figure 4.143 The effect of the foamed magnesia cement floor heating insulation layer.

guarantee for the quality of the foamed magnesium cement. Practice has proved that after taking the above measures, when the density is less than 300 kg/m^3, its compressive strength can reach above 2 MPa. If such indicators can be stably achieved in large-scale construction, the excellent performance of the foamed magnesium cement will be obvious.

The appearance quality of the foamed magnesium cement floor heating insulation layer is excellent. The density is uniform both above and below, with good stability and excellent insulation effect. The surface is flat without cracks, and the surface strength is high (Fig. 4.143). The construction method of foamed magnesium cement floor heating insulation is superior to chemical heat insulation materials. It has high compressive strength, light weight, good sound insulation and heat insulation performance, good moisture-proof and antiseepage performance, short construction period, quick effect, low cost, and easy quality control, which meets the requirements of energy conservation and environmental protection. This technology not only solves the construction quality problems of low-floor heating insulation layers in buildings, but is also more significant in terms of energy-saving, environmental protection, construction speed, reduction of cost, reduction of construction energy consumption, and reduction of labor intensity, achieving good social and economic benefits.

4.21 Application of magnesium phosphate

Magnesium phosphate cement (MPC) is composed of calcined magnesium oxide, phosphate salts, and retarders. It reacts with water at room temperature, releasing a large amount of heat and hardening rapidly. MPC has some characteristics similar to cement, while also possessing the property of chemical bonding to ceramics.

MPC has many superior properties compared to traditional cement materials: rapid setting and hardening, high early strength; good volume compatibility, high bonding strength; when used as a repair material, it has excellent corrosion resistance and wear resistance. MPC can also solidify various wastes, and has a higher efficiency in solidifying solid wastes containing heavy metals, which is beneficial to environmental protection.

Given the above characteristics of MPC, several common application forms are introduced here.

1. Rapid repair material.
   MPC has a fast setting and hardening process, high early strength, good volume stability, and a high bonding strength with the repaired material. Compared to ordinary silicate cement, it has better performance. During repair, MPC can well fill the pores of silicate cement, making the repair interface more stable and forming a higher bonding strength. These properties enable MPC to have wide applications in rapid repair of defense projects, highways, and main urban roads.
2. Solidify harmful and radioactive waste.
   The solidification mechanism of magnesium phosphate is to first convert the polluting elements into low-solubility phosphates, and then encapsulate them in the MPC matrix. Compared with ordinary silicate cement, the solidification effect of MPC on radioactive waste is more significant.

Through research, it has been found that MPC has a good effect in solidifying medium and low-level radioactive waste, and can be used as a good medium and low-level radioactive waste solidification substrate.

3. As a clinical bone substitute material.
   Magnesium phosphate bone cement is a new type of inorganic bone cement material due to its excellent biocompatibility, plasticity, high mechanical strength, and osteogenic potential. It has been introduced into the field of bone defect repair, meeting the requirements of clinicians for the plasticity, stability, and bone induction of the material.
4. As a casting sand binder.
   In the casting industry, the molding material is the foundation that determines the quality of the castings. And the quality of the molding material depends on the type of binder. The phosphate-based casting sand binder is a new type of inorganic binder that has been developed in recent years. Compared with organic resin binders and inorganic sodium silicate binders, it has excellent mechanical properties and thermal stability. It has the characteristics of complete disintegration, low gas emission, nontoxic and environmentally friendly, and relatively low cost.

MPC can harden rapidly within minutes to tens of minutes and possesses very high bonding strength, earning it the nickname "inorganic glue" in the engineering field. It can cure both in air and in water, maintaining long-term strength development in aqueous environments. Consequently, it is highly suitable as a self-hardening casting sand binder.

In summary, MPC has excellent properties such as short setting time, high early strength, and good corrosion resistance, which makes it widely used in the construction industry. With the continuous improvement of China's urbanization level, the demand for new buildings is decreasing, and as time goes by, the repair of buildings will be a huge demand gap. MPC can be used as a good repair material, and it can also be used as an emergency repair material in emergency rescue and disaster relief. MPC is a new type of air-hardening cementitious material, and due to its excellent properties, it has received extensive research.

Appendix 1

Correspondence table of molasses degree, density, and concentration of magnesium chloride and magnesium sulfate

In the production of magnesium-based materials, the specific gravity is commonly used to indicate the concentration of brine. In fact, the specific gravity reflects the density of the brine, and the impurities in the brine also contribute to a certain amount of specific gravity, which results in the brine formed from inferior brine having a higher specific gravity, while the concentration of magnesium chloride in the brine cannot meet the production requirements, and the quality of the finished plates is greatly compromised. When exploring related technologies of magnesium-based materials, people often calculate the material ratio based on the chloride magnesium content in the brine, that is, the molar quantity. To facilitate the conversion between molar quantity and specific gravity and other data among industry colleagues, a reference table comparing common brine ratios with specific gravity, density, etc. has been specially compiled. See Appendix 1.1.

Appendix 1.1 Reference table comparing common brine ratios with specific gravity, density, etc

Ratio of marinated slices to water	Bé (degree of alcohol)	Density/ (g/cm^3)	Specific gravity	Molar concentration/ (mol/L)	Molar ratio of water to magnesium chloride	Salinity/ ‰	Chloride/ %
0.6	20	1.16	1.16	2.06	26.1	186.73	139.00
0.65	20.5	1.1647	1.1647	2.17	24.5	196.16	146.02
0.7	21.5	1.1741	1.1741	2.28	23.3	205.03	152.63
0.75	22.5	1.1837	1.1837	2.40	22.1	213.40	158.85
0.8	23.5	1.1934	1.1934	2.51	21.2	221.31	164.74
0.85	24.5	1.2033	1.2033	2.61	20.3	228.78	170.30
0.9	25	1.2083	1.2083	2.71	19.5	235.86	175.58
1	26	1.2185	1.2185	2.88	18.2	248.97	185.33
1.1	27	1.2288	1.2288	3.04	17.2	260.82	194.16
1.2	28	1.2393	1.2393	3.19	16.3	271.60	202.19
1.3	29	1.25	1.25	3.34	15.5	281.44	209.50
1.38	30	1.2609	1.2609	3.46	15.0	288.72	214.92

Note: 1. This table was drawn based on the data obtained after preparing brine using sea salt flakes and distilled water at a room temperature of 20°C. The actual experiments and production activities are subject to factors such as temperature, batch of salt flakes, water quality conditions, and measurement instrument errors, so the obtained data may deviate from this table.

2. Due to the limitations of instrument accuracy and operational errors, this table is only applicable as a reference for experiments and production activities with general measurement accuracy requirements.

When preparing 26°Bé brine, the halogen sheet and water can be mixed in a 1:1 ratio. After the halogen sheet has completely dissolved, the density of the solution is approximately 1.2185 g/cm^3. Each liter of the brine contains about 2.88 mol/L of magnesium chloride, and the salinity of the brine is 248.97‰, while the chlorine content is 185.3304‰. In 26°Bé brine, the molar ratio of water to magnesium chloride is 18.2. At this molar ratio, it satisfies the formation of the 5·1·8 phase in the standard consistency slurry. When the brine reaches 20°Bé, the molar ratio of water to magnesium chloride is 26.1. The amount of 5·1·8 phase formation in the standard consistency slurry decreases at this time, affecting the strength of the product. When the brine reaches 28°Bé, in this system, there will be a large amount of $MgCl_2$ remaining. One should be vigilant of the risk of back brining.

In the production of sulfur-magnesium products, the quality control of the magnesium sulfate solution is often achieved through the Baumé scale data. When preparing the magnesium sulfate solution using heptahydrate magnesium sulfate, the dissolution of heptahydrate magnesium sulfate absorbs heat, which significantly prolongs the time required for preparing the magnesium sulfate solution. Production personnel often cannot obtain the Baumé scale data of the magnesium sulfate solution in time, and thus it is difficult to precisely control the material ratio. To facilitate colleagues in the industry to prepare the magnesium sulfate solution and to convert between molar quantities and Baumé scale data, a reference comparison table of common magnesium sulfate solution ratios and Baumé scale, density, etc. has been specially compiled. See Appendix 1.2.

When preparinga 27°Bé magnesium sulfate solution, the heptahydrate magnesium sulfate and water can be mixed in a ratio of 0.7:1. After the heptahydrate magnesium sulfate is completely dissolved, the density of the solution is approximately 1.229 g/cm^3. Each liter of heptahydrate magnesium sulfate solution contains approximately 1.95 mol/L of magnesium sulfate, and the salinity of the heptahydrate magnesium sulfate solution is 201.09‰. In the 27°Bé magnesium sulfate solution, the molar ratio of water to magnesium sulfate is 28.3. At this molar ratio, the formation of the 5·1·7 phase in the sulfur-oxygen-magnesium system is favorable. When the magnesium sulfate solution reaches 23°Bé, the molar ratio of water to magnesium sulfate is 36.6, and the product is mainly composed of $Mg(OH)_2$, with the strength of the product significantly decreasing. When the magnesium sulfate solution reaches 32°Bé, the molar ratio of water to magnesium sulfate is 22.2, the proportion of the 5·1·7 phase in the product further increases, and the strength of the product also increases.

Appendix 1.2 Reference table for common magnesium sulfate solution ratios, specific gravity, density, etc

Ratio of magnesium sulfate heptahydrate to water	Bé (degree of alcohol)	Density/ (g/cm^3)	Specific gravity	Molar concentration/ (mol/L)	Molar ratio of water to magnesium sulfate	Chloride/ %
0.5	23	1.189	1.189	1.53	36.6	162.79
0.6	25	1.208	1.208	1.75	31.7	183.14
0.7	27	1.229	1.229	1.95	28.3	201.09
0.8	29	1.25	1.25	2.14	25.7	217.05
0.9	31	1.272	1.272	2.32	23.7	231.33
1	32	1.283	1.283	2.47	22.2	244.18

Note: 1. This table was drawn based on the data obtained after preparing the solution using first-grade anhydrous magnesium sulfate and distilled water at a room temperature of 20°C. The actual experiments and production processes are subject to factors such as temperature, batch of anhydrous magnesium sulfate, water quality conditions, and measurement instrument errors. Therefore, the obtained data may deviate from this table.
2. During storage and transportation of anhydrous magnesium sulfate, it is prone to lose moisture and powderize. The powderized anhydrous magnesium sulfate does not affect its use. However, when using the powderized anhydrous magnesium sulfate and referring to this table to prepare a solution with a specified Baumé degree, the amount of anhydrous magnesium sulfate added should be appropriately reduced.
3. Due to the limitations of instrument accuracy and operational errors, this table is only applicable as a reference for experiments and production activities with general measurement accuracy requirements.

Appendix 2

Correspondence table of molar ratio and mass ratio

The production ratio of magnesium-based products should be in line with the production environment. Under different temperature and humidity conditions, products made with different ratios will have certain differences in the final quality. The ratio of magnesium oxide to magnesium chloride and magnesium sulfate is generally referred to as the molar ratio, but in actual production, the proportion of materials is often calculated by weight. Whenever the environment changes and the ratio needs to be dynamically adjusted, the calculation is rather cumbersome. To facilitate colleagues in the industry to adjust the ratio, the following table is specially drawn showing the corresponding molar ratio and weight ratio of chlorinated magnesium oxide and sulfuric magnesium oxide.

Appendix 2.1, Appendix 2.2, and Appendix 2.3 list the proportion of different grades of magnesium oxide and brine, as well as the quality data of the adjusting agents that can be selected under different molar ratios. The mixing ratio of different brix-adjusting agents can be referred in Appendix 1.1 and Appendix 1.2.

Appendix 2.1 Reference corresponding table of molar ratio and weight ratio (chlorinated magnesium oxide)

(1) When the activity of magnesium oxide is 60%

Molar ratio K	Magnesium oxide		Brine (containing 45% magnesium chloride)		
	Liveness	Quality/kg	Bouyant force	Concentration	Mass/kg
4	60%	10	21	0.185	19.256
			22	0.190	18.75
			23	0.197	18.08
			24	0.204	17.46
			25	0.213	16.725
			26	0.225	15.83
			27	0.236	15.095
			28	0.245	14.54
			29	0.253	14.08
			30	0.262	13.597
5	60%	10	21	0.185	15.4
			22	0.190	15
			23	0.197	14.467
			24	0.204	13.97
			25	0.213	13.38
			26	0.225	12.67
			27	0.236	12.076
			28	0.245	11.63
			29	0.253	11.26
			30	0.262	10.878
6	60%	10	21	0.185	12.838
			22	0.190	12.5
			23	0.197	12.056
			24	0.204	11.64
			25	0.213	11.15
			26	0.225	10.556
			27	0.236	10.063
			28	0.245	9.694
			29	0.253	9.387
			30	0.262	9.065
7	60%	10	21	0.185	11.004
			22	0.190	10.714
			23	0.197	10.333
			24	0.204	9.979
			25	0.213	9.557
			26	0.225	9.048
			27	0.236	8.626
			28	0.245	8.309
			29	0.253	8.046
			30	0.262	7.77

8	60%	10	21	0.185	9.628
			22	0.190	9.375
			23	0.197	9.042
			24	0.204	8.732
			25	0.213	8.363
			26	0.225	7.917
			27	0.236	7.548
			28	0.245	7.27
			29	0.253	7.04
			30	0.262	6.799
9	60%	10	21	0.185	8.56
			22	0.190	8.33
			23	0.197	8.04
			24	0.204	7.76
			25	0.213	7.43
			26	0.225	7.04
			27	0.236	6.71
			28	0.245	6.46
			29	0.253	6.258
			30	0.262	6.04

When the fixed molar ratio is 6, for every 10 kg of magnesium oxide with an active content of 60%, 11.15 kg of 25°Bé brine can be used for blending, and the concentration of the brine is 21.3%.

(2) When the activity of magnesium oxide is 65%

Molar ratio K	Magnesium oxide		Brine (containing 45% magnesium chloride)		
	Liveness	Quality/kg	Bouyant force	Concentration	Mass/kg
4	65%	10	21	0.185	20.86
			22	0.190	20.31
			23	0.197	19.59
			24	0.204	18.92
			25	0.213	18.12
			26	0.225	17.15
			27	0.236	16.35
			28	0.245	15.75
			29	0.253	15.25
			30	0.262	14.73
5	65%	10	21	0.185	16.69
			22	0.190	16.25
			23	0.197	15.67
			24	0.204	15.13
			25	0.213	14.50
			26	0.225	13.72

			27	0.236	13.08
			28	0.245	12.60
			29	0.253	12.20
			30	0.262	11.78
6	65%	10	21	0.185	13.90
			22	0.190	13.54
			23	0.197	13.06
			24	0.204	12.61
			25	0.213	12.08
			26	0.225	11.44
			27	0.236	10.90
			28	0.245	10.50
			29	0.253	10.17
			30	0.262	9.82
7	65%	10	21	0.185	11.92
			22	0.190	11.61
			23	0.197	11.19
			24	0.204	10.81
			25	0.213	10.35
			26	0.225	9.80
			27	0.236	9.34
			28	0.245	9.00
			29	0.253	8.72
			30	0.262	8.42
8	65%	10	21	0.185	10.43
			22	0.190	10.16
			23	0.197	9.80
			24	0.204	9.46
			25	0.213	9.06
			26	0.225	8.58
			27	0.236	8.18
			28	0.245	7.88
			29	0.253	7.63
			30	0.262	7.37
9	65%	10	21	0.185	9.27
			22	0.190	9.03
			23	0.197	8.71
			24	0.204	8.41
			25	0.213	8.05
			26	0.225	7.62
			27	0.236	7.27
			28	0.245	7.00
			29	0.253	6.78
			30	0.262	6.55

When the fixed molar ratio is 6, for every 10 kg of magnesium oxide with an active content of 65%, 12.08 kg of 25°Bé brine with a concentration of 21.3% can be used for blending.

Appendix 2.2 Reference table for molar ratio and mass ratio (sulfur-magnesium oxide)

(1) When the activity of magnesium oxide is 60%

Molar ratio K	Magnesium oxide		Magnesium sulfate solution (with a purity of 99% magnesium sulfate)		
	Liveness	Quality/kg	Bouyant force	Concentration	Mass/kg
5	60%	10	25	0.192	18.61
			26	0.200	17.87
			27	0.217	16.47
			28	0.222	16.10
			29	0.228	15.67
			30	0.235	15.21
6	60%	10	25	0.192	15.51
			26	0.200	14.89
			27	0.217	13.72
			28	0.222	13.41
			29	0.228	13.06
			30	0.235	12.67
7	60%	10	25	0.192	13.29
			26	0.200	12.76
			27	0.217	11.76
			28	0.222	11.50
			29	0.228	11.19
			30	0.235	10.86
8	60%	10	25	0.192	11.63
			26	0.200	11.17
			27	0.217	10.29
			28	0.222	10.06
			29	0.228	9.79
			30	0.235	9.50
9	60%	10	25	0.192	10.34
			26	0.200	9.93
			27	0.217	9.15
			28	0.222	8.94
			29	0.228	8.71
			30	0.235	8.45
10	60%	10	25	0.192	9.31
			26	0.200	8.93
			27	0.217	8.23
			28	0.222	8.05
			29	0.228	7.86
			30	0.235	7.60
11	60%	10	25	0.192	8.46
			26	0.200	8.12
			27	0.217	7.48
			28	0.222	7.32

			29	0.228	7.12
			30	0.235	6.91
12	60%	10	25	0.192	7.75
			26	0.200	7.44
			27	0.217	6.86
			28	0.222	6.71
			29	0.228	6.53
			30	0.235	6.34

When the fixed molar ratio is 6, for every 10 kg of magnesium oxide with an active content of 60%, 13.72 kg of 27°Bé magnesium sulfate solution can be used for blending, with the concentration of the magnesium sulfate solution being 21.7%.

(2) When the activity of magnesium oxide is 65%

Molar ratio K	Magnesium oxide		Magnesium sulfate solution (with a purity of 99% magnesium sulfate)		
	Liveness	Quality/kg	Bouyant force	Concentration	Mass/kg
5	65%	10	25	0.192	20.16
			26	0.200	19.35
			27	0.217	17.84
			28	0.222	17.44
			29	0.228	16.98
			30	0.235	16.47
6	65%	10	25	0.192	16.80
			26	0.200	16.13
			27	0.217	14.87
			28	0.222	14.53
			29	0.228	14.15
			30	0.235	13.73
7	65%	10	25	0.192	14.40
			26	0.200	13.82
			27	0.217	12.74
			28	0.222	12.45
			29	0.228	12.13
			30	0.235	11.77
8	65%	10	25	0.192	12.60
			26	0.200	12.10
			27	0.217	11.15
			28	0.222	10.90
			29	0.228	10.61
			30	0.235	10.30
9	65%	10	25	0.192	11.20
			26	0.200	10.75
			27	0.217	9.91

			28 29 30	0.222 0.228 0.235	9.69 9.43 9.15
10	65%	10	25 26 27 28 29 30	0.192 0.200 0.217 0.222 0.228 0.235	10.08 9.68 8.92 8.72 8.49 8.24
11	65%	10	25 26 27 28 29 30	0.192 0.200 0.217 0.222 0.228 0.235	9.16 8.80 8.11 7.93 7.72 7.49
12	65%	10	25 26 27 28 29 30	0.192 0.200 0.217 0.222 0.228 0.235	8.40 8.06 7.43 7.27 7.07 6.86

When the fixed molar ratio is 6, for every 10 kg of magnesium oxide with an active content of 65%, 14.87 kg of 27°Bé magnesium sulfate solution can be used for blending, with the concentration of the magnesium sulfate solution being 21.7%.

Appendix 2.3 Reference table for molar ratio and mass ratio (sulfur-chlorine compound)

Sulfur-chlorine compound solution ratio	The molar ratio of different substances			
	MgO	$MgCl_2$	$MgSO_4$	H_2O
A = 60 B = 60	1.00	0.09	0.07	3.48
A = 72 B = 48	1.00	0.11	0.06	3.52
A = 84 B = 36	1.00	0.13	0.04	3.52
A = 96 B = 24	1.00	0.14	0.03	3.52
A = 108 B = 12	1.00	0.16	0.01	3.52

A = 120 B = 0	1.00	0.18	0.00	3.52

Note: The mass of magnesium oxide in the table is calculated based on 100 kg, and the content of active magnesium oxide is calculated as 60%; the dosage of the sulfur-chlorine composite solution is 120 kg, among which the brine is 25°Bé and the magnesium sulfate solution is 27°Bé; the total mass of the magnesium sulfate composite solution is represented by L, the dosage of the brine is represented by A, and the dosage of the magnesium sulfate solution is represented by B; the data of the molar ratio of different substances is based on the molar quantity of MgO.

When using 100 kg of magnesium oxide with an active content of 60% for production, a slurry can be prepared by adding 84 kg of 25°Bé brine and 36 kg of 27°Bé magnesium sulfate solution as the raw materials. At this time, the mass ratio of brine to magnesium sulfate is 7:3, the molar ratio of $MgCl_2$ to MgO in the system is 0.13, the molar ratio of $MgSO_4$ to MgO is 0.04, and the molar ratio of H_2O to MgO is 3.52.

Appendix 3

List of common magnesium product formulas

1. Fireproof door core board (the brief production process introduction is given only for the fireproof door core board, and other products will not be elaborated further) (see Table A3.1, Table A3.2, and Table A3.3).

Production process:

(1) All materials are weighed and prepared, and brine is pre-prepared.

(2) Pour the materials into the mixer in the prescribed sequence. First, pour the brine. If there is GX-1# modifier, add it to the brine first. Then add the magnesium oxide powder. Add GX-4# or GX-8# while stirring, then add short fibers. Use the foaming machine to foam, add an appropriate amount of foam to the mixer and stir evenly, then pour it into the mold.

(3) Curing requirements: place the mold containing the slurry that is about to solidify into the prepared curing room (continuously monitor the temperature, humidity, and moisture changes).

(4) Demolding and curing: generally, based on the environmental conditions, remove the mold after 8–12 hours of curing, and then continue to cure for 5–7 days.

Production process:

(1) All materials have been weighed, and the magnesium sulfate solution has been prepared.

(2) Pour the materials into the mixer in the prescribed sequence. First, pour the magnesium sulfate solution, then add the GX-15# modifier to the magnesium sulfate solution, followed by adding the magnesium oxide powder. Add GX-13# while stirring, then add the short fibers. Next, use the foaming machine to foam, add an appropriate amount of foam to the mixer and stir evenly, then pour it into the mold.

(3) Curing requirements: place the mold containing the slurry that is about to solidify in the prepared curing room (continuously monitor the temperature, humidity, and moisture changes).

(4) Demolding and curing: generally, based on the environmental conditions, the mold can be removed after 8–12 hours of curing, and then further curing for 5–7 days is required.

Production process:

(1) All materials have been weighed and the sulfur-chlorine composite solution has been prepared.

Table A3.1 Reference mix ratio of chlorinated magnesium fireproof door core board slurry (unit: kg).

Magnesium oxide	Brine (25°Bé)	Fiber	GX-1# (or GX-8#)	GX-4#	GX-7#
100	100–110	0.5–1	0.5–1	0.6–0.8	1:80 dilution with water

Note: Production temperature: 20°C; magnesium oxide powder: 85 mesh, with activity of 65%.

Table A3.2 Reference slurry ratio for magnesium oxide fireproof door core board (unit: kg).

Light-burned magnesium oxide powder	Magnesium sulfate solution (30°Bé)	Fiber	Pulverized material	GX-15#	GX-13#	GX-7#
100	160	1–1.5	30	0.5–1	1	1:80 dilution with water

Note: Production temperature: 20°C; magnesium oxide powder: 85 mesh, with activity of 65%.

Table A3.3 Reference blending ratio for sulfur-chlorine composite fireproof door core board material (unit: kg).

Light-burned magnesium oxide powder	Prepare a composite solution (27°Bé) by mixing 8 parts of magnesium chloride with 2 parts of magnesium sulfate	Fiber	Special modified agent	GX-13#	GX-7#
100	120	1–1.5	0.5–1	1	1:80 dilution with water

Note: Production temperature: 20°C; magnesium oxide powder: 85 mesh, with activity of 65%.

(2) Pour the materials into the mixer in the prescribed sequence. First, pour the sulfur-chlorine composite solution, then add the special modifier to the sulfur-chlorine composite solution, followed by adding magnesium oxide powder, adding GX-13# while stirring, then add short fibers; then foam with the foaming machine, add an appropriate amount of foam to the mixer and stir evenly, and pour it into the mold.

(3) Curing requirements: place the mold containing the slurry that is about to solidify into the prepared curing room (continuously monitor the temperature, humidity, and moisture changes).

(4) Demolding and curing: generally, based on the environmental conditions, remove the mold after 8–12 hours of curing, and then continue to cure for 5–7 days.

2. Brick machine trays (refer to Table A3.4).
3. Wave-shaped colored tiles (refer to Table A3.5).
4. Piglet electric heating plates (refer to Table A3.6 and Table A3.7).
   (1) Surface layer.
   (2) Core layer.
5. Springboard (refer to Table A3.8).
6. Construction formwork (refer to Table A3.9).
7. Fireproof sealing plates (refer to Table A3.10).
8. Hand-woven frame of the greenhouse (refer to Table A3.11).
9. Enema-type large intestine framework (refer to Table A3.12).
10. Kiwi fruit support (refer to Table A3.13).
11. Fireproof board (refer to Table A3.14).
12. Sound barrier (refer to Table A3.15).
13. Chimney (outer square inner circle) (refer to Table A3.16).
14. Sandwich partition panels (silicate calcium panels) (refer to Table A3.17).
    Foam to 700–800 kg/m^3 and add EPS to 500 kg/m^3.
15. Exterior wall insulation board (homogeneous board): 250 kg/m^3 (refer to Table A3.18).
    Foam to 500 kg/m^3 and add EPS to 250 kg/m^3.
16. Building formwork (sprayed molding) (refer to Table A3.19).
17. Art fencing (refer to Table A3.20).
18. Partition board (fireproof board with core): 800 mm×250 mm×100 mm, 500–600 kg/m^3 (refer to Table A3.21).

Table A3.4 Reference material ratio for brick machine trays.

Materials	MgO	$MgSO_4$(28°Bé)	GX-15#	GX-13#	Saw powder	Bamboo comb	Glass fiber cloth (coated)
Quantity	7.5 kg	12.75 kg	37.5 g	225 g	4.2 kg	1 layer	4 layers

Table A3.5 Reference mix ratio for wave-shaped colored tile mortar.

Materials	MgO	$MgSO_4$(28°Bé)	GX-15#	Saw powder	p-p	Glass fiber cloth
Quantity	8 kg	11 kg	30 g	3.2 kg	30 g	2 layers

Table A3.6 Reference mix ratio for the surface layer material of piglet electric heating plates.

Materials	MgO	$MgSO_4$(28°Bé)	GX-15#	Talcum powder	Iron oxide red
Quantity	1.3 kg	1.7 kg	8 g	1 kg	Appropriate amount

Table A3.7 Reference material ratio for the core layer of the electric heating plate for piglets.

Materials	MgO	$MgCl_2$(25°Bé)	GX-15#	Saw powder	Perlite	Glass fiber cloth (coated with plastic)
Quantity	3.7 kg	4.8 kg	24 g	0.55 kg	0.37 kg	3 layers

Table A3.8 Reference blending ratio for springboard material slurry.

Materials	MgO	$MgSO_4$(28°Bé)	Saw powder	GX-15#	GX-13#	Glass fiber cloth (coated with plastic)	Bamboo comb
Quantity	4 kg	6 kg	120 g	24 g	200 g	4 layers	1 layer

Table A3.9 Reference mix ratio for construction formwork material.

Materials	MgO	$MgSO_4$(28°Bé)	Saw powder	GX-15#	p-p	Glass fiber cloth	Bamboo comb
Quantity	12 kg	19.5 kg	3.9 kg	360 g	36 g	3 layers	1 layer

Table A3.10 Reference slurry ratio for fireproof sealing plates.

Materials	MgO	$MgSO_4$(28°Bé)	GX-15#	GX-13#	Talcum powder	Glass fiber cloth
Quantity	7 kg	10 kg	42 g	56 g	3.5–4.9 kg	4 layers

Table A3.11 Reference slurry ratio for roll-type greenhouse frame.

Materials	MgO	$MgSO_4$(26°Bé)	GX-15#	GX-13#	Glass fiber cloth (90 cm)
Quantity	5 kg	4.4 kg	35 g	40 g	3 layers

Table A3.12 Reference slurry ratio for enema-type large intestine framework.

Materials	MgO	$MgCl_2$(26°Bé)	GX-8#	GX-12#
Quantity	10 kg	10 kg	70 g	90 g

Table A3.13 Reference slurry ratio for kiwi fruit stands.

Materials	MgO	$MgSO_4$(28°Bé)	GX-15#	Quartz sand	Water-cooled sand	Bamboo fibers	Glass fiber cloth
Quantity	200 kg	180 kg	2 kg	50 kg	60 kg	10–12 pieces	1 layer

Table A3.14 Reference blending ratio for fireproof panels.

Materials	MgO	$MgSO_4$(28°Bé)	GX-15#	Talcum powder	Saw powder
Quantity	6 kg	7.2 kg	36 g	1.2 kg	1.2 kg

Table A3.15 Reference slurry ratio for sound barrier.

Materials	MgO	$MgSO_4$(28°Bé)	GX-15#	GX-13#	Mountain flour	Pulverized fuel ash	Saw powder
Quantity	100 kg	120–130 kg	1–1.2 kg	0.8 kg	45 kg	25 kg	20–10 kg

Table A3.16 Reference slurry ratio for the chimney.

Materials	MgO	$MgSO_4$(30°Bé)	GX-7#	pp Fiber	Mountain flour	Pulverized fuel ash	Saw powder	Glass fiber cloth
Quantity	50 kg	55 kg	100 times	0.5 kg	5 kg	5 kg	4 kg	1 layer

Table A3.17 Reference mix ratios for sandwich partition panels.

Materials	MgO	$MgSO_4$(30°Bé)	GX-15#	GX-13#	GX-7#
Quantity	8 kg	8 kg	64 g	40 g	80 times

Table A3.18 Reference slurry ratio for exterior wall insulation boards.

Materials	MgO	$MgSO_4$(30°Bé)	GX-15#	GX-13#	GX-7#	p-p	EPS
Quantity	13 kg	13 kg	52 g	65 g	80 times	65 g	Appropriate amount

Table A3.19 Reference slurry ratio for building formwork.

Materials	MgO	$MgSO_4$(30°Bé)	GX-15#	GX-13#	Saw powder
Quantity	15 kg	21 kg	120 g	30 g	4.15 kg

Table A3.20 Reference mix ratio for art fencing materials.

Materials	MgO	$MgSO_4$(28°Bé)	GX-15#	GX-13#	River sand	Pebble
Quantity	400 g	400 g	2 g	2.4 g	1:2.25	1:1.25

Table A3.21 Reference mix ratio of the mortar for the fireproof board core.

Materials	MgO	MgSO4(30°Bé)	GX-15#	GX-13#	GX-7#	Fiber	Ceramsite
Quantity	6 kg	6 kg	0.9%	0.5%	70 times	0.06 kg	Appropriate amount

Index

Note: Page numbers followed by "*f*" and "*t*" refer to figures and tables, respectively.

M

P

U

W

X

Printed and bound by CPI Group (UK) Ltd, Croydon, CR0 4YY
22/09/2026
02220638-0004